Spaces of Spirituality

Spirituality is, too often, subsumed under the heading of religion and treated as much the same kind of thing. Yet spirituality extends far beyond the spaces of religion. The spiritual makes geography strange, challenging the relationship between the known and the unknown, between the real and the ideal, and prompting exciting possibilities for charting the ineffable spaces of the divine which lie somehow beyond geography. In setting itself that task, this book pushes the boundaries of geographies of religion to bring into direct focus questions of spirituality. By seeing religion through the lens of practice rather than as a set of beliefs, geographies of religion can be interpreted much more widely, bringing a whole range of other spiritual practices and spaces to light. The book is split into three sections, each contextualised with an editors' introduction, to explore the spaces of spiritual practice, the spiritual production of space, and spiritual transformations. This book intends to open to up new questions and approaches through the theme of spirituality, pushing the boundaries on current topics and introducing innovative new ideas, including esoteric or radical spiritual practices. This landmark book not only captures a significant moment in geographies of spirituality, but acts as a catalyst for future work.

Nadia Bartolini is an associate research fellow at the University of Exeter. Her work has looked at how tangible heritage is incorporated in contemporary urban planning in Rome. Prior to undertaking her PhD, she worked in Indigenous research and policy in the Canadian Federal Government. Her research focuses on issues surrounding urban cultures, heritage and the built environment. She has published on spiritualities that lie outside mainstream religions in London, Manchester and Stoke-on-Trent, in particular how spiritual values are transmitted through communities and across generations.

Sara MacKian is currently Senior Lecturer in Health and Wellbeing at The Open University. Her research to date has been driven by a curiosity for how people and organisations interact around issues of health, wellbeing and meaning making. More recently she has developed research around alternative spiritualities in contemporary society, based on a fascination with the relationship between the real and the imaginary, the body and the spirit, this world and the otherworldly. She is author of *Everyday Spirituality* (2012).

Steve Pile teaches Geography at The Open University. He has published on issues concerning place and the politics of identity. Steve is author of *Real Cities* (2005) and *The Body and The City* (1996), which both develop a psychoanalytic approach to geography. It is through these projects that he became interested in alternative spiritualities and their relationship to contemporary modernity. His many collaborative projects include the recent collection, *Psychoanalytic Geographies*, edited with Paul Kingsbury.

Routledge Research in Culture, Space and Identity
Series editor: Dr. Jon Anderson, School of Planning and Geography, Cardiff University, UK

The *Routledge Research in Culture, Space and Identity Series* offers a forum for original and innovative research within cultural geography and connected fields. Titles within the series are empirically and theoretically informed and explore a range of dynamic and captivating topics. This series provides a forum for cutting edge research and new theoretical perspectives that reflect the wealth of research currently being undertaken. This series is aimed at upper-level undergraduates, research students and academics, appealing to geographers as well as the broader social sciences, arts and humanities.

For a full list of titles in this series, please visit www.routledge.com/Routledge-Research-in-Culture-Space-and-Identity/book-series/CSI

Surfing Spaces
Jon Anderson

Violence in Place, Cultural and Environmental Wounding
Amanda Kearney

Arts in Place: The Arts, the Urban and Social Practice
Cara Courage

Explorations in Place Attachment
Jeffrey S. Smith

Geographies of Digital Culture
Edited by Tilo Felgenhauer and Karsten Gäbler

The Nocturnal City
Robert Shaw

Geographies of Making, Craft and Creativity
Edited by Laura Price and Harriet Hawkins

Spaces of Spirituality
Edited by Nadia Bartolini, Sara MacKian and Steve Pile

Spaces of Spirituality

**Edited by Nadia Bartolini,
Sara MacKian and Steve Pile**

Routledge
Taylor & Francis Group
LONDON AND NEW YORK

First published 2018
by Routledge
2 Park Square, Milton Park, Abingdon, Oxon OX14 4RN

and by Routledge
711 Third Avenue, New York, NY 10017

Routledge is an imprint of the Taylor & Francis Group, an informa business

British Library Cataloguing-in-Publication Data
A catalogue record for this book is available from the British Library

Library of Congress Cataloging-in-Publication Data
A catalog record for this book has been requested

ISBN: 978-1-138-22606-7 (hbk)
ISBN: 978-1-315-39842-6 (ebk)

Typeset in Times New Roman
by Apex CoVantage, LLC

Contents

SECTION 2
The spiritual production of space

SECTION 3
Spiritual transformations

Figures

Contributors

Kim Beecheno is a teaching fellow in Gender and Social Policy at King's College London. Her PhD focused on the ways in which Christianity (Catholicism and Pentecostalism) is addressed in both secular and faith-based centres for female survivors of domestic violence in Brazil. Her research interests cover intersections between gender, religion, feminism and violence, with a particular focus on Latin America.

Kath Browne is a Professor in Geographies of Sexualities and Genders at Maynooth University. Her research interests lie in sexualities, genders and spatialities. She works with Catherine Nash and Andrew Gorman Murray on understanding transnational resistances to LGBT equalities. She is the lead researcher on the 'Making Lives Liveable: Rethinking Social Exclusion' research project and has worked on LGBT equalities, lesbian geographies, gender transgressions and women's spaces. Kath has authored a number of journal publications, co-wrote with Leela Bakshi *Ordinary in Brighton: LGBT, activisms and the City* (Ashgate, 2013), and *Queer Spiritual Spaces* (Ashgate, 2010), and has co-edited a number of books, most recently, *The Routledge Companion to Geographies of Sex and Sexualities (Routledge, 2016)* and *Lesbian Geographies (Routledge, 2015)*.

Louisa Cadman (Sheffield Hallam University) has worked within the field of Foucauldian and poststructural geographies, with a particular interest in questions of power and resistance in relation to health care and mental health.

Claire Dwyer is a Reader in Human Geography at University College London, where she is also co-director of the Migration Research Unit. Her research focuses on geographies of ethnicity, religion and multiculturalism and she is currently researching suburban religious landscapes in West London. She is the co-author of *New Geographies of Race and Racism, Transnational Spaces, Geographies of New Femininities, Qualitative Methods for Geographers,* and *Geographies of Children and Young People: Identities and Subjectivities.*

Julian Holloway is Senior Lecturer in Human Geography in the Division of Geography and Environmental Management at Manchester Metropolitan University. He has published on the geographies of religion, spirituality and the occult, with a particular reference to spectrality, haunting and monstrosity. His more recent work interrogates the geographies of sound and the sonic apprehension of space and place. Each of these research topics are connected by a theoretical interest in embodiment, practice, affect and materiality.

Peter Hopkins is Professor of Social Geography at Newcastle University. A key focus of his research to date has been upon the geographies of religion, faith and spiritualities which has included work with a diversity of religious faiths, including Muslim, Christian, Sikh and other religious and non-religious young people.

Patricia 'Iolana holds a PhD in Literature, Theology and the Arts from the University of Glasgow, and is an ordained Pagan Minister and Interfaith Activist. She specializes in personal experiences with Goddess, and approaches this from a psychodynamic methodology called Depth Thealogy. Her publications include: *Literature of the Sacred Feminine: Great Mother Archetypes and the Re-emergence of the Goddess in Western Traditions* (2009), *Goddess Thealogy: An International Journal for the Study of the Divine Feminine 1(1)* (2011), *She Rises, Vol 2* (2016), *Goddess 2.0: Advancing A New Path Forward* (2016), and the forthcoming *Feminine States of Consciousness* (2018).

Tariq Jazeel teaches Human Geography at University College London in the UK. His research explores cultural and aesthetic constitutions of the political, and his work is broadly positioned at the intersections of critical geography, South Asian studies and postcolonial theory. He is the author of *Sacred Modernity: nature, environment and the postcolonial geographies of Sri Lankan nationhood* (2013), and co-editor of *Spatialising Politics: culture and geography in postcolonial Sri Lanka* (2009, with Cathrine Brun). He is an editor of the journal *Antipode*, and is on the Editorial Collective of *Social Text*.

Jennifer Lea (University of Exeter) has worked on geographies of the body, with a particular interest in spiritual practices and wellbeing. She has recently developed an interest in post-natal depression and new motherhood.

Catherine J. Nash is a Professor in the Department of Geography and Tourism Studies at Brock University, Canada. Her research focus is on sexuality, gender and urban places. Her current research interests include changing urban sexual and gendered landscapes in Toronto; a focus on digital technologies and sexuality in everyday life; new LGBT mobilities; and international resistances to LGBT equalities in Canada, GB and Australia. Her books include *Queer Methods and Methodologies* (2010) with K. Browne and *An Introduction to Human Geography* (Canadian Edition) (2015) with E. Fouberg, A. Murphy and H. de Blij.

Elizabeth Olson is Associate Professor of Geography and Global Studies at the University of North Carolina – Chapel Hill. She has worked in areas related to inequality and religion, and her current research examines the historical and contemporary ethics and politics of care through the experiences of young people who are engaged in informal caregiving in the United States.

Chris Philo, Professor of Geography at the University of Glasgow since 1995, is fascinated by all things geographical. His specialist interests have been the historical and social geographies of 'madness', 'asylums' and psychiatry (or 'MAP', as he sometimes short-hands it), including a concern for the spaces of mental health, ill-health, treatment, care and recuperation in the present. One dimension of the latter has been a concern for spaces of spiritual health, understood in a broadly existential rather than narrowly 'religious' fashion, and hence Chris was delighted to work with Jen Lea and Louisa Cadman on an AHRC-funded project addressing what we termed 'a new urban spiritual' – which specifically embraced the spaces, times and practices of yoga and meditation in the city. That project, which officially ran 2010–2012 but continues to shape our thinking and writing today, was the basis for the chapter in the present volume.

professor dusky purples has worked and lectured internationally as a reader and spiritual guide for precarious, misplaced, and tenured akademiks alike. Her areas of experience-based expertise include tarot, astrology, and crystalline storytelling. Though professor purples has spent much of the last decade on extended sabbatical, she is always available for consultations, conferences, and private events. You can reach her directly at prof[dot]duskypurples[at]gmail[dot]com. She's within reach and ready to help you read, all you have to do is Ask Her How!

Alison Rockbrand is currently undertaking a PhD at the University of Exeter, where she studies contemporary esoteric theatre and performance art from the insider's perspective. She has worked with numerous esoteric traditions including witchcraft, chaos magic, western left hand path traditions, demonology, satanism, and is an active member of the pagan community. Her theatre company, Travesty Theatre based in Montreal, Canada (2000–2007) was dedicated to experimental performance, including the use of vaudeville and cabaret, Antonin Artaud's Theatre of Cruelty, collective creation, physical theatre, clown and dance. After moving to the UK in 2005 she directed Paul Green's occult play *Babalon*, which was subsequently recorded for radio broadcast and released with a publication of his plays. She has written for Oracle Occult magazine, Silkmilk magazine, Women's Voices in Magick and is the author of a grimoire. Along with Cryptozoologist Richard Freeman, she co-hosts Exe Files Paranormal Radio on Phonic FM in Exeter.

Richard Scriven is a lecturer in Human Geography at the School of Geography and Archaeology, National University of Ireland Galway. His research examines contemporary pilgrimage as an embodied practice, a form of political action and a spatial therapeutic process.

Olivia Sheringham is a lecturer in Human Geography at Queen Mary University of London. Her research interests span migration and religion, creolization and identity formation, and geographies of home and the city. She is currently working with Alison Blunt and Casper Laing Ebbensgaard on a project examining home, city and migration in East London, and was recently involved in a collaborative project called Globe with artist Janetka Platun. Her publications include *Transnational religious spaces: Faith and the Brazilian migration experiences* (Palgrave Macmillan, 2013), *Encountering Difference* (with Robin Cohen, Polity, 2016), and several peer-reviewed articles in journals including *Ethnic and Racial Studies, Diaspora: a Journal of Transnational Studies* and *Portuguese Studies*.

Lia Dong Shimada is Senior Research Fellow for the Susanna Wesley Foundation, based at the University of Roehampton. She was awarded a PhD in Geography from University College London in 2010 and a Masters degree in Theology and Religious Studies from King's College London in 2014. Lia is a mediator and facilitator specialising in diversity, conflict and organisational practice in faith communities; from 2010 to 2013 she implemented the national diversity strategy for the British Methodist Church. Lia is the general editor of the forthcoming book *Migration and Faith Communities* (Jessica Kingsley Publishers).

James Thurgill is an Associate Professor and cultural geographer at the University of Tokyo, Japan. James received his PhD in Cultural Geography from Royal Holloway, University of London, where he completed his doctoral thesis on *Enchanted Geographies: experiences of place in contemporary British landscape mysticism*. James' work is concerned with spectrality and the phenomenology of absence, spatial narrative, folklore and affective geographies. His most recent work examines literary geographies of absence in the writings of M. R. James and Lafcadio Hearn.

Justin K. H. Tse (謝堅恆) is a Visiting Assistant Professor in the Asian American Studies Program at Northwestern University. He is the lead editor of *Theological Reflections on the Hong Kong Umbrella Movement* (Palgrave, 2016) and is working on a book manuscript titled *Religious Politics in Pacific Space: Grounding Cantonese Protestant Theologies in Secular Civil Societies*. His publications can be found in *Population, Space, and Place*, *Global Networks*, *Progress in Human Geography*, *Chinese America: History and Perspective*, *Ching Feng*, *Review of Religion in Chinese Society*, *Bulletin for the Study of Religion*, *Relegens Thréskeia*, and *Syndicate*.

Karin Tusting is Senior Lecturer at the Department of Linguistics and English Language, Lancaster University. Her research interests include linguistic ethnography, workplace literacies, digital literacies communities, language and identity. Her most recent research project, 'The Dynamics of Knowledge Creation', studies academics' writing practices, http://wp.lancs.ac.uk/acadswriting/.

Giselle Vincett is a sociologist of religion with a special interest in marginalized groups, and especially in poverty (or social and economic exclusion) in Britain and Europe. She is currently Mercator Fellow at the University of Leipzig.

Annabelle Wilkins is a postdoctoral researcher in Geography and Environment at the University of Southampton. Her research focuses on the relationships between home, work, migration and cities, and she is currently Research Fellow on an ERC project examining the social, spatial and economic dimensions of home-based self-employment. Her publications include an article in the journal *Gender, Place and Culture*, and a book (in progress) entitled *Migration, Work and Home-Making in the City: Dwelling and Belonging among Vietnamese Communities in London*, to be published by Routledge.

David Gordon Wilson is a former partner in a City of London law firm, who completed a PhD in Religious Studies at Edinburgh in 2011 and has served as a committee member of the British Association for the Study of Religions; he currently owns a rare and second-hand bookshop in London.

Linda Woodhead MBE is Professor of Sociology of Religion in the Department of Politics, Philosophy and Religion in Lancaster University and Director of the Institute for Social Futures. She holds honorary doctorates from the Universities of Uppsala and Zurich and is a Fellow of the Academy of Social Sciences. Here books include *That Was the Church That Was: How the Church of England Lost the English People* (with Andrew Brown 2016), A *Sociology of Prayer* (with Giuseppe Giordan, 2015), *Christianity: A Very Short Introduction* (2nd revised edition, 2014), *Everyday Lived Islam in Europe* (with Nathal Dessing and Nadia Jeldtoft, 2013), *Religion and Change in Modern Britain* (with Rebecca Catto, 2012), *A Sociology of Religious Emotions* (with Ole Riis, 2010), *Religions in the Modern World* (2009), and *The Spiritual Revolution* (with Paul Heelas, 2005).

1 Spaces of spirituality

An introduction

Nadia Bartolini, Sara MacKian and Steve Pile

1 Placing spirituality

Writing in 1993, Martha Henderson asked the question 'what is spiritual geography?' Her question was prompted by the publication of two books, both of which used spiritual geography in their subtitle: *Beliefs and Holy Places* (1992) by James Griffin and *Dakota* (1993) by Kathleen Norris. Henderson's response to the question focuses upon the geography of the spiritual. She argues that spirituality is a subject that can be approached by people working in many different disciplines and, indeed, working in transdisciplinary modes. In this view, the spiritual is one aspect of the relationship between people, place and the earth. Fundamentally, spirituality is connected to the earth and is, therefore, a part of the human ecology and history of particular sites and places. Thus, spiritual geography taps into the long-standing connection that people have with places. Places, as she puts it, 'momentarily trap and illuminate [the] supernatural ability of humans to adapt, create and re-create their surroundings' (page 470). Supernatural? She does not explain. However, Henderson is trying to grasp something unquantifiable: the relationship between the known and the unknown, between the real and the ideal. So, the spiritual is about more than religious beliefs and practices and the creation (and recreation) of sacred sites and spaces. It is not, therefore, an analogue for other ways of thinking about human life, such as the industrial or the biological or the behavioural. Introducing the spiritual starts to interfere with commonplace understandings of place by pushing them into a consideration of the 'supernatural': the unknown, the unknowable, the ineffable and the numinous. The spiritual makes geography strange. Indeed, the spiritual is itself a strange territory: not just uncharted, but calling into question what can be charted. So, this book is not an attempt to provide a cartography of the spiritual, as this would be to disavow the way that the spiritual interferes with geography. Instead, we wish to explore the many different ways that space and spirituality can be entangled, in ways that are surprising, challenging and (hopefully) provocative.

While Henderson wishes to approach spirituality through place, Julian Holloway and Oliver Valins argue that spirituality can be explored at a range of different spatial scales, from, for example, the body to the global (2002, page 5; see also Bartolini *et al.*, 2017). This might imply that spirituality operates within, and confirms the operability of, nested hierarchies of scale. Yet, Holloway's and Valins' aim is to draw out the different ways that spirituality and space are entangled through notions of scale. Rather than spirituality simply being in evidence at different scales, spirituality is seen as productive of those scales. Thus, for example, the body is itself understood and lived in different and distinct ways through spirituality. Indeed, spirituality is woven through everyday life. Moreover, as Jennifer Lea, Chris Philo and Louisa Cadman (Chapter 9) argue, the weaving of spiritual forms through everyday life does not necessarily reveal itself in dramatic or obvious ways. They focus upon the 'small stuff' of spirituality: the 'micro-instances' of other ways of being in the world. They explore the significance of stillness in spiritual life and how forms of stillness can then infuse everyday life. Thus, practices learned in yoga sessions can then be used to 'pause' or 'still' everyday situations or be used to cope with the ordinary stresses of life. Often, this goes unnoticed even by the people doing it. And, even if they do, it is unremarkable and easily forgotten. Yet, these unnoticed micro-instances of spirituality are part of how everyday life is conducted, sustained and endured.

This makes it hard, perhaps impossible, to disentangle the spiritual from the production of space in general. Holloway and Valins observe:

> 'Religious and spiritual matters form an important context through which the majority of the world's population live their lives, forge a sense (indeed an ethics) of self, and make and perform different geographies'.
>
> (page 6)

Everyday life is infused with practices that carry religious and spiritual connotations, often unthinkingly: this is especially clear in the types of foods that are eaten or the clothes in which people feel comfortable (or uncomfortable) or the festivals that people observe; but also in people's celebration of births and marriages – and how they cope with death and bereavement. More than this, Holloway and Valins argue,

> 'religion is a crucial component [of] the construction of even the most "secular" societies. Through, for example, systems of ethics and morality, architecture, systems of patriarchy and the construction of law, government or the (increasing) role of the voluntary sector'.
>
> (page 6)

Religion, and by extension spirituality, do not stand outside of modernity in such a view, but are fundamental to how it is constructed: whether through the implicit moral codes that govern people's conduct or through its explicit laws and their execution. Seemingly secular decisions about what is right or fair or just are informed by value systems derived as much from implicit religious and spiritual assumptions as from the explicit formation of principles by other means. Personal and shared values, then, are hard to shake from religious and spiritual precepts – and this is political, too: for it shapes how people think about and treat others; how they construct and live through collectives, of all kinds; and the values that they seek to operationalise in their everyday lives.

Yet, it is hard (impossible?) to make universalisable statements about religion and spirituality, at any scale. There are intense debates about whether Western societies are post-secular or not, post-Christian or not; about the rise and distribution of religious fundamentalisms across the planet, though especially where it becomes evident in acts of violence, terrorism and war; about whether new forms of spirituality are alternative or mainstream, meaningful or merely a product of fashion (think, for example, about the appearance and spread of the Jedi religion in the West); about the meaning and significance of different kinds of clothing; about the decline or intensification of religious observances; or about the coexistence of laws founded on different religious or spiritual principles. Taken together, these debates challenge the relentless and ubiquitous assumption that social processes are driven by economic, political and cultural logics that have nothing to do with religion and spirituality.

While the trajectories of the social, in different places, at different scales, cannot be universalised, what we can say is that religion and spirituality remains complicit in the production of space and scale. Yet, this is to take religion and spirituality as a singularity: one kind of thing that makes and unmakes geography in distinctive, perhaps even unique, ways. Too often, spirituality is marshalled under, or alongside, the heading of religion and treated as much the same kind of thing. For sure, religions are formed by practices, performances and sacred spaces that are all designed to evoke the divine – and this marks religion out as different from other social practices. Indeed, as the divine is beyond geography, religions work hard to produce a consistency of the numinous across different places, through highly formalised practices, performances and sacred spaces. Yet, not only are religions constantly splitting, mutating and syncretizing, creating contact with the divine is not the only purpose of religious practices, performances and sacred spaces. They are also spiritual, concerned with the nature of spirit, both human and nonhuman. Spirituality, and its forms, extend far beyond religion. Spirituality is therefore not, in this book, a synonym for, nor coextensive with, religion.

Setting spirituality inside, alongside and aside from religion allows us to pose new challenges for understanding the production of space. These are less to do with the structure of beliefs and practices, and more to do with how beliefs and practices play out in, or intersect with, everyday life. Significantly, thinking through spirituality sets new puzzles and challenges for understanding the production of space.

2 Understanding space and spirituality

Paul Cloke and J. D. Dewsbury, writing in 2009, take up the challenge of thinking through the relationship between spirituality and space. To do so, they introduce the idea of 'spiritual landscapes'. Spiritual landscapes, for them, are constituted by the relationship between 'bodily existence, felt practice and faith in things' (page 696). Helpfully, they distinguish spirituality from religion. First, they argue that spirituality can be experienced in a wide variety of religious and non-religious forms. To establish this, they use the examples of experiences with nature, of meditation, and of ghosts. Second, they argue that formal religion can be practiced and experienced in un-spiritual ways, but especially where religious institutions and practices become interwoven with practices of domination and exploitation. So, while ways of understanding religion – through, for example, ritual, beliefs and faith are useful for understanding spirituality, they are not enough. Cloke and Dewesbury argue for an experiential approach to spiritual landscapes, focused on people's personal experiences, their ways of inhabiting, and engaging with, the world.

Thus, the expression 'spiritual landscapes', for Cloke and Dewesbury, is a way to understand the relationship between the spiritual and the spatial. The spiritual, for them, is a disposition that involves both faith and an openness to the possibility of 'other-worldliness'. Notice that this shifts the terrain of spirituality away from the divine, as such, onto a much broader set of possibilities for 'other-worldliness'. In a sense, this idea harks back to Henderson's use of the supernatural to evoke the ineffable. The spiritual implies some kind of world beyond the visible and the material. Key, for Cloke and Dewesbury, is that this other-worldliness is a possibility: faith is associated with this possibility and, indeed, comes to be defined by it. Faith, for example, in Heaven. Or Hell. Or the after-life. Spirituality, however, is not limited to its faith in the possibility of other worlds. It is practised and performed – and becomes manifest in its performances and practices. Such performances and practices include, for example, prayer and meditation, retreats and pilgrimages, singing and chanting, art and music, but also more profane activities such as ghost hunting and dark tourism.

In Chapter 17, Alison Rockbrand explores the performance of esoteric theatre. She shows how esoteric rituals are drawn into theatrical performances that include the audience, enabling them to take part in a spiritual journey. Esoteric theatre creates a transitional space that enables the transmutation of the self that can be carried into the world. Esotericism, thus, undermines the boundaries between worlds (see also Goodricke-Clarke, 2008). Spirituality, through its performances and practices, proliferates through everyday life, often in ways that go unrecognised and unacknowledged, sometimes in ways that are easily dismissed and disavowed. Significantly, Rockbrand also points to the therapeutic and healing aspects of spiritual practices (see also Lea, 2008; and, Williams, 2010, 2016).

Arguably, more than through specific beliefs or rules, it is through bodily practices that people come to live out their spirituality (see Mills, 2012; and, Olson *et al.*, 2013). As Holloway has shown (2003, 2006), the body is active in the production of sacred and spiritual spaces. As importantly, it is through the body that the sacred and spirituality come to make sense. Indeed, religious and spiritual practices organise the senses and distribute them in specific ways. This can run counter to the privileging of the visual in Western cultures as spiritual practices intervene in the whole body and reorganise the senses, through practices such as yoga, meditation, praying, hymns, chanting, festivals, scents, candles and special foodstuffs. All this suggests that ways that we recognise the spiritual and the spirituality of ordinary life need to be expanded, so as to see better new ways that spirituality is being expressed and experienced.

As importantly, Cloke and Dewesbury (2009) evoke the entanglement between space and spirituality through the idea of 'landscape'. For them, landscape is about embodied ways of being-in-the-world. Significantly, these landscapes are, like spirituality, built out of practices and performances as well as lived experiences. There is, then, no spirituality, no landscape, no spiritual landscape that simply has its own pre-formed intrinsic qualities. Landscape, then, suggests an indeterminate range of possible spatial relationships and geographical outcomes, one of which may indeed be 'a landscape', but equally it might be a sacred space or a roadside shrine or a haunted house (see also Olson, Hopkins and Kong, 2012). Dwyer (Chapter 7) shows how religious architecture has altered the landscape along Highway 99 in Vancouver. The juxtaposition of religious buildings creates its own effects. The road becomes a metaphor for the journey towards the divine. Rather than competing, or undermining claims to the one true path to the divine, the highway becomes a part of the practice of reaching the divine. As importantly, it becomes evidence of the divine, with everyday miracles and religious observances now set side by side (and within easy reach, if you have a car).

Following Lily Kong (2001), Cloke and Dewesbury reaffirm that the task is to consider the ways in which spaces become entangled with spirituality such that they become sacred. Sacred spaces are a product of the rituals, performances and practices that make space sacred. This may sound circular, but it indicates that spirituality can be understood through the ways that it produces spaces and spatial practices for itself. Consequently, churches and yoga retreats, pilgrimages and festivals, all highlight the particular spiritual practices through which spirituality is itself constituted (see for example Rose, 2010; and, Conradson, 2012). Thus, spirituality is not simply a matter of personal beliefs, as it is spatially performed and constituted. Indeed, the spaces and spatial practices of spirituality are revealing both of their underlying beliefs and also of how those beliefs sustain ways of inhabiting and producing the world (see Lea, 2009; or, Finlayson, 2012). This is easily witnessed: for example, in architectural plans for sacred spaces, in plans to travel to sacred sites, in the transformation of space during festivals, and in political activism of all kinds.

Even so, as Richard Scriven shows in Chapter 5, the spatial practices of spirituality are often quite marginal, both socially and spatially, yet can be thoroughly transformative. Indeed, the seeming marginality of pilgrimage can disguise its personal effects and affects. What is, then, less easily witnessed is the entanglement between spirituality, personal experience and other ways of being in the world. Indeed, the pilgrimage itself can often appear, especially in its mass forms, as if it is only performed, undertaken only so that it can be seen to be undertaken. Often, spirituality can be seen the same way. In Scriven's hands, thinking through the entanglement between the performative and the experiential becomes a way to understand the significance of spirituality and spiritual transformation.

This performative and experiential understanding of spirituality – and indeed also of space – can unsettle the distinction between modernity and religion. On the surface, modernity might appear secular, profane, scientific and rational, while, on the other hand, religion may appear a legacy of pre-modern beliefs in superstition, in the supernatural, in animism and in magic. Yet, in this account, spirituality and modernity would appear entwined, imbricated, embroiled through their constitution of thoroughly modern sacred spaces and spatial practices. Thus, Tariq Jazeel (Chapter 4) unpicks the relationship between the sacred and the modern. He argues that the entwinement of Buddhism and modernity produced, what he calls, a tropical modern architectural space. This architecture, significantly, is both modern and spiritual. That is, as he says, that the space performs a secular modernity, upon which modern Sri Lanka relies, but the space is also recognizable as having a Buddhist structure of feeling. Rather than seeing the modern and the sacred as in opposition in this architectural space, it must be understood through its duality: both modern and sacred.

Understanding the sacred and the profane requires us to see them as relationally constituted, but also practised, performed and experienced; not separate, but entangled. In Chapter 10, Elizabeth Olson, Peter Hopkins and Giselle Vincett explore the ways that young people engage with religion by testing the limits of belief and spirituality through, what might be seen as, sacrilege. Olson, Hopkins and Vincett argue that the dichotomy of modernity and religion is actively unhelpful in understanding how people negotiate their personal and social lives through religion. Youthful spirituality does not, they argue, fall neatly into the category of religion. The dichotomy between the sacred and the profane has fallen. And not just this dichotomy. Thus, an attendance to spirituality also questions the relationship between the material and the immaterial. Taking spirituality into account radically alters how matter is understood by valorising the immaterial, whether this is in the form of deities or the divine, or body and soul. This critique of dichotomies might appear abstract, but it has the effect of broadening the possible ways that spirituality might be expressed, experienced and discovered. Spirituality is no longer confined to religious rituals and sacred spaces, but rather spread through everyday life – indeed, the implication is that spirituality is just as easily found in ordinary, mundane life as in (for example) churches and church services (McGuire, 2008). So, exploring the spaces of spirituality means being prepared to find spirituality in unexpected places, expressed in ways that may not at first sight appear to be spiritual at all (MacKian, 2012). This book, consequently, takes the opportunity to look awry at space and spirituality.

3 Religion in, and out of, place

We have argued that taking spirituality as a starting point undermines various dichotomies, one of which is between the sacred and the profane. This argument enables us to approach religion in a slightly different way. In our view, religion is a form through which spirituality is expressed, performed and experienced. It is not spirituality's only form, but it is nonetheless a form that we must include in any exploration of spirituality. Just as we attend to the specific forms that spirituality takes, so we must resist the temptation to see religion as a universal. That is, we must see religion as situated in its distinctive contexts. Thus, however much a religion has its own internal dynamics and trajectories, it is imbricated in and/or constitutive of wider social processes. Moreover, as Lily Kong has shown, religion has its own geographies – and these really matter (2010).

Writing in the wake of the terrorist attack on the World Trade Centre on 11 September 2001 and a seemingly endless procession of bombings since, Kong (2010) perceptively observed that the rise of Islamic extremism has altered both geopolitical imaginations and understandings of the place of religion. Religion has, in this view, shifted and altered how we understand the world. It is not just the ways that events elsewhere suddenly appear in unexpected places, as when suicide bombers or refugees appear in markets, hotels, beaches, streets and tower blocks, but that such occurrences draw new geographies of connection and dislocation. Religion is redrawing the map of the world: revealing connections, contesting and subverting borders, making new territories (both virtual and real), rendering safe places unsafe, yet creating new kinds of sanctuary, community and humanity. All these need new maps to be drawn, yet this need is barely acknowledged.

Kong points to the flows of people that have created greater religious diversity in many places. She observes that:

> 'new sources of migrants, new religions, new conflicts, new territories and new networks have all become the subject of analyses. [. . .] Different sites of religious practice beyond the "officially sacred", different sensuous sacred geographies, different religions in different historical and place-specific contexts, different geographical scales of analysis, and different constituents of population have all gained research attention'.
>
> (2010, page 756)

Since being written, as Kong anticipates, these trends have only intensified. However, Kong's larger point is that religion is not simply a dimension of personal and social life, it increasingly provides the framework through which personal and social life is understood and experienced. In this light, it is religion that provides the nation-state with a lens through which to understand who is likely to be dangerous or subversive or require special treatment. Thus, religion does not just enter debates about flows of migrants and refugee crises, it frames them in ways that allow states to identify wanted and unwanted migrants, good and bad refugees. This can be witnessed as easily in US President Donald Trump's attempts to restrict travel from six predominantly Muslim nations (in 2017) as in widely expressed fears that Islamist terrorists would use the refugee crisis in the Mediterranean to access the European Union (in 2016 and after). Religion is now a means through which social and political life are being organised.

The paradox is that religious practice is increasingly 'disorganised': that is, being conducted outside of the formal structures that are intended to organise them. Kong has highlighted how religious practice is to be found in unofficial as well as official sites. So, beyond mosques and synagogues, religion is conducted in living rooms, schools, museums, online, on streets, by roads, in banks and in boardrooms (2010, page 756). The sacred site is supplemented by spiritual places. Perhaps increasingly so. Shrines are not only spontaneously set up in all kinds of places, they can also take the most prosaic of forms. An example would be the Ghost Bikes of New York, where white bicycles memorialise a cyclist that has been killed in a road traffic accident. Perhaps beginning in St Louis in 2003, the Ghost Bikes are now in evidence in London and Berlin, Toronto and Seattle. The Ghosts Bikes are not just an act of memorialisation, nor just a political intervention designed to highlight the lack of road safety for cyclists, they also sacralise space – by invoking the idea of the ghost: that is, the persistence of spirit after death. Thus, the Ghost Bikes are a blend of grief, politics and spirit that deliberately punctuate space, but also make space spiritual. The Ghost Bike is but one of a myriad of possible examples: informal shrines, whether to memorialise the dead or to offer lucky charms or to mark significant events, are common around the world.

Similarly, in the wake of the Grenfell Tower fire in London in 2017, it was through nearby churches, mosques and gurudwaras that the first practical help was organised. The improvised collection and distribution network centred on St Clement's Church. Indeed, the seemingly spontaneous expressions of compassion were often couched in spiritual terms, including the creation of shrines, the use of candles and the invocation of God and angels. On a wall of condolence, where heartfelt messages had been left as well as requests for information about missing people, in large colourful letters was written, underneath a heart made out of twine, 'Pray For Our Community'; in smaller letters, just above the word community, 'our loss is heaven's gain' (see www.itv.com/news/2017-06-16/grenfell-tower-tragedy-shames-us-all/). Such instances not only erode the distinction between sacred and secular space, they also undermine the separation of different kinds of religious spaces from one another. More than this, it suggests that spirituality can lie beyond the formal spaces and practices of religion.

Indeed, religious and sacred spaces can themselves be opened up to reinterpretation along alternative religious, spiritual or occult lines of thought. James Thurgill (Chapter 14) looks at the case of Glastonbury Abbey in Somerset. Glastonbury Abbey was founded in the CE7th and, by the CE14th, grew to become one of the richest and most significant monasteries in England. The Abbey was dissolved in 1539 and fell into decline. And now stands as a ruin. Yet, since the CE12th, Glastonbury Abbey has been strongly connected to Arthurian mythology, so said to be the location of Arthur and Guinevere's tomb. The Abbey is also connected to Christian legends: not only is it said to be founded by Joseph of Arimathea in CE1st (living on through the hawthorn), it is also claimed that he was the last custodian of the Holy Grail. It is said that Joseph is buried beneath Glastonbury Tor, at the entrance to the underworld. Thurgill shows that the relationship between place and spirituality creates opportunities for these to be reimagined and for spirituality to be mobilised in unexpected ways. Thus, his investigation of the sacrality of place shows the exact opposite of what we might expect from sacred space: the meaning of place is never immutable, coherent or singular.

One of the interesting features of contemporary Western societies is the rise of the so called 'no religion' category (Woodhead, 2016). For example, in the UK, in the 2011 Census about 40% of respondents described themselves as having 'no religion'. In the same census, all the major religions (Christianity, Islam, Buddhism, Sikhism) showed a decline in both absolute numbers and in proportion of the total population. Given what we have said, we cannot assume that 'no religion' means no spirituality; as importantly, nor does it necessarily mean no religion in any form. Put another way, the rise of 'no religion' cannot simply be read as being a rise in secularity. Indeed, other evidence suggests that agnosticism has remained relatively stable – and represents maybe only 5–10% of people in the West (see Bartolini *et al*., 2017). Through seeing religion beyond its 'official' forms, and by expanding the sensuous registers through which the religious can be expressed and experienced, it is possible to see how it is that people might think of themselves as not being religious while at the same time having more and less deep-seated spiritual beliefs (see also Gökariksel, 2009).

4 Faith, community and identity

Instead of religion just being organised by a sense of the divine, religion can also be a means through which meanings, affects and identities are expressed. We can usefully extend this idea. Thus, we need to think not just about what people believe in, but what spirituality smells like, what it eats, how it looks, what it sounds like, what spaces it creates for itself, what its bodily regimes, comportments and conducts are, and its defining affects and emotions. We could take any religion or form of spirituality as an example (see Finlayson and Mesev, 2014; and, Sanderson, 2012). However, let us think of how Roman Catholicism defines itself through a specific combination of smells, foods, songs, chants, clothes, rituals, and affectual and emotional performances – that are all heavily circumscribed and over-determined with meaning, affect and identity. Religious and spiritual meanings, affects and identities create a means through which people can reaffirm their faith, create a sense of self, and also form wider communal bonds. This can be especially significant in places where people feel marginalised or excluded from 'normal' or 'dominant' society. Thus, religious and spiritual identities can provide a way to negotiate, challenge or ignore dominant cultural forms, by providing alternative forms of affiliation and community.

Consequently, much research explores the relationship between faith, community and identity. A key aspect of this work is the ways in which migrants express their faith in the new contexts in which they find themselves, not always by choice (see, for example, Dwyer, 2000; and, Sheringham, 2010; or, Olson and Silvey, 2006; Aitchison *et al*., 2007; and, Hopkins and Gale, 2009). As we have discussed, there are visible forms such as buildings and practices, but migrants can also use their spirituality as a means of providing mutual aid in the form of money, shelter, knowledge and work. Olivia Sheringham and Annabelle Wilkins (Chapter 11) show how, for Brazilian and Vietnamese migrants to London, religion permeates every aspect of their experience. Often, migrant experiences are precarious and uncertain, involving movement from, between and to hazardous and hostile situations. Religion can provide not only a means through which to connect to others (see also Holloway, 2012), but also a source of memory, a way of establishing home and a spiritual reconnection with the world – as deities and spirits travel along with migrants. Alongside this, faith communities can become the focus for challenges to the state around injustices, rights and the recognition of difference. Indeed, much work has explored how faith groups act politically in different settings (see for example Sutherland, 2014). Similarly, inter-faith aid can be as significant as faith-centred support. Such inter-faith mutual aid can be a prompt for faith communities to act together politically, especially around support for refugees and migrant rights activism.

Of course, negotiating cultural life elsewhere does not always go well. Interestingly, then, it is religion that affords some people the ability to express their social marginalisation and exclusion. Indeed, religion can be a means to struggle against mistreatment and injustice more broadly. In Chapter 12, Kim Beecheno reveals women's resistance to domestic violence. In Latin America, it is often observed how religion has become politicised in the struggle against poverty. However, less recognised has been the use of religion against the widespread physical and sexual violence within the favelas. Thus, drawing upon religious precepts, perhaps backed by the authority of the Church, can be a source of empowerment and provide new ways of negotiating different forms of violence. With relative ease, religion can become a primary point of identification for the marginalised and excluded over other forms of social division, such as race, class, gender and sexuality. Perhaps a better way to say this is that religion provides a means of organising the harms of race, class, gender and sexuality into a coherent whole, a singularity that can then be mobilised by a politics of identity centred upon religion (see also Hackworth, 2012). This is profoundly place-based. And it is about connection across space, as processes of politicisation and identification stretch around the world.

There is a paradox, here. Religions, on the one hand, seem to be a repository of fixed beliefs, rituals and practices. Yet, on the other hand, they are continually shaped by the worlds in which they find themselves. Perhaps, because of this, religions remain relentlessly mobile. Indeed, such mobile forms as the mission and the pilgrimage proliferate continuously. While the pilgrimage can be undertaken by the devout and be a proper expression of one's faith through the trial of the journey itself, it can just as easily be undertaken as a way to discover faith and spirituality. Instead of the journey being the product of faith, the journey can be a means through which to discover it. Patricia 'Iolana (Chapter 16) shows how the Goddess Movement deploys Jungian ideas to create a spiritual journey. This journey enables a deep connection between the individual and the world, along a path known as individuation. At the end of this path is an archetype, *anima*, which in some ways lies at the intersection of the collective unconscious, sexuality, mind, history and the numinous. Here is the Goddess, both a universal principle and earthly. She combines opposites. This is an important principle, for it allows us to see both sides of the coin at once.

So, religion has been used to stabilise identity and to sustain fixed political positions. Browne and Nash (Chapter 3) explain how religion has been used as a means to resist the legalisation of same-sex marriage in the West, but especially Ireland. They show how Christianity has struggled with same sex relationships, both within the clergy and in broader society. Indeed, as the Roman Catholic and Protestant Churches are both international, these difficulties are compounded by the differing attitudes of their congregations across different social settings. As they show, dramatic splits have emerged between more progressive and more conservative elements within Christianity. Interestingly, these processes of identification can produce the splitting, or strategic use, of identities, so that people express themselves differently in different social and cultural settings. People can "shape shift" from one context to another, deploying religious and non-religious identities flexibly depending on a variety of factors: such as fitting into dominant cultural norms or seeking to avoid conflict and tension. Because of this, religious identity cannot be assumed to be a given (see Hopkins *et al.*, 2010). Instead, people are constantly negotiating religion. Religion, in this view, is profoundly schizophrenic: constantly working to stabilise identity, while at the same time working within contexts not of its own choosing. This schizophrenia has a geography.

5 Religion is (not) a territory

Religions have territorialised the world in various ways. Most obviously, they have created bounded and scaled territories, using physical structures to mark the centre of those territories. The church or mosque, the pilgrimage site, the pilgrimage, the mission and missionaries, and the like, all seek to implement, maintain and extend the territory of a religion. Further, religions have become entangled with the nation-state, whether this is in Henry VIII's England or through religious leadership in countries such as Iran or in other events such as the partition of India and Pakistan. Further, religions establish a distinction between sacred and non-sacred space. This introduces a paradox: religion produces a hierarchy of territories over which it has dominion, yet this tends to ensure that only specific spaces within that territory are sacred.

Alongside its territorial practices, religion also relies on geographies of connection-across-space to establish its influence, such as through the development of mega-churches (see Warf and Winsburg, 2010). An example of this is what is known in Christian traditions as evangelism, which not only seeks to extend the territory of Christianity, but also creates and utilizes forms of influence in order to do so. Justin Tse (Chapter 2) explores Christian evangelism in two University settings. He shows that evangelism does not necessarily produce a coherent position. Instead, it is constantly undergoing reappraisal in different settings. Consequently, the territorial and connective forms of evangelism need to be critically assessed in the settings that they emerge. Religious territories and connections can just as easily overlap with other religions as not. Indeed, religions can interact in unexpected ways, sometimes producing what are now often described as inter-faith communities (see also Stevenson *et al.*, 2010).

The complex interaction between territorialised and connective religious geographies has been revealed by Claire Dwyer, David Gilbert and Bindi Shah's work in London's suburbs (2012). They note the diversity of prominent religious buildings in West London: a Sikh Gurdwara, a Russian Orthodox Christian Cathedral, a Hindu temple, a Mosque and a Jain temple. Scattered amongst these are less obvious, and sometimes quite hidden, places of worship (see Heng, 2016). This diversity produces a map of overlapping faith territories. Meshed with the hierarchically organised spaces of the Church of England – the parish, the diocese – are the spaces produced by temples, synagogues, mosques, centres, foundations, cultural societies, missions and the like. It is not simply that this diversity of faith communities undermines the seeming homogeneity of suburban life; the overlapping territories of faith produce inter-faith interactions that themselves are generative of new religious practices (Mills, 2012). Indeed, arguably, it is this interaction between faiths that enables people to identify spiritually with more than one faith, whilst at the same time detaching themselves from the idea of organised religion as a singular source of identity: 'no religion', in this sense, would be the consequence of multi-/inter-faith interaction. 'No religion', in this framing, would become a thoroughly modern way to be religious: always producing new syncretic forms of faith, belief and spirituality.

In Chapter 15, David Wilson offers a personal insight into the syncretism of Spiritualist practices by drawing out its relationship to wider shamanic traditions. He frames the Spiritualist practice of spirit communication by comparing it with mediumship in Siberian and North American traditions. Significantly, Spiritualism has drawn heavily on Native American spirit guides since the 1920s. In the UK, Spiritualists are especially familiar with Silver Birch, who spoke through the medium Maurice Barbanell. Wilson makes a plea for examining the processes through which religious traditions emerge and sustain themselves. Significantly, this involves looking at how they draw upon, and internalise, ideas and practices from wider, related traditions. From this perspective, religions' syncretism can disguise the ways they contain a diversity of ideas, some of which may appear esoteric or indeed antagonistic.

Similarly, according to Dwyer, Gilbert and Shah, religion must be considered in its social contexts. Thus, the "super diversity" of faith communities in London is a product of migration patterns into and through the city (since always). So, one way to read the relationship between migration and the city is to witness its religious buildings and practices. However, migration does not simply happen over space, it also produces space. For Dwyer, Gilbert and Shah, migration into London's suburbs has produced what can be called, variously, 'semi-detached faith', 'edge-city faith' and 'ethnoburb faith', drawing on the experience of faith and suburban life across modern Western societies. In part, such terms suggest that faith communities produce new religious forms and practices in suburbs (Wilford, 2012): that is, that there is something distinctive about faith in the suburbs, as Dwyer herself argues in Chapter 7. Yet, there is also a hint that faith has modified the suburb in some way (e.g. Connell, 2005; Hackworth and Stein, 2012). Indeed, there is a hint that the circulation, overlapping and mixing of faiths is doing something to suburban life that has yet to be fully understood. Dwyer, Gilbert and Shah rise to the challenge of understanding the place of faith in the production of suburban life.

The diversity of faith communities is of course not unique to London. All cities, arguably, evidence this diversity. However, faith communities are not just territories, they are also networks of connections that trace themselves across the world in specific ways. People within faith communities can use the wider sense of a religion to locate themselves in the world. Thus, geographers have shown how religious identities can be used by people to 'locate' themselves in places other than those where they live. These stretched or detached identities can enable people to adopt a wide variety of stances in relation to the social setting in which they find themselves. It is easy to think that migration is about people moving from one place to another (whether voluntarily or not), but more moves than people. Indeed, the world is mobile, too. People can migrate without moving: the world of social media enables people to 'travel' without going anywhere. These worlds are not simply mediatised social relations, as they become a means through which people identify and express religiosity and spirituality.

The mobility of religion is still marked by practices such as pilgrimage and by roles such as the missionary or the volunteer (e.g. Maddrell and Scriven, 2016; or, Baillie Smith *et al.*, 2013; and Cloke *et al.*, 2007). Moreover, priests and ministers are not static, but often on the move between different posts within the organisation. Significantly, in Methodism and Spiritualism, churches do not have a standing minister. In Chapter 6, Lia Shimada explores the mobility of ministers in the British Methodist Church. She argues that mobility is woven into the fabric of Methodist Christianity, causing it to constantly engage with different geographical and social contexts. Mobility, in a Methodist context, is normal. This mobility, as Shimada reveals, can create opportunities for spiritual renewal, but also tensions and conflict. Importantly, this continual mobility installs a geography of connection (which becomes something specific in Methodist thought, as you will see), which continually shapes both ministers, their spaces and the spaces of spirituality.

The missionary, of course, has long been associated with colonialism and Empire (see Kong, 2010, page 760). It is clear that religions and Empires have parallel territorial strategies. There is first contact, then a deliberate effort to occupy and control people by the creation of territories. Previously existing territorial arrangements are overwritten as Empire and its religion organise space according to its expectations and requirements. The territorial impositions of Empire and Religion have produced long lasting, and sometimes seemingly permanent, ongoing, harm. The roots of many conflicts are spawn of this relationship between Empire and Religion, whether we think of Northern Ireland or Palestine, First Peoples or Latest Peoples. However, religion has also produced a means through which colonial processes have been recorded, contested and subverted. Thus, the territory does not simply stabilise a religion, ensuring its timeless expression of faith and devotion. Territorialisation also becomes a means of syncretism, mutation and adaptation. Indeed, paradoxically, the means through which religions seek to stabilise themselves, as they seek to create territories and extend their territorial reach, can necessitate adaptation and change.

Religions organise themselves in ways that produce geographical scales: the parish, the diocese, the nation, international. Yet, they are not limited to these scalar, territorialised logics of organisation. Religion can just as easily be organised through every modern form of communication, making avid use of television, the internet, social media and the like. Religion makes geography; religion makes connections; religion moves. Yet, in this process of producing spaces for itself, religion also changes. More than this, we can glimpse the production of spaces for spirituality beyond religion. And the creation of these spaces for spirituality is profoundly political, as it underpins the logics of community and care through which people act in the world (see also Williams, Cloke and Thomas, 2012).

6 Acting in the world

There has been much talk of the decline of religion (see Beaumont and Baker, 2011, for an overview; also, Cloke and Beaumont, 2013; and, Tse, 2014). The decline both in participation in formal religion such as regular attendance at divine services, especially in (but not only) Christianity and also in people identifying themselves as belonging to a particular religion in census counts has enabled many to argue that religion is dying (on its pews). This decline is termed secularisation, because it is assumed that these trends are evidence of people turning away from religion altogether. On the other hand, others have pointed out that the state has increasingly had to take account not only of religious views, but of the sheer diversity of religious views within modern Western nation-states. Thus, Britain has been described as a "post-Christian" society, most famously by former Archbishop of Canterbury Rowan Williams (in 2014). This term is not intended to suggest that Britain is no longer Christian, but that Britain contains within it many significant religions such that it can no longer claim to be *only* Christian. An argument has developed that suggests that Western societies are becoming less and less secular, as nations and states have to take more and more account of the increasingly diverse expressions of religion in their midst (see Bowman and Valk, 2012). Indeed, former British Prime Minster David Cameron responded to the suggestion that Britain is a post-Christian society by claiming that Christianity suffuses every part of British cultural life, providing it with its moral compass.

An alternative counter-argument to the idea that Western societies are relentlessly secularising, is that people are increasingly turning to forms of spirituality that provide an alternative to, or supplement in a variety of ways, formal religion. This argument has been most clearly put forward by Heelas and Woodhead (2005), based on fieldwork in Kendal (in the north of England). In Chapter 8, Karin Tusting and Linda Woodhead revisit the main findings of the original Kendal study. In a rigorous re-examination of the original study, they show how markedly Christianity has declined as a force that shapes spirituality and morality. Indeed, they argue, Christianity has almost become 'alternative' to mainstream cultural life. Yet, this does not mean spirituality has become marginal. Forms of religion and spirituality that seemed 'alternative' a decade ago have become more familiar, more a part of mainstream cultural life. The decline of formal religion, in this account, is not evidence of secularisation, but rather of the proliferation of spiritual forms through every part of society (see, for example, Saunders, 2012).

Religion and spirituality continue to provide a focal point for acting on the state, or of requiring that the state act in different ways (Jamoul and Wills, 2008; Williams, Cloke and Thomas, 2012). Religion and spirituality, in this respect, provide a moral order that can be used to act on, act alongside, act in opposition to, the normal codes of conduct of governments and nations. And this in turn causes governments and nations to act in response. Barely a day goes by when some form of dispute around religion hits the news. It is not just that a list is easily assembled. But that this list is ever-growing. You may wish to create your own list, but let us think of how governments have sought to regulate the wearing of religious symbols and dress in airports, classrooms, banks and public (in streets or on the beach). Or the ways in which particular religious groups get called to account for themselves after terrorist acts (which are by no means limited to Islamic and Christian so-called extremists). Each terrorist act forces – or enables – governments to restrict freedoms of various kinds for people as a whole, and for so-called target groups in particular.

Julian Holloway (Chapter 13) engages with the idea that there is an entanglement between states, geopolitics and religion. This suggests that geopolitics has a religious dimension, which can be more and less explicit (see, also, Dittmar and Sturm, 2010; and, Sturm, 2013). An instance, of this, is the ongoing 'problem' of allowing Turkey, as an Islamic state, to join the Christian European Union. While the difficulties of allowing Turkey to join the European Union are usually couched in terms of economic convergence and human rights, these difficulties often track back to a fundamental disjuncture between Christianity and Islam. This finds expression, most recently, in fears that allowing Turkey to join the EU would open Europe's borders to Islamic terrorists and an unstoppable flood of Muslim refugees from the Middle East. Holloway reminds us that conflicts between states can be fought on spiritual, and indeed occult, grounds, too. He reveals a strong connection between geopolitics and occult thinking in World War 2. (Hitler was famously interested in the occult, while Churchill was covertly sympathetic to Spiritualism, leading both to consider the possibility of otherworldly influences in fighting World War 2.) Further, Holloway shows the intersection between the British colonial imagination and occultism, especially through Theosophy. We might also be reminded of how Nancy Reagan used her astrologer, Joan Quigley, to protect Ronald Reagan while in the White House. Governments are not – governmentality is not – cauterised from religious, spiritual and occult thinking.

Government, here, is not simply the nation-sate, but can include transnational governmental organisations as well as forms of government at more local levels. This is neither coherent nor integrated. An example is the Sanctuary City movement. Sanctuary Cities are to be found in cities, local governments, in towns, in universities. This spatial disorganisation can bring it into opposition with state logics at different levels. Thus, the Sanctuary City movement in the US came into direct conflict with the migration policies espoused by President Trump. Importantly, the religious and spiritual idea of sanctuary becomes a means to resist not only the idea that political space is universal, but also the ways in which the state reserves the right to act universally. The sanctuary then is a limit on the state and its capacity to assume the cooperation of people, organisations and the like. This limit is moral, guiding people to think of their connection to other people (especially people in distress) differently, but also spatial, creating a demarcated space which is welcoming and generous. Thus, the Sanctuary City movement creates ruptures in the supposedly smooth spaces of the state, of the nation and of government.

The Sanctuary City movement deliberately draws on an idea of sanctuary that is located in many faiths, including (but not limited to) Christianity, Islam, Hinduism, Judaism and Buddhism. Although not an exclusively religious or spiritual engagement with the politics of refugees and asylum, the idea of sanctuary provides a ready idea through which to mobilise political acts, through protest, activism in elections, enacting sanctuary practices and the like. Religious and spiritual ideas are not always as overt in political acts as in the Sanctuary City movement, however. Instead, the mobilisation of religious and spiritual values – whether by conservative or radical elements – can sometimes require a little excavating. Arguably, however, such ideas are never very far from the surface, as homilies such as 'turn the other cheek' or 'an eye for an eye', for example, get drawn into modes of civility (see also Cloke, 2011; or, Sutherland, 2014).

While states, elections, censuses provide a means to register the stuff of life lived religiously and spiritually, in fact acting in the world is far more mundane, prosaic and ordinary. Once every 10 years, people get to declare themselves as this or that or nothing in their census return. This tells us little – very little – about what people do in the meantime. It is the ordinary observances of religious and spiritual thinking and being that are least easy to grasp, as they can be so fleeting and therefore seemingly irrelevant. How many times a day might someone say 'oh my God' or mutter 'Jesus Christ' under their breath, yet think of themselves as entirely without religion? Does it matter that people pray, believe in the power of prayer, yet might not believe in God? Angels are intriguing examples of how people can slide around religious, spiritual and non-religious thinking. Angels can be invoked as a messenger of God, as a messenger (without the message coming from God), as a figure watching over people, as a guide to future action. The angel, in the form of a divine messenger, is central to Christian beliefs. Yet, the angel's many forms enable it to escape an exclusively Christian understanding. The angel permeates everyday life: the white feather that says an angel is watching over you; the loved one that is now with the angels . . . or has become an angel; the pub/restaurant/bar down the road. These are the moments that do not appear on the census, yet are remarkably important in people's everyday lives.

7 Spaces of spirituality

A word about Geography. This book operates at different scales and in different kinds of spaces and places. It is led by its case study material, connecting the stuff of spirituality to specific times and spaces. This, we feel, is in keeping with the idea that it is unwise to make universalising statements about spirituality (or religion). This is not a global geography, designed to speak from or about everywhere at once. Instead, we have sought case studies that render spaces and places a little stranger, a little less familiar, than we first thought. And, for this reason, many of our case studies are in the West: a strange and curious place that passes itself off, too easily, as the familiar and the normal. Even so, we hope that this impulse to make the world a little stranger, a little less familiar, can be productively carried elsewhere. Not as a way of exoticising or romanticising the world, but as a way of seeing the extraordinary construction of the ordinary everyday (Figure 1.1).

We have divided the book into three sections: the first considers the spaces of spiritual practices; the second examines the production of spiritual spaces in everyday life; and, the third explores spiritual transformations in and of the world. Each section has its own introduction to help readers see themes emerging from the chapters.

This book is not an attempt to close down, or to organise, debates on the relationship between spirituality and space. It is rather to suggest that we are only at the beginning of this journey. We are at a moment when a broadened definition of religion and spirituality can reveal how much more important religion and spirituality are, both in determining the fate of larger social processes – from governmentality to geopolitics, from migration to understanding labour contracts – but

Figure 1.1 The Hanley Church Bar and Restaurant in 2015.

The Hanley Church Bar and Restaurant in Stoke-on-Trent was formerly a Spiritualist Church. Before that, it was a Methodist Church (built 1860). Every Tuesday night, the restaurant holds very popular tarot card evenings: partly to maintain a spiritual link to the building's past, but also so you can find out what the future holds in store for you, and enjoy a three-course meal. Reproduced with kind permission of Daniele Sambo.

also in shaping the ordinary, everyday lives. We tried to suggest, above, that there are many ways (and many more ways) to evoke these issues. Thus, in this book, the authors show us what can be achieved, with an expanded and generative notion of spirituality. So, we are delighted to end this book with the words of professor dusky purples (Chapter 18). Professor purples entreats us to ask her how to read the world differently, offering us a programme through which we might prepare and orient ourselves, offering us new ways to read and think. Perhaps, dear reader, you might wish to begin with this chapter?

Acknowledgements

This volume is a direct response to issues arising from our research project, "Spirited Stoke: Spiritualism in the Everyday Life of Stoke-on-Trent" (2015–2016), funded by the AHRC Grant AH/L015447/1. During this project, we approached Faye Leerink at Routledge, who was immediately enthusiastic and supportive of the idea; her enthusiasm and support has never flagged. At Routledge, we have been guided through the process by Priscilla Corbett and Ruth Anderson. We hope we have not been too trying of their patience! Finally, of course, we must thank the authors: Thank You!

This book is for Ellie, Pippa and Ben.

References

Aitchison, C., Hopkins, P. and Kwan, M. P., 2007, *Geographies of Muslim identities: diaspora, gender and belonging* (Aldershot: Ashgate).

Baillie Smith, M., Laurie, N., Hopkins, P. and Olson, E., 2013, 'International volunteering, faith and subjectivity: negotiating cosmopolitanism, citizenship and development' *Geoforum* 45, pp. 126–135.

Bartolini, N., Chris, R., MacKian, S. and Pile, S., 2017, 'The place of spirit: modernity and the geographies of spirituality' *Progress in Human Geography* 41(3), pp. 338–354.

Beaumont, J. and Baker, C., 2011, 'Introduction: the rise of the postsecular city' in J. Beaumont and C. Baker, eds, *Postsecular cities: space, theory and practice* (London: Continuum), pp. 1–11.

Bowman, M. and Valk, U., eds, 2012, *Vernacular religion in everyday life: expressions of belief* (London: Equinox Publishing).

Cloke, P., 2011, 'Emerging geographies of evil? Theo-ethics and postsecular possibilities' *Cultural Geographies* 18, pp. 475–493.

Cloke, P. and Beaumont, J., 2013, 'Geographies of postsecular rapprochement in the city' *Progress in Human Geography* 37(1), pp. 27–51.

Cloke, P. and Dewesbury, J-D., 2009, 'Spiritual landscapes: existence, performance and immanence' *Social and Cultural Geography* 10(6), pp. 695–711.

Cloke, P., May, J. and Johnsen, S., 2007, 'Ethical citizenship? Volunteers and the ethics of providing services for homeless people' *Geoforum* 38, pp. 1089–1101.

Connell, J., 2005, 'Hillsong: a megachurch in the Sydney suburbs' *Australian Geographer* 36(3), pp. 315–332.

Conradson, D., 2012, 'Somewhere between religion and spirituality? Places of retreat in contemporary Britain' in P. Hopkins, L. Kong and E. Olson, eds, *Religion and place: landscape, place and piety* (New York: Springer), pp. 185–202.

Dittmar, J. and Sturm, T., eds, 2010, *Mapping the end of times: American evangelical geopolitics and apocalyptic visions* (Farnham: Ashgate).

Dwyer, C., 2000, 'Negotiating diasporic identities: Young British South Asian Muslim women' *Women's Studies International Forum* 23(4), pp. 475–486.

Dwyer, C., Gilbert, D. and Shah, B., 2012, 'Faith and Suburbia: secularisation, modernity and the changing geographies of religion in London's suburbs' *Transactions of the Institute of British Geographers* 38, pp. 403–419.

Finlayson, C., 2012, 'Spaces of faith: incorporating emotion and spirituality in geographic studies' *Environment and Planning A* 44, pp. 1763–1778.

Finlayson, C. and Mesev, V., 2014, 'Emotional encounters in sacred spaces: the case of the Church of Jesus Christ of Latter-day Saints' *The Professional Geographer* 66, pp. 436–442.

Gökariksel, B., 2009, 'Beyond the officially sacred: religion, secularism, and the body in the production of subjectivity' *Social and Cultural Geography* 10, pp. 657–674.

Goodricke-Clarke, N., 2008, *The Western esoteric traditions: a historical introduction* (Oxford: Oxford University Press).

Hackworth, J., 2012, *Faith based: religious neoliberalism and the politics of welfare in the United States* (Athens and London: University of Georgia Press).

Hackworth, J. and Stein, K., 2012, 'The collision of faith and economic development in Toronto's inner suburban industrial districts' *Urban Affairs Review* 48(1), pp. 35–61

Heelas, P. and Woodhead, L. with Seel, B., Szerszynski, B. and Tusting, K., 2005, *The spiritual revolution: why religion is giving way to spirituality* (Oxford: Basil Blackwell).

Henderson, M., 1993, 'What is spiritual geography?' *Geographical Review* 83(4), pp. 469–471.

Heng, T., 2016, 'Making 'unofficial' sacred space: spirit mediums and house temples in Singapore' *Geographical Review* 106, pp. 251–234.

Holloway, J., 2003, 'Make believe: spiritual practice, embodiment and sacred space' *Environment and Planning A* 35, pp. 1961–1974.

Holloway, J., 2006, 'Enchanted spaces: the séance, affect, and geographies of religion' *Annals of the Association of American Geographers* 96(1), pp. 182–187.

Holloway, J., 2012, 'The space that faith makes: towards a (hopeful) ethos of engagement' in P. Hopkins, L. Kong and E. Olson, eds, *Religion and place: landscape, place and piety* (New York: Springer), pp. 203–218.

Holloway, J. and Valins, O., 2002, 'Editorial: placing religion and spirituality in geography' *Social and Cultural Geography* 3(1), pp. 5–9.

Hopkins, P. and Gale, R., 2009, *Muslims in Britain: race, place and identities*. Edinburgh: Edinburgh University Press.

Hopkins, P., Olson, E., Pain, R. and Vincett, G., 2010, 'Mapping intergenerationalities: the formation of youthful religiosities' *Transactions of the Institute of British Geographers* 36, pp. 314–327.

Jamoul, L. and Wills, J., 2008, 'Faith in politics' *Urban Studies* 45(10), pp. 2035–2056.

King, L., 2001, 'Mapping 'new' geographies of religion: politics and poetics in modernity' *Progress in Human Geography* 25(2), pp. 211–233.

Kong, L., 2010, 'Global shifts, theoretical shifts: changing geographies of religion' *Progress in Human Geography* 34(6), pp. 755–776.

Lea, J., 2008, 'Retreating to nature: rethinking 'therapeutic landscapes' *Area* 40(1), pp. 90–98

Lea, J., 2009, 'Liberation or limitation? Understanding Iyengar Yoga as a practice of the self' *Body and Society* 15(3), pp. 71–92.

MacKian, S., 2012, *Everyday spirituality: social and spatial worlds of enchantment* (Basingstoke: Palgrave Macmillan).

Maddrell, A. and Scriven, R., 2016, 'Celtic pilgrimage, past and present: from historical geography to contemporary embodied practices' *Social and Cultural Geography* 17(2), pp. 300–321.

McGuire, M., 2008, *Lived religion: faith and practice in everyday life* (Oxford: Oxford University Press).

Mills, S., 2012, 'Duty to God/my Dharma/Allah/Waheguru: diverse youthful religiosities and the politics and performance of informal worship' *Social and Cultural Geography* 13(5), pp. 481–499.

Olson, E., Hopkins, P. and Kong, L., 2012, 'Introduction – religion and place: landscape, politics, and piety' in P. Hopkins, L. Kong and E. Olson, eds, *Religion and place: landscape, place and piety* (New York: Springer), pp. 1–20.

Olson, E., Hopkins, P., Pain, R. and Vincett, G., 2013, 'Retheorizing the postsecular present: embodiment, spatial transcendence, and the challenge to authenticity among young Christians in Glasgow, Scotland' *Annals of the Association of American Geographers* 103(6), pp. 1421–1436.

Olson, E. and Silvey, R., 2006, 'Transnational geographies: rescaling development, migration, and religion' *Environment and Planning A* 38(5), pp. 805–808.

Rose, M., 2010, 'Pilgrims: an ethnography of sacredness' *Cultural Geographies* 17(4), pp. 507–524.

Sanderson, E. R., 2012, 'Emotional engagement in the context of development and spirituality research' *Emotion, Space and Society* 5, pp. 122–130.

Saunders, R., 2012, 'Pagan places: towards a religiogeography of neopaganism' *Progress in Human Geography* 37(6), pp. 786–810.

Sheringham, O., 2010, 'Creating "Alternative Geographies": religion, transnationalism and everyday life' *Geography Compass* 4(11), pp. 1678–1694.

Stevenson, D., Dunn, K., Possamai, A. and Piracha, A., 2010, 'Religious belief across "Post-Secular" Sydney: the multiple trends in (de)secularisation' *Australian Geographer* 41(3), pp. 323–350.

Sturm, T., 2013, 'The future of religious geopolitics: towards a research and theory agenda' *Area* 45(2), pp. 134–140.

Sutherland, C., 2014, 'Political discourse and praxis in the Glasgow Church' *Political Geography* 38(1), pp. 23–32.

Tse, J., 2014, 'Grounded theologies: 'religion' and the 'secular' in human geography' *Progress in Human Geography* 38(2), pp. 201–220.

Warf, B. and Winsburg, M., 2010, 'Geographies of megachurches in the United States' *Journal of Cultural Geography* 27(1), pp. 33–51.

Wilford, J., 2012, *Sacred subdivisions: the postsuburban transformation of American Evangelicalism* (New York: New York University Press).

Williams, A., 2010, 'Spiritual therapeutic landscapes and healing: a case study of St. Anne de Beaupre, Quebec, Canada' *Social Science and Medicine* 70, pp. 1633–1640.

Williams, A., 2016, 'Spiritual landscapes of pentecostal worship, belief and embodiment in a therapeutic community: new critical perspectives' *Emotion, Space and Society* 19, pp. 45–55.

Williams, A., Cloke, P. and Thomas, S., 2012, 'Co-constituting neoliberalism: faith-based organisations, co-option, and resistance in the UK' *Environment and Planning A* 44, pp. 1479–1501.

Woodhead, L., 2016, 'The rise of "No Religion" in Britain: the emergence of a new cultural majority' *Journal of the British Academy* 4, pp. 245–261.

Section 1

Spaces of spiritual practices

Nadia Bartolini

> 'It's not easy to be spiritual all of the time; you have to work at it'.

This quote is from a December 2014 diary entry of someone who is hesitant to define herself within a particular religious affiliation, but who believes in life after the death of the body. Vivian strives to find meaning and make sense of her life. This spiritual quest resonates with many people, and does not necessarily align with religious categories and specific doctrines. Yet, her practices can be understood as being 'spiritual' and mapped onto her quotidian life: she meditates with calming music, she attends a church where she feels welcome, she gets healing, she sends good thoughts to friends and family, she reads biographies and various history books to understand societal patterns, and she finds ways to give back to the community through volunteering and other caring activities.

Part of unfolding spirituality in contemporary life is to appreciate the complexities involved in defining what we mean by 'spiritual'. Vivian was brought up in a religious household, but a number of events and experiences in her life led her to question her religion. So today, she prefers to talk about her spirituality. But as the quote suggests, being spiritual is not an easy task. For her, it requires work.

Vivian's situation might strike a chord with many people. The baggage transmitted from our upbringings may, at some point in our lives, lead us to seriously question our innermost thoughts and sense of identity. What spirituality does, then, is capture an essence that moves beyond the confines of organized religion. This does not negate religion altogether; rather, it provides a means to contextualize how beliefs, and associated tensions, emerge.

In the first section of this book, all five chapters explore practices through a set of organized religious belief systems, such as Christianity and Buddhism. Yet, within these clearly defined categorizations lie uncertainties that mask the tensions inherent in trying to grasp something that is, as Justin Tse suggests, between the spiritual and the supernatural. To do this, Tse's chapter explores how these intangible, otherworldly forces can find 'feet on the ground' by using propositional truths. As such, by using a propositional approach, tensions within Evangelical intellectuals in America find commonalities through the spiritual and supernatural plane. As the Evangelicals navigate these 'earthly' controversies, and establish political and social relevancy, Tse shows how the mapping of their supernatural worlds transcend institutional frameworks.

In some cases, tensions between representatives of a religious group are inherent in the face of modern life. Kath Browne and Catherine Nash's chapter investigates the 2015 Irish referendum on same-sex marriage when over 62% voted in favour of amending the definition of marriage to include same-sex couples. This particular case touches on sexuality and marriage, two issues that juxtapose the modern and the traditional. If sexuality is being addressed more openly today – generally, as well as in Catholicism – it nonetheless has a tendency of pushing moral and spiritual boundaries. The chapter points to the Catholic Church's nebulous place in contemporary Ireland, which resonates with other traditionally Catholic territories debating value systems (such as Quebec). Along the same lines as Tse, the authors demonstrate through the use of digital archive data collection the difference in opinions from representatives. Importantly, their study shows how the tension is not directed at LGBT rights, but rather toward issues relating to what could be described as foundational to Irish Catholicism: marriage, family and Irish identity.

This begs the question, once again, of how tradition – and the heritage of particular religious groups in a given location – can 'fit' with modern life. If Browne and Nash capture the ways that the Irish state and the Catholic Church are entwined, Tariq Jazeel's chapter is subtler in deploying how the religious and the secular are understood in the built environment.

Drawing on Raymond Williams' structures of feeling, Jazeel considers how the architect Geoffrey Bawa's parliament complex in Sri Lanka blurs the boundaries between spaces of secularity and of religiosity. However, this does not preclude the structures from being devoid of elements that push and pull these terms closer and wider apart. Instead, Jazeel argues that "sacred modernity in the Sri Lankan context should be understood on its own terms" (p. 59), and by doing so, he moves away from the post-secular binaries that make the secular and the religious fixed and separate categories where the religious can only be seen as residing outside of the secular.

These chapters query the idea of the modern as is it understood from the Enlightenment, and in effect, take seriously the proposition that Charles Baudelaire stated in 1863 when considering the work of painter Constantin Guys:

> 'Modernity is the transitory, the fugitive, the contingent, the half of art of which the other half is the eternal and the immutable'.

Similar to Baudelaire, the authors in this first section consider the transitory with the immutable, the elusive with the permanent as aspects of modernity that go hand in hand. Whilst the appearance may be conflicting, these ideas are also accepting of how notions of tradition – and the heritages, identities and nationalisms that can ensue – can be bound up in wanting to appear 'progressive' and address inequalities. In so doing, the authors also merge spiritual matters with practices that are embedded in politics. Rather than attempt to sever the territories of politics and religion, the chapters work to unravel how spiritual matters resurface amidst debates in ordinary life, from social media and academic appointments to architectural design and state affairs. In this sense, spirituality does not reside outside the realm of everyday life, even when there are attempts to develop and keep spiritual life separate from the noise of contemporary society.

Moving, embodying, circulating

Richard Scriven's chapter deals precisely with the desire to retreat from the chaos of modern life to seek contemplation as pilgrims travel to Lough Derg in Ireland. As Scriven points out, it is a place where pilgrims "withdraw from the rest of the world" (p. 69). Here, the pilgrimage evokes a journey, a traveling and a spiritual journey. Both journeys are important to consider as they consist in engaging the body and soul. By traveling away from modern life to embrace the rituals of this lonely yet 'very spiritual place' (as described by a TripAdvisor reviewer in August 2016), pilgrims willingly accept ridding themselves of the comforts of their daily lives, such as their mobile phones and their shoes. This stripping away of material possessions is significant as it is emblematic of releasing the soul so that the spirit does not get distracted by trivialities. Hence, it is this willingness to achieve a more spiritual plane – one that casts a hope of inner transformation – that enables pilgrims to endure bodily pain, as reported by another TripAdvisor reviewer in July 2015:

> 'This was my 10th pilgrimage to this very holy and special place, it is always a very challenging experience, with lack of sleep, being barefoot and fasting for three days, but it is worth every minute of the hardships experienced'.

Whilst the concept of pilgrimage might allude to going back in time, or escaping responsibilities, there is a challenge sought, a certain meaning in the doing where the practising of 'hardships' is something desired time and again. Perhaps it is akin to what Jane Bennett attests to when considering enchantment:

> 'To be enchanted is, in the moment of its activation, to assent wholeheartedly to life – not to this or that particular condition or aspect of it but to the experience of living itself'.
>
> (2001: 159–160)

Capturing this moment of awe does not necessarily correspond with the imagery of bliss and contentment. It can also be associated with bodily pain. This is not unlike the tensions present in modernity. These terms can indeed encompass a range of dichotomies that seem disparate, yet in practice, make sense.

Along the same vein, vocational callings could be seen as the spiritual complement to the chaos of contemporary lifestyles. Callings are considered unique, special, supernatural. So, when Methodists in the UK are sent off to serve a particular community, one could imagine the pleasure and dedication of the ministers. Yet, Lia Shimada's chapter specifically attests to how spiritual endeavours are combined with the mundane, practical geographies of moving every 5–7 years. The 'connexional' structure of governance of the Methodist ministers shapes the itinerant nature of their spiritual work in ways that one might not have previously considered. The physical movement and the spiritual devotion are entwined with the psychological adjustments of 'fitting in' and enabling oneself (and sometimes their families) to cope with the disruption of a transient lifestyle. One respondent in Shimada's chapter explains this state of flux through his faith by referring to the pilgrimage, while another sees it as 'an expression of social justice' (p. 91). Here, both supernatural and earthly worlds collide, where meaning is constructed through embodied trials and spiritual compassion.

What is important to remember is that all the chapters in this first section produced work from a spatial perspective. As spiritual practices are exposed through everyday, contemporary life and from the point of view of their geographies, we gain a better understanding of the specificities of place, as well as how tensions and new relationships emerge.

References

Baudelaire, C. 1863. Le peintre de la vie moderne, IV La modernité. *Le Figaro*. URL: www.uni-due.de/lyriktheorie/texte/1863_baudelaire.html, accessed 14/04/2017.

Bennett, J. 2001. *The Enchantment of Modern Life: Attachments, Crossings, and Ethics*. Princeton University Press: Princeton, NJ.

2 Spiritual propositions

The American evangelical intelligentsia and the supernatural order

Justin K. H. Tse

Introduction: mapping the spiritual geographies of an evangelical intelligentsia

The spiritual geographies of the theological movement known as 'evangelical Christianity' are seldom taken seriously, especially among its intellectual elites and their critics. Typically conceived as a 'conservative' version of Protestant Christianity – the strand of Christian faith that historically broke with the Roman Catholic Church around the dawn of modernity – 'evangelical' Protestants tend to emphasize the literal interpretation of the Bible because its pages reveal the good news – the *gospel*, the *evangel* (the root of the word 'evangelical') – of salvation from an afterlife of damnation and a present experience of divine alienation through faith in the death and resurrection of Jesus Christ, considered to be divinity in human form. Although such an understanding of the Christian Gospel emphasizes individual faith with few implications for institutional membership, what might be called 'spiritual geographies' are often taken in contrast to the 'sacred archipelagos' of evangelicalism's seemingly organized structures, deep-pocketed networks, and scripted realities floating in a sea of secularity (Wilford 2012; Bartolini *et al.* 2017). However, I hope to demonstrate in this chapter that there are evangelical ways of unfolding spiritual geographies that transcend their institutional structures that have not yet been fully explored. In other words, there is a mismatch between the perceived institutional edifices of evangelical Protestantism and the individual spiritualities fostered by its doctrine, and my aim is to explore the spiritual geographies fostered by this disconnect.

In this account of evangelical spiritual geographies, I want to examine the rhetoric of what I call the *evangelical intelligentsia*, the journalists, academics, clergy, and other public intellectuals who either speak as self-identifying evangelicals or as empathetic fellow-travellers offering what Wilford (2012: 7) calls a 'brother's account' of the movement (Worthen 2014; Sutton 2014; Strachan 2015). What is intriguing is that both this intelligentsia and their non-evangelical intellectual critics tend themselves to be shy about discussing their spiritual geographies. Critics of evangelicalism often take the movement to be ideological cover for more cynical materialistic endeavours, a marriage of Christ and 'cowboy capitalism' (Connolly 2008), with obsessions about amassing megachurch territory (Connell

2005), imposing their moral values and neoliberal economic ideologies on secular civil society (Hackworth 2013; Han 2011), and bringing about the end of the world (Dittmer and Sturm 2010) – precisely the opposite of a liberating spiritual geography. The response of some in the intelligentsia has been intriguing: their claim is that evangelicalism is barely coherent at present, as it is constantly undergoing re-definition (Bebbington 1988; Miller 1999; Larsen 2007; Hunter 2010).

I want to perform a counter-reading of these intellectuals, then, to show that evangelicalism as a term primarily describes an orientation toward the supernatural, a map of spiritual geographies. I perform a close reading of published online material about two case studies in evangelicalism where seeming ideological disagreements among the evangelical intelligentsia end up displaying how much they actually agree about how to map spiritual geographies: InterVarsity Christian Fellowship's (IVCF) (2016) controversial theological position paper about human sexuality and the 2015–6 dismissal proceedings against tenured political scientist Larycia Hawkins at Wheaton College for quoting Pope Francis's statement that Christians and Muslims 'worship the same God' (Gleim 2016). My central argument is that evangelical Protestants tend to orient themselves toward the supernatural by seeking to understand God, spirits, and spiritual reality by way of propositional sentences that create ideological worlds. What I therefore take seriously is that evangelicals believe in a different plane of existence from the natural one, which can be described as *spiritual* and *supernatural*, but the question is – *how do they articulate these spiritual and supernatural worlds in ideological ways?*

Articulated this way, the spiritual geographies of evangelicals defy some of the common expectations about what holds evangelicals together as *evangelicals*. By examining the controversy at IVCF over its theological position paper on human sexuality, I hope to show that the ideological world created by evangelical propositions about the supernatural may not be based on the inerrancy of the Bible, but instead on a biblical narration of a spiritual world that circumscribes everyday action. I will then use Wheaton College's attempt to fire Larycia Hawkins to highlight the difference between institutional spiritual geographies and intellectual convictions about the supernatural that defy the boundaries of an evangelical institution. Through these episodes, I hope that this paper contributes to the study of spaces of spirituality by showing that evangelical Christianities are not necessarily always institutional impositions of religion but can also lead to ways of being that transcend institutions as well.

Gospel sexuality: moving beyond inerrancy with InterVarsity Christian Fellowship's theological position paper on sexuality

On 6 October 2016, *TIME* Magazine reported that IVCF was asking employees who did not agree with its new theological statement on sexuality to self-disclose their positions so that IVCF could begin with them the process of 'involuntary termination' by 11 November. Specifically at issue, *TIME* argued, was any position that differed with IVCF's proposition 'that any sexual activity outside of a husband and wife is immoral' (Dias 2016). Interviewing former IVCF staff worker Bianca Louie, the *TIME* reporter found that she and ten others had formed a queer collective within IVCF and that the word on the street was that IVCF was purging its staff members of those who privately held that same-sex marriage did not contradict biblical teaching.

What made this debate distinctively evangelical was that both sides relied on the veracity of spiritual propositions to make their case. What makes it interesting is that the conversation seems to have progressed beyond an obsession about the inability of biblical truth to be in error. Historically, struggles over inerrancy have animated evangelical hostilities over gender and sexuality from the 1970s to the 1990s. Over this period, some of the founders of the Evangelical Women's Caucus (EWC) disclosed that they were in fact lesbians, leading to widespread distrust among others in the intelligentsia regarding what was being called *evangelical feminism*, a reading of the Bible that emphasized the equality of women and men in creation (Cochran 2005; Ingersoll 2005). What was striking about that era of backlash, the anthropologist Andrea Smith (2008) has noted, is that it was not really about sex, but about biblical inerrancy, the question of whether every proposition in the Bible is scientifically true.

On the surface, it might seem that IVCF is shutting down the latest iteration of this intra-evangelical debate with an inerrancy argument. Indeed, there are still struggles over inerrancy within evangelicalism; for example, several high-profile faculty were recently fired from Westminster Theological Seminary because their writings on Scripture did not neatly conform to inerrancy standards especially around the Genesis creation stories (Pulliam [Bailey] 2008; Withrow 2014). Moreover, reports of the IVCF sexuality policy rollout revealed that such a seeming emphasis on inerrancy engendered some very scandalous practices. *Religion Dispatches* published a piece shortly after the *TIME* article detailing some of the misuse of such theologies of sexuality by some IVCF staff. Particularly jarring was the opening hook, the story of Michael Vasquez, an IVCF staff worker at the University of Utah who had reported in 2013 to his supervisor that he was gay. Vasquez told the press that what happened next was as traumatizing as it was bizarre: his supervisor at first met with him to pray for his homosexual orientation to be taken away, and when nothing else seemed to work, instructed him to watch straight pornography. With what was now being called a 'purge' at IVCF, *Religion Dispatches* reporter Deborah Jian Lee (an ex-evangelical herself) suggested that such self-reporting would not only result in emotional trauma, but also a process of termination that would be less than gracious (DJ Lee 2016).

However, the framework of inerrancy is arguably insufficient for capturing the spiritual geographies in the offending document in question. Titled 'A Theological Summary of Human Sexuality,' the authors of the paper pitch their position on sexuality by appealing to how evangelicals understand sexuality as part of the larger spiritual reality of God's grace: 'As men and women created in the image of God, relationships with family, friends, and spouses bring us the deepest joy of human experience. God's common grace is given to all people (Matthew 5:45) and evident in every sector of life. He designed the sexual relationship between a husband and wife to be enjoyed as a meaningful experience.' This 'theological foundation – grounded in the character of God' becomes the matrix whereby 'human sexuality' – 'that particular aspect of God's creation gift where, in marriage, we engage in physical sexual intimacy that is personal, self-giving, and spiritual in nature' – can be understood in the 'theological categories' of '(1) creation, (2) fall, (3) redemption, and (4) restoration' (InterVarsity Christian Fellowship/USA 2016: 1).

The document's authors then explicate each point by drawing from the Bible a picture of a larger spiritual reality of divine grace and redemption, in effect mapping IVCF's perception of spiritual geographies. The creation of God, they argue, is premised on a self-giving love that became translated through the words of Jesus in the New Testament through the Greek word *agape*, an orientation of self-sacrifice and other-directed love required of all Christians whether they are single or married. Deviations from this spiritual geography of *agape* love and grace-filled creation are described as 'The Fall: Not the Way It Is Supposed to Be':

> We live in a world where the common experience of sexuality is broken and distorted to some extent, sometimes to the extremes of manipulation, abuse, and violence. There is a striking difference between 'knowing' one's spouse and using, abusing, or neglecting one's spouse. We have a sense that it is not intended to be this way. How did we drift so far from the Creator's grand design for human relationships? How did we move from self-sacrifice to self-gratification? How did we move from meaningful sexual intimacy to casual sex?
>
> (InterVarsity Christian Fellowship/USA 2016: 6)

The authors then list a variety of ways that the original creation of sexuality has become spiritually broken, including in sexual abuse, divorce, premarital sex, lust, adultery, and pornography. However, the longest section is on same-sex relationships, which is broken down into 'attraction,' 'identity,' and 'behaviour' (InterVarsity Christian Fellowship 2016: 12–14). Regretting that 'many Christians have not loved same-sex-attracted people as we ought' – a failure in its own right to live out the matrix of grace prescribed by God's original creation – the document insists that 'God's intention for sexual expression is to be between a husband and wife in marriage,' which means that 'every other sexual practice is outside of God's plan and therefore is a distortion of God's loving design for humanity' (InterVarsity Christian Fellowship 2016: 12). IVCF posits a tension in spiritual prescriptions of God's creation: to follow the matrix of love with regard to those who are attracted to persons of the same sex and identify as such while insisting that that same grid of grace means that practicing same-sex sexual acts deviate from the original created order. The narrative of redemption, then, posits that in such a world that has deviated from the order of creation, God sent his Son to die and rise again so that all creation might be restored to the original matrix of self-giving *agape* and other-directed sacrifice.

At face value, these spiritual geographies may seem standard for evangelicals: they are derived from biblical exegesis, they adopt a propositional approach to spiritual geographies, they propose a moral order that protests against the fallenness of creation (2016: 13). But IVCF does not rely on the blunt instrument of biblical inerrancy. Resonating with the work of evangelical theologian David Fitch (2011), IVCF describes an emphasis on the narrative arc of the Gospel that can be read through Scripture but does not have to be beholden to its every jot and tittle, one that moves from creation to fall to redemption to restoration. What is being proposed is a mapping of evangelical spiritual realities within the big supernatural picture of the Gospel in Scripture, not the scientific veracity of every jot and tittle of Scripture. No wonder, then, that IVCF vice president Greg Jao says that the policy is 'about the authority of Scripture, which leads us to read Scripture in a certain way' through this matrix of redemption but never uses the word *inerrancy*. The propositions form a story, allowing Jao to point out the lived tensions in this spiritual geography:

> I remember talking with a student who says that it was at InterVarsity that I was first loved and cared for deeply enough that I could admit to myself after years of denial that I had same-sex attraction, and it was at InterVarsity that I encountered Jesus in the Scriptures and gave my life to him, and it's in InterVarsity that I feel that I can lay my sexual identity before Jesus and let him guide me, which in her case, she says, 'I'm choosing chastity because that's what Jesus calls me to, and I'm doing it with joy.'
>
> (CBN News 2016)

Here, Jao maps InterVarsity's orientation toward the spiritual order. Like Fitch, the central proposition is that the order of creation is founded on the love that opened this student up toward self-discovery, but because those same sets of spiritual propositions proscribe the sexual behaviour that would be part of her identity, she sacrifices her sexual orientation to maintain this evangelical *spiritual* orientation.

Discounting as it may be of nonheteronormative sexual practices, the IVCF position paper on sexuality illumines how IVCF is trying to insist on its reading of a spiritual geography, one created by grace and premised on *agape* self-giving love. Yet it is that same matrix, with the same propositions, that its opponents within the same intellectual circle contest this mapping. As Bianca Louie told the *TIME* reporter who broke the story:

> I think one of the hardest parts has been feeling really dismissed by InterVarsity. . . . The queer collective went through a very biblical, very spiritual process, with the Holy Spirit, to get to where we are. I think a lot of people think those who are affirming [same-sex marriage] reject the Bible, but we have landed where we have because of Scripture, which is what InterVarsity taught us to do.
>
> (Dias 2016)

Read via Fitch (2011), Louie's comments are in fact far more conservative than IVCF. IVCF is attempting to elevate the conversation beyond inerrancy and toward an explicit discussion of the spiritual geographies proposed by the redemptive arc of Scripture. Not only does Louie appeal to a process of spiritual discernment – one that is presumably premised on the same sort of self-giving love that enables the formation of a community like IVCF – but she emphasizes that it is the very words of Scripture, not only its narrative matrix, that have informed her understanding of how sexuality should be mapped as part of an evangelical spiritual geography. Louie, in other words, is claiming the evangelical high ground on both narrative and inerrancy. So too, Vasquez – the University of Utah staff worker traumatized by attempts to pray the gay away and use pornography while at it – continues to share a similar theological understanding of space:

> Just doing life with college students is enough to bring change to campus. . . . Just to acknowledge the inherent dignity in LGBT students on campus will transform their lives and their experience of the Kingdom. So, whether it's through a formal ministry or not, my desire is simply to see students encounter Jesus, no different as it was when I was on InterVarsity staff.
>
> (DJ Lee 2016)

The end goal here is that this spiritual geography is circumscribed by a personal relationship with Jesus Christ, whose self-giving love is worth the encounter of students regardless of Vasquez's institutional affiliation. It is, after all, not IVCF but Jesus himself who goes beyond creating the world to redeeming it from its fallenness, which means that he personally encounters students whether inside or outside of IVCF to make their world from 'what it is not supposed to be' to what it is supposed to be. At heart, then, the debate over IVCF's articulation of the supernatural world presents a map of spiritual geographies that can be exported even outside of the institution. What remains consistent between the institution and its dissenters, though, is an insistence on propositions in this narrative account of the Gospel.

Dismissing Larycia Hawkins: race and the spiritual limits of interreligious solidarity

A similar analysis can be applied to the ideological fissures among the evangelical intelligentsia on the question of race. In 2015, the dismissal proceedings at Wheaton College in the Chicago area against tenured political scientist Larycia Hawkins divided evangelicals. News of this case rippled across the evangelical intellectual world because Wheaton is an evangelical liberal arts college that became such a bastion for the neo-evangelical movement in the 1940s that some even dared to call it the 'evangelical Harvard.' Indeed, the spiritual geographies from which the dismissal originated are seldom noted, as what is remarked upon more often is the fact that Hawkins was the only African American woman to be teaching at Wheaton – and as further point of fact, there were only five black faculty in total, including her. Her firing evoked the sociological analysis delivered by Emerson and Smith (2001) that American evangelicalism remained divided by race in an unconscious way – not because white evangelicals tended to be consciously racist, but because their propensity toward individual spiritual practice often did not account for the structural reasons behind poverty and marginalization. Of course, as Deborah Jian Lee (2015) has recently reported, there are also more nefarious accounts of open racism behind closed doors: Lee highlights how the historian Randall Balmer (2006) discovered in a closed-door meeting with evangelical political operatives of the Religious Right that their movement for moral values touted the story of abortion as the reason for which they mobilized, but in fact they got together to oppose the desegregation of schools. As such studies and stories have widely circulated among the evangelical intelligentsia, Hawkins's dismissal raised questions about race and gender within evangelicalism itself, especially as earlier that same year, IVCF had endorsed the Black Lives Matter movement at their annual conference in Urbana, Illinois (InterVarsity Christian Fellowship 2015).

Hawkins's dismissal was triggered by a Facebook status update on 10 December 2015 in which she advocated her 'embodied solidarity' with Muslims in the face of American Islamophobia, as 'theoretical solidarity is not solidarity at all.' Hawkins outlined her plan for her 2015 'Advent Worship' as she prepared for the coming of Jesus Christ during the Christmas season: 'I will wear the hijab to work at Wheaton College, to play in Chi-town, in the airport and on the airplane to my home state that initiated one of the first anti-Sharia laws (read: unconstitutional and Islamophobic), and at church.' For Hawkins's readers, this was an invitation to 'all women into the narrative that is embodied, hijab-wearing solidarity with our Muslim sisters – for whatever reason,' and she lists off the following as invitees to her embodied solidarity: Muslims who do not 'wear the veil normally,' atheists and agnostics who find 'religion silly or inexplicable,' Catholic and Protestant Christians 'like me,' and those who already cover their head in worship 'but not a hijab.' All of this was prefaced by a theological justification: 'I stand in religious solidarity with Muslims because they, like me, a Christian are people of the book. And as Pope Francis stated last week, we worship the same God' (Gleim 2016).

Hawkins was immediately placed on administrative leave, which eventually snowballed into dismissal proceedings even though she had tenure. Although Hawkins taught political science (not theology) and even though she had tenure (which should have made firing her difficult), all Wheaton faculty are required to sign a statement of faith with propositions from which they cannot deviate, highlighting again an evangelical propositionalist approach to spiritual geographies. The theological debate between Hawkins and Wheaton thus became a sparring match over that statement's propositions. Following a letter sent to her on 15 December questioning her commitment to exclusively worshipping the Christian God in an evangelical Protestant way, Hawkins responded with a rigorous point-by-point treatise defending her faithful adherence to Wheaton's statement of faith. A close reading of Hawkins's reply (as I shall show) reveals that the school not only questioned her statement that Christians and Muslims worship the same God, but also whether she was sufficiently Protestant as opposed to being Catholic because she had invoked Pope Francis and called for Protestants and Catholics to share in the 'embodied solidarity' of wearing the hijab.

In her response, Hawkins does three things. First, she reviews the evangelical scholarly literature on whether Christians and Muslims worship the same God (which means that this has been a matter of scholarly contention for some time), finds both yes and no answers because Muslims deny the Christian conceptions of the three-personedness of God and the deity of Christ but affirm with Jews and Christians that there is only one God, and concludes that 'my statement is not a statement on soteriology or Trinitarian theology, but one of embodied piety. When I say that "we worship the same God," I am saying what Stackhouse [one scholar in the evangelical intelligentsia's discussion of Muslim-Christian relations] points out, namely that "when pious Muslims pray, they are addressing the One True God, and that God is, simply, God."' Second, she addresses the contention around calling Muslims her 'brothers and sisters' by affirming the common creation story among Jews, Christians, and Muslims that all are descended from the same common humanity bearing God's image, so this statement is 'in full agreement with the Wheaton College statement of faith, identifying each person as an image-bearer of God.' Third, she argues that Wheaton's objections to her Catholic sympathies because of Protestant-Catholic disputes around the Eucharist and the Immaculate Conception of the Virgin Mary are misplaced because there are disagreements about these theological conceptions among Protestants themselves (Nazworth 2016).

Still, that Wheaton College filed dismissal proceedings against Hawkins in January 2016 suggested that her rigorous propositional reply explicating her understanding of spiritual geographies did not sufficiently overlap with the school's institutional understanding of the supernatural map. As a *New York Times Magazine* report describes it, tensions had always been there between Hawkins and the Wheaton administration, especially with Wheaton's provost Stanton Jones; one memorable account in the article has Jones accusing Hawkins of endorsing Marxism in her laudatory account of black liberation theology in a faith-and-work integration paper required for all tenure-track faculty (Graham 2016: 53). Dissatisfied with Hawkins's theological rebuttal on the Muslim question, Jones recommended her dismissal in early January 2016, triggering campus-wide protests that put Hawkins's face on the front cover of the *Wheaton Record* for its 14 January 2016 issue and resulted in the faculty pushing back *en masse* on the administration for dismissing their colleague. In the face of a disciplinary hearing in February, Hawkins held her own press conference at the United Methodist Church's Chicago Temple flanked by an interreligious group of faith leaders in Chicago. With Wheaton sufficiently embarrassed, the school suddenly pulled back from firing Hawkins, just as Hawkins decided to leave the school; the result was a 'reconciliation' farewell event described as emotional for all parties involved, during which Wheaton administrators, faculty, students, and staff said good-bye and conveyed their regrets to Hawkins. Hawkins currently works as the Abd el-Kader Visiting Faculty Fellow at the University of Virginia's Institute for Advanced Studies in Culture.

While it may be tempting to regard this case solely as one of institutional racism and sexism – and given Hawkins's prior clashes with the administration, it cannot be denied that race was a factor in her struggle – what is perhaps more fascinating here is how much the spiritual geographies outlined by both Hawkins and Wheaton are approached purely propositionally. At issue is a set of propositions about the spiritual world, Wheaton's statement of faith. In twelve sentences, the statement's propositions provide 'a summary of biblical doctrine that is consonant with evangelical Christianity,' reaffirming the 'salient features of the historic Christian creeds, thereby identifying the College not only with the Scriptures but also with the reformers and the evangelical movement of recent years' and defining 'the biblical perspective which informs a Wheaton education' by casting 'light on the study of nature and man, as well as on man's culture' (Wheaton College 1992). For all of the contention around Hawkins's supposed interreligious inclusivity and Catholic sympathies, her approach to 'Advent Worship' in her Facebook post is based on the proposition that Muslims are part of a common humanity with whom everyone should stand in 'embodied solidarity' during their persecution. The citation of Pope Francis is not an entry into sacramental communion with him, but an affirmation of his proposition that Christians and Muslims worship the same God. In so doing, Hawkins's propositions describe a spiritual order in which what is shared is a common humanity that transcends religious identity politics and institutional conformity. That Wheaton saw this as threatening indicates that their propositions present a supernatural world in which religious identities matter in order to safeguard their institution from incoherence. The fissure between Hawkins and Wheaton, then, is over what evangelical spiritual geographies are for: a spiritual identity politics (as for Wheaton) or a supernatural pathway to a common humanity regardless of institutional affiliation (as for Hawkins)?

In other words, the Larycia Hawkins case highlights difference between these two spiritual worlds dividing the evangelical intelligentsia, much as they agree that supernatural truths should be approached propositionally. As I noted above, the conflicts between Hawkins and Jones do not originate with this assertion of 'embodied solidarity' with Muslims, but from Hawkins's usage of black liberation theology in her required faith-and-work integration paper. It was not just Hawkins's identity as the only black woman on Wheaton's faculty that made trouble for her; it turns out that her understanding of the spiritual world was constituted by a sensibility that racialization is a form of oppression that needs to be named for the sake of supernatural restoration. For Jones, the naming of such oppressive racial and class projects evoked a kind of Marxism from which he had to defend his spiritual world. In so doing, it is in fact Jones's spiritual geographies, not Hawkins's, that are constituted by identity politics, a need in this case to be clear about what evangelicals as an institutional group believe about the nature of God and his relationship to the world. Framed this way, it becomes clearer that neither Emerson and Smith's (2001) sociological analysis about evangelical colourblindness as a symptom of evangelicals' individualistic spirituality nor Balmer's (2006) revelation that the Religious Right was a coalition to preserve racial segregation applies here. Instead, this is a new fight among the evangelical intelligentsia over the constitution of the supernatural order, while using the same propositional language, with regard to whether Hawkins's evangelical spiritual map, propositionally evangelical as it is, should be bounded to an evangelical institution like Wheaton's, for a major part of an evangelical *institution*'s mapping of the supernatural order is to preserve its own institutional identity, whereas Hawkins is advancing a spiritual geography that transcends institutional boundaries.

Conclusion

In this chapter, I hope to have shown that despite the evangelical intelligentsia's attempt to nuance the increasingly bizarre political images associated with evangelicals in the popular imaginary, what is more interesting about evangelical Protestantism in the United States is the intelligentsia's own spiritual geographies. Evangelicalism is premised on an *orientation* to spiritual geographies known as *propositionalism*. My contention is that the fractures among an evangelical intellectual class can also be chalked up to approaches to spiritual geographies that include politics. An analysis of these recent debates shows that the contentions within this intelligentsia have been intellectually productive, as many have moved beyond characterizing their stances with the old fundamentalist and evangelical ideological tropes of biblical inerrancy, suspicion toward Catholic teachings, and individualizing practice. Instead, new propositions are being used, and therefore new debates are being had about the worlds being built through evangelical convictions. While some of these debates are being had at the institutional level, the stakes over which these contentions are being had is over geographies of the spiritual that transcend institutions through individual practices of faith.

Such debates add to a broader understanding of spiritual geographies because evangelicalism is often taken to fall under the category of geographies of *religion*, an institutionalized form of theology that may or may not point to supernatural realities (Wilford 2012; Bartolini *et al.* 2017). In this chapter, I hope to have demonstrated that evangelicalism is better characterized not as a set of institutions, but as a *network of intellectuals* who sometimes reinforce and sometimes undermine their own institutions, and at the heart of their debate is the constitution of a supernatural order that is not easily institutionally boxed in. As a popular evangelical catchphrase goes, 'I don't have a religion; I have a relationship!' The intelligentsia might cringe at such folksiness, but as I have shown, their debates are articulating what that relationship is by positing that one's personal relationship with a spiritual world is premised on the veracity of propositions, not necessarily by institutional affiliation. It is this discursive distinction that marks this intellectual circle as evangelical, showing that what appears to be their fragmentation may in fact be their greatest marker of spiritual coherence.

References

Bailey, S.P. [Pulliam]. 2008, 1 Apr. Westminster Theological Suspension: Peter Enns's book *Inspiration and Incarnation* created a two-year theological battle that resulted in his suspension. *Christianity Today*. Accessed 14 Feb. 2017, from www.christianitytoday.com/ct/2008/aprilweb-only/114-24.0.html.

Balmer, R. 2006. *Thy Kingdom Come: An Evangelical's Lament; How the Religious Right Distorts the Faith and Threatens America*. New York: Basic.

Bartolini, N., Chris, R., MacKian, S., and Pile, S. 2017. The place of spirit: modernity and the geographies of spirituality. *Progress in Human Geography*, 1–17. DOI: 10.1177/0309132516644512.

Bebbington, D.W. 1988. *Evangelicalism in Modern Britain: A History From the 1730s to the 1980s*. London: Routledge.

CBN News. 2016, 10 Oct. InterVarsity explains the real reason for its new gay marriage policy. *YouTube*. Accessed 6 February 2017, from https://youtu.be/H9r3j5MTsUc.

Cochran, P. 2005. *Evangelical Feminism*. New York: New York University Press.

Connell, J. 2005. Hillsong: a megachurch in the suburbs. *Australian Geographer*, 36(3), 315–332.

Connolly, W.E. 2008. *Christianity and Capitalism, American Style*. Durham, NC and London: Duke University Press.

Dias, E. 2016, 6 Oct. Top evangelical college group to dismiss employees who support gay marriage. *TIME*. Accessed 6 Feb. 2017, from http://time.com/4521944/intervarsity-fellowship-gay-marriage/.

Dittmer, J. and Sturm, T., eds. 2010. *Mapping the End Times: American Evangelical Geopolitics and Apocalyptic Visions*. Surrey, UK and Burlington, VT: Ashgate.

Emerson, M.O. and Smith, C. 2001. *Divided by Faith: Evangelical Religion and the Problem of Race in America*. New York: Oxford University Press.

Fitch, D. 2011. *The End of Evangelicalism: Discerning a New Faithfulness for Mission: Towards a New Evangelical Political Theology*. Eugene, OR: Wipf and Stock.

Gleim, J. 2016, 22 Jan. Larycia Hawkins' actual violation of Wheaton's statement of faith hints at the school's true intentions. *Religion Dispatches*. Accessed 10 Feb. 2017, from http://religiondispatches.org/larycia-hawkins-actual-violation-of-wheatons-statement-of-faith-hints-at-the-schools-true-intentions/.

Gökariksel, B. and Secor, A. 2014. The veil, desire, and the gaze: turning the inside out. *Signs: Journal of Women in Culture and Society*, 40(1), 177–200.

Graham, R. 2016, 16 Oct. Acts of faith. *New York Times Magazine*, 48–53, 60–61.

Hackworth, J. 2013. *Faith Based: Religious Neoliberalism and the Politics of Welfare in the United States*. Athens, GA: University of Georgia Press.

Hunter, J.D. 2010. *To Change the World: the Irony, Tragedy, and Possibility of Christianity in the Late Modern World*. New York: Oxford University Press.

Han, J.J.H. 2011. "'If you don't work, you don't eat': evangelizing development in Africa." In J. Sook, ed., *New Millennium South Korea: Neoliberal Capitalism and Transnational Movements*. London: Routledge. 142–158.

Ingersoll, J. 2005. *Evangelical Christian Women*. New York: New York University Press.

InterVarsity Christian Fellowship/USA. 2015. InterVarsity and #BlackLivesMatter. *InterVarsity Christian Fellowship/USA*. Accessed 10 Feb. 2017, from http://intervarsity.org/news/intervarsity-and-blacklivesmatter.

InterVarsity Christian Fellowship/USA. 2016. A theological summary of human sexuality. *Scribd*. Accessed 10 Feb. 2017, from www.scribd.com/document/326684433/InterVarsity-Christian-Fellowship-Theology-of-Human-Sexuality-Paper.

Larsen, T. 2007. Defining and locating evangelicalism. Ch. 1, in T. Larsen and D.J. Treier, eds., *The Cambridge Companion to Evangelical Theology*. Cambridge: Cambridge University Press. 1–14.

Lee, D.J. 2015. *Rescuing Jesus: How People of Color, Women and Queer Christians Are Reclaiming Evangelicalism*. Boston, MA: Beacon.

Lee, D.J. 2016, 18 Oct. Inside InterVarsity's purge: trauma and termination at the premier evangelical student org. *Religion Dispatches*. Accessed 6 Feb. 2017, from http://religiondispatches.org/inside-intervarsitys-purge-trauma-and-termination-at-the-premier-evangelical-student-org/.

Miller, D.E. 1999. *Reinventing American Protestantism: Christianity in the New Millennium*. Berkeley, Los Angeles, CA and London: University of California Press.

Nazworth, N. 2016, 7 Jan. Larycia Hawkins' theological statement following 'same God' controversy. *The Christian Post*. Accessed 10 Feb. 2017, from www.christianpost.com/news/wheaton-college-larycia-hawkins-theological-statement-muslims-christians-worship-same-god-controversy-154375/.

Smith, A. 2008. *Native Americans and the Christian Right: The Gendered Politics of Unlikely Alliances*. Durham, NC and London: Duke University Press.

Strachan, O. 2015. *Awakening the Evangelical Mind: An Intellectual History of the Neo-Evangelical Movement*. Grand Rapids, MI: Zondervan.

Sutton, M.A. 2014. *American Apocalypse: A History of Modern Evangelicalism*. Cambridge, MA: The Belknap Press of Harvard University Press.

Wheaton College. 1992, 17 Oct. Statement of faith and educational purpose. *Wheaton College*. Accessed 14 Feb. 2017, from www.wheaton.edu/About-Wheaton/Statement-of-Faith-and-Educational-Purpose.

Wilford, J. 2012. *Sacred Subdivisions: The Postsuburban Transformation of American Evangelicalism*. New York: New York University Press.

Withrow, B.G. 2014, 29 Jul. How Westminster Theological Seminary came to define fundamentalism for me. *Huffington Post*. Accessed 14 Feb. 2017, from http://new.www.huffingtonpost.com/brandon-g-withrow/how-westminster-theologic_b_5624650.html.

Worthen, M. 2014. *Apostles of Reason: The Crisis of Authority in American Evangelicalism*. New York: Oxford University Press.

3 Resisting marriage equalities

The complexities of religious opposition to same sex marriage

Kath Browne and Catherine Jean Nash

Introduction

There can be little doubt that religious orthodoxies play a significant role in asserting that progressive lesbian, gay, bisexual, trans and Queer (LGBTQ)[1] legislation and understandings of homosexuality, and increasingly trans lives, are 'against God'. Biblical references are used to evidence both the so-called depravity of homosexuality ('man should not lie with another man')[2] and also the 'naturalness' of God appointed man-woman marriage as the basis for procreation and healthy families (see Browne and Nash, 2014; Nash and Browne, 2015). These views are highly visible in many contemporary debates (including education), but one of the most prominent is around same sex marriage. Various groups and organisations across the globe, and perhaps most visibly in the Global North (including the USA, the UK, Ireland and Australia), have sought to resist the implementation of same sex marriage in part through recourse to religious (mainly specific forms of Christian) ideologies. We term this form of ideology and associated action heteroactivisms, as it seeks to reiterate heteronormative orders (that is male/female relationships within normative genders that are also classed and racialised). In the UK, the passage of same sex marriage legislation in 2013 included provisions expressly barring the Church of England from performing same sex marriages, in order to assuage worries that churches would be 'forced' to perform same sex marriages against their will.

Although it is often assumed that religion is diametrically opposed to lesbian, gay, bisexual and trans rights, the situation is far more complex than this easy assumption allows. Indeed, where we reconsider 'religions' and spiritualties beyond traditional and majority religions, there is evidence that alternative forms of sexual and gender identities and lives receive some form of acceptance. For some spiritual communities, inclusions of LGBT people (and other marginalised groups) are core to their spiritual practices and identities. It is clear that when exploring the everyday practicing of spiritualties and religions a diverse array of inclusions/exclusions and reworkings are apparent (see for example Browne *et al.*, 2010; Hunt, 2016; Rodgers, 1995; Yip, 2008). As Andersson *et al.* (2011) and Vanderbeck *et al.* (2011) show, even in churches where there is vocal opposition to LGBT rights, members of the congregation often voice more complex understandings and engagements with these issues. Queer spiritual spaces can also be found in religions whose underpinning theologies and public pronouncements might be described as homophobic, biphobic and transphobic (e.g. Browne *et al.*, 2010). Yet, it is clear that religious ideologies can pervade state and legislative discourses and play a key role in law making in contexts where there is ostensibly a separation of church and state (Johnson and Vanderbeck, 2014).

In this chapter, we explore the 2015 Irish referendum on same sex marriage, where over 62% of voters voted in favour of amending the definition of marriage to include same sex couples. Our goal is to contribute to the literature that refuses to dichotomously pit religions, even those vocally opposing LGBTQ rights generally and same sex marriage in particular, against LGBTQ rights. To do so, we examine the diversity of views voiced within the Roman Catholic Church in Ireland in the lead-up to the referendum. Despite the Pope's stance against same sex marriage (as leader of the worldwide Roman Catholic Church), and the Roman Catholic Church's official line that supported a No vote in the marriage referendum, clergy in Ireland were found both publically and privately on both sides of the debate which, as Mulhall (2015) argues, demonstrates the interventions were 'by no means unanimous'.

The chapter begins by outlining the specific context of the same sex marriage debate in Ireland focusing on the position of the Irish Roman Catholic Church in relation to key elements of the debate. It then examines the 'two sides' of the argument presented by Catholic clergy, beginning with the rationale for demanding a No vote, before examining the ways in which Parish priests in particular supported a Yes vote. The chapter contends that these debates showcase the complex and multifarious relationships between sexualities, spirituality and space. Exploring these can, we hope, open up new dialogues between those often-entrenched positions that pit religious freedom against sexual equalities.

Same sex marriage and Catholic Ireland

Twenty-two years after Ireland decriminalised homosexuality, a national vote was held on a constitutional amendment to the definition of marriage. Any amendment to Ireland's Constitution requires a referendum, and there have been various referenda on social issues such as divorce and abortion. Because of this constitutional requirement, Ireland was the first country to hold a national referendum on same sex marriage. On 22 May 2015, the Irish were asked to vote on the question: 'Do you approve of the proposal to amend the Constitution contained in the under-mentioned Bill?' The proposed amendment would alter the constitutional statement on marriage to read: 'Marriage may be contracted in accordance with law by two persons without distinction as to their sex'. All the main political parties supported a Yes vote. Some 1,949,725 people or 60.52% of electorate voted in the referendum, with 1,201,607 people or 62.07% voting Yes, and 734,300 people or 37.93% voting No (Elections Ireland, 2015). Reports indicated that many people living and travelling abroad also returned to Ireland to vote (see O'Leary, 2015; Mullhall, 2015; Silvera, 2015).

The Roman Catholic Church (herein 'the Church') was the main opponent to the proposed constitutional revision to the definition of marriage. Ireland is understood as a 'Catholic country' with some 3.86 million people identifying as 'Catholic' in the 2011 Census, constituting 84.2% of the population. Of this number, 92% were of Irish descent (Hyland, 2012; Irish Census, 2011). Despite these figures, adherence to the major precepts of Catholicism, including attending Mass and other rituals, is in decline with a marked increase in secularism and liberalism (Breen and Reynolds, 2011; and, Girvin, 1996). Nevertheless, control of the majority of Irish primary (96%) and secondary schools (51%) remains very much within the hands of the religious orders (Mulhall, 2015). Indeed, as Andersen (2010) notes, young people between the ages of 18–29 are embedded in the culture of Catholicism, but are less institutionalised in their practices and spiritual beliefs than previous generations. Inglis (2007) contends that individual identification with Catholicism as a religious heritage is taking the place of more orthodox adherence to rules and practices, with Irish Catholics mixing Catholicism with other religious and spiritual beliefs.

Paedophile scandals have rocked the Catholic Church in Ireland since the beginning of the 21st century (see for example Böhm *et al.*, 2014; Crowe, 2008; and Pilgrim, 2011). Accusations of child sex abuse by Catholic priests and laity have been met with woefully inadequate responses (Dunne, 2004), causing an ongoing and painful scandal within the Church and amongst faithful Catholics (Hogan, 2011; Savage and Smith, 2003). In Ireland, three government-backed inquires in the 2000s issued detailed reports documenting the specifics of these scandals and the largely ineffective responses by the Church (Murphy *et al.*, 2005; Commission to Inquire into Child Abuse, 2009; Murphy *et al.*, 2009). The causes of sex abuse are multifaceted and certainly not limited to the Irish context, indeed as Terry (2015) notes, accusations of sexual improprieties and abuse are an ongoing, global issue in the Church. However, research highlights the specific factors that created the conditions of possibility for the form this abuse took in the Irish context, particularly emphasising the dominant positioning of the Church as Ireland gained independence from Great Britain as a key factor (alongside poverty and social exclusions) (see Garrett, 2013; McLoone-Richards, 2012). These scandals, together with the secularisation of Irish society through modernisations and economic development, weakened the Church's positioning within Irish social and political contexts. The ability to harness the electorate from the pulpit has diminished, which has been attributed to the Church's lack of control of the media (Donnelly and Inglis, 2010). The increasing separation of the Church and the Irish state gained momentum as clergy, and the institutions of the Church itself, has lost some influence and is arguably ceasing to play a central role in contemporary Irish society.

The Church's failure to prevent the constitutional amendment on marriage might be understood as incontrovertible evidence of the Church's declining importance in Ireland. However, as we argue here, this overstates the matter, as this fails to take into account the multiple and often-conflicting positions of Catholic clergy during the same sex marriage debates nor gives sufficient weight to the potential influence of supportive clergy on the overall result. In what follows, we provide a more detailed consideration of the arguments both for and against the constitutional redefinition of marriage made by Catholic clergy in the period leading up to the referendum. In doing so, we suggest that the Catholic Church's opposition was not as monolithic as one might expect and that it is in the details that a better understanding of the influence and role of the Church in these debates can be gleaned.

Vote No: marriage, children, and family

In examining the archives of the Irish press,[3] we found 48 instances of members of the Catholic clergy from across the country directly opposing same sex marriage, evidenced through direct quotes in the Irish press or as reported in various news articles. These results are in keeping with Mulhall (2015), who asserts that the interventions of the Catholic Church in the same sex marriage debates were 'muted'. This suggests significant opposition within the Church hierarchy to the passing of the proposed constitutional amendment, and although the arguments are diverse, they often collectively draw on familiar and linked perspectives, that are more accurately named as heteroactivism rather than homophobia or transphobia. In this section, we will outline some of the key points made in support of a No vote. Interestingly, and in contrast to those on the 'Yes' side, those quoted as being against the constitutional change tended to be more senior in the hierarchy of the Church.[4]

The style and framing of oppositional arguments is important to consider as this reflects what those on the vote 'No' side thought to be their strongest arguments. Most broadly, and despite the Church's strong opposition to same sex marriage, the Church attempted to strike a conciliatory note by arguing even those voting in support of same sex marriage should not be considered 'wrong':

> People have to make their own mature decision, be it yes or be it no. I would hate for people to be voting no for bad reasons, for bigoted reasons, for nasty reasons, for bullying reasons. People have to make up their own minds and I'm quite happy that people can do that in front of God, be it yes or be it no. I don't doubt that there are many people who are practicing Churchgoers of whatever Church background who will in conscience vote Yes, and that's entirely up to them. I'm not going to say they're wrong.
>
> (Donal McKeown, Bishop of Derry, quoted in Baklinski, 2015)

As this quote illustrates, Donal McKeown sought not to overtly direct laity on how to vote, but to frame the decision-making process as a 'vote of conscience' in 'front of God'. At times this conscience vote was presented as a 'choice' although it was clearly intimated that any right thinking person, in good conscience, could only choose to vote 'No'. The Church's more circumspect and less dictatorial approach reflects its diminished social position arising from the child abuse scandals that seriously undermined its ability to preach political directives from the pulpit (see Donnelly and Inglis, 2010, a point picked up by those advocating a Yes vote, see below). Such an approach reflects a more heteroactivist stance that struggles to be defined as 'homophobic', but nonetheless seeks to reiterate heteronormative orders.

This seemingly softer and more indirect appeal was made not only to the heterosexual Catholic population but also directly to 'gay and lesbian people . . . together with their parents and family members' (Bishop John Fleming, quoted in O'Brien and McGarry, 2015). By encompassing families as well as those who are most affected by the legislation, the Church sought to appeal to lesbians and gay men to consider 'the good of society' ahead of their personal or individual goals. This attempt to connect with the families of those most affected by the vote mirrors the Yes campaign's focus on personal relationships to encourage solidarity with the aspirations of lesbians and gay men to marry. It also seeks to reiterate an individual (Christian?) sacrifice for the good of the Church and broader society. For example, John Fleming, Bishop of Killala sought to persuade lesbians and gays, as well as those close to them, to vote against the amendment by claiming that:

> The Church's vision for marriage and the family is based on faith and reason. It is shared by other faith traditions and by people who have no religious belief. [The proposed amendment] not only redefines marriage in the Constitution but it also, as a result, changes the understanding of the family as outlined in the Constitution. Everyone, including gay and lesbian people, together with their parents and family members, must think carefully on all the issues involved and vote accordingly.
>
> (John Fleming, Bishop of Killala, quoted in O'Brien and McGarry, 2015)

Although suggesting the matter is one for serious reflection, Fleming's argument is clear: the state and the Church are so closely linked that an amendment to the constitutional (state) definition of marriage would have wide reaching (and negative) effects on broader understandings of the 'family' and marriage. The Church in Ireland is historically and intricately entwined with the state in ways that make arguments about their mutual interests more compelling than might be the case elsewhere.

Throughout the debates, familiar heteroactivist arguments regarding the nature and constitution of the 'family' appear under the guise of religion and care for society (see Nash and Browne, 2015; Browne and Nash, 2013, 2014). One key (and recurring) argument against same sex marriage and LGBTQ equalities is the claim that the fundamental 'nature' of marriage is rooted in male-female relationships positioned at the heart of a stable and healthy society:

> Society values the complementary roles of mothers and fathers in the generation and upbringing of children. The differences between a man and a woman are not accidental to marriage but are fundamentally part of it.
>
> (John Kirby, Bishop of the Diocese of Clonfert, quoted in MacDonald, 2015b)

In this claim, the Church suggests that society's values and Christian values, as advanced through the Irish Roman Catholic Church, are thoroughly intertwined such that both require the traditional, heterosexual family be 'protected' for the common good. The loss of this referendum dramatically highlights how the Church is increasingly distanced from the interests of the state and has lost considerable influence over state social and political policies.

As we have argued elsewhere (Browne and Nash, 2013, 2014; Nash and Browne, 2015), and as is certainly the case here, another key heteroactivist argument mounted by the Church is that man and woman are not 'accidental' to marriage. Further, the male-female complementarity is essential for the healthy rearing of children, thereby providing for the present and future stability of society. This reasserts the role of the Church and Christianity, as the moral compass for Irish societal codes and norms:

> We make our position clear not just from a faith point of view but also because we believe it is good for children, that it's good for family and it's good for society to preserve the uniqueness of marriage as we have traditionally understood it.
>
> (Eamon Martin, Archbishop of Armagh and Primate of All Ireland, quoted in MacCormaic, 2015)

By linking Church, state and society in this way, the heteroactivist narrative creates a 'we' that needs to protect those 'traditions' that ensure society's future. These traditions are evoked in ways that see marriage as a timeless and placeless entity. It is the 'foundation' of society and thus any threat to marriage is a threat to society (and its members):

> . . . common sense alone tells us that every child should have its 'mammy and daddy'. This has been the way since the dawn of civilization in every culture and on every continent. . . . The referendum on 22 May is seeking to change the very meaning of marriage. It is like removing concrete foundations under a house and saying that any material will do. In what has turned out to be a desperately one-sided public debate I hope you will think long and hard about your decision.
>
> (Phonsie Cullinan, Bishop of Diocese of Waterford and Lismore, quoted in Towey and Duncan, 2015)

Phonsie Cullinan's plea to 'think long and hard' about this decision, and the equation of man/woman marriage with the foundations of society, not only places marriage at the centre of society, it seeks to re-establish the specific Christian views of the orthodox Catholic clergy as central to the debate. Second, the appeal to 'common sense' returns the focus of the debate to questions of 'natural' procreation and genetics. Diarmuid Martin, Archbishop of Dublin, went so far as to argue that, '[e]ven if it were possible to clone a child, that child would still bear the genetic imprint of a male and a female. Genetic parentage is not irrelevant' (quoted in MacDonald, 2015a). This came in apology for the 'offense' that Kevin Doran caused by suggesting that those who have children by other means 'are not parents. They may have children, but you see this is the point, people who have children are not necessarily parents' (Kevin Doran, Bishop of Elphin, quoted in MacDonald, 2015c).

Not surprisingly, denying the multiple ways people become 'parents' had the effect of drawing into the debates a broad range of different types of families, including single parent families, adopted families, blended families and many other non-(hetero)normative families. The Church's official stance on same sex marriage also reasserted the Church's 'traditional' view of marriage against a range of contemporary forces visible in a secular and diverse Ireland.

Not only were children seen as an important part of marriage, marriage itself was understood by definition to entail the ability to procreate:

> The reality is that those who wish to change the Constitution are not actually looking for marriage equality. They are looking for a different kind of relationship which would be called marriage; a relationship which includes some elements of marriage, such as love and commitment, but excludes one of the two essential aspects of marriage, which is the openness of their sexual relationship to procreation. This is only possible if we change the meaning of marriage and remove that aspect of openness to procreation. Part of the challenge for us as a society, of course, is that we (and that includes many practising Catholics) have to a greater or lesser extent given up on the idea that sexual intercourse and an openness to procreation are essentially linked. That makes it more difficult to get our heads around why there might be any problem about changing the meaning of marriage. There is nothing wrong with being nice to them, but that is not what the referendum is about.
>
> (Kevin Doran, Bishop of Elphin, quoted in McGarry, 2015)

Doran makes a number of related points about the purpose of marriage, its 'true' definition and the relationship between the Church and gays and lesbians. His logic seeks to extend the Church's doctrine beyond 'Christians' through an appeal to what is framed as a universal and inevitable 'truth' based in 'reason'. In this 'reality', there are marriages that are 'real' because they are based on love and a sexual relationship that can result in children and those that are based in love but do not include the 'essential aspect' of the possibility of procreation. Despite attempts to move away from just Christians, the argument is of course based in Christianity and this position reflects the Church's prohibition against contraception and its doctrinal concerns about the purported disconnection between procreation and marriage. Such a view once again asserts a particular form of Christianity that saw the Church's traditional Catholic stance on procreation remain at odds with contemporary gender equalities and women's freedoms in a modernising Ireland.

Doran suggests there is a need to 'be nice to them', that is, to gays and lesbians, but that the question of marriage and family is a much broader issue. This claim that the Church is 'caring' (or at least not 'mean-spirited') was a central rhetorical point within Church arguments in conjunction with the Church's overall conciliatory tone described above. Emphasising love as a central Christian value, this heteroactivist framing sought to soften the hurtful and exclusionary language against LGBT relationships and was often deployed before a critique of the 'unnaturalness' of same sex relationships and claims about the negative impact of the genderless definition of marriage. However, it could also be perceived as reflecting the paternalistic voice of the Church – one that was tough but 'concerned' and 'caring' in their desire to exclude same sex attractions, love and relationships from the definition of marriage:

> We are not being mean-spirited towards those who have same-sex attractions. On the contrary, we regard marriage as the central and crucial social relationship, which is of natural law and plays an indispensable part in human life. Our view of Christian marriage, properly explained and understood, is not in any way disrespectful of people who experience same-sex attraction. As a Church, we believe every person is equal in the sight of God and should always be treated with love, dignity and respect. There is no denying the fact that marriage faces difficulties throughout the Western world today. These pressures impinge on all, but particularly on children. Following the Referendum on Children's Rights our laws now enshrine the principle that, in all decisions relating to a child, the welfare of the child must be paramount. A society that identifies the two parties in marriage as spouse I and spouse II has lost sight of a deep truth of human nature. Are we going to be the first generation in human history to say that mothers and fathers don't matter anymore in the upbringing of children? Children have a right to grow up in a family with a father and a mother capable of creating a suitable environment for the child's development and emotional maturity. This referendum is not and should not be about judging the various family types which have always existed as a reality in Ireland. Married parents and single parents deserve as much support as possible as they live out the challenging vocation of parenthood. [. . .] Despite what we are led to believe this referendum is not about same-sex relationships or about equality, but about the family.
>
> (Michael Neary, Archdiocese of Tuam, quoted in MacDonald, 2015b)

We quote Michael Neary in depth here to demonstrate how this heteroactivist narrative moves from one of 'love, dignity and respect', towards a conceptualisation of family that excludes same sex couples and suggests that children will suffer from having same sex parents. Delinking family from same sex relationships and parenting means that the referendum was not about 'same sex relationships, or about equality' (Drennan, 2015), but instead about family, that is, the only sort of arrangement that is truly a 'family' – the heterosexual, married couple. This view was shown time and again to be socially and legally inaccurate. The changes to the Constitution proposed through the referendum had little effect on the legal status of same sex parents and no effect on laws around adoption, custody or access to fertility treatments. However, the 'No' campaign continually evoked the rights of children, and the figure of the child as potentially being damaged by the proposed constitutional amendment. Here, the Christian doctrine of love couches this message, all are 'equal in the sight of God', but family is only created through men and women, mothers and fathers (even if these are single parents and unmarried).

As we have noted elsewhere, heteroactivists also sought to minimise the potential impact of excluding gays and lesbians from marriage by retrospectively supporting civil partnerships rather than redefining marriage (see Browne and Nash, 2014). In the alternative some sought to reframe the debate not as a human rights issue but to suggest that while all people are equal, marriage is a 'unique' and special institution. This is supported by the European Court of Human Rights' 2014 decision declaring that same sex marriage is not a human right. For example, Philip Boyce (Bishop of Raphoe, quoted in Harkin, 2015a) argued that, 'equality and human rights should be afforded to everyone, but it should be done without sacrificing the institution of marriage and the family'. Such an argument sought to reposition the Church not as 'behind the times' but as a truly caring and innovating institution, seeking to find respectful solutions to difficult issues while essentially maintaining the status quo:

> A pluralist society can be creative in finding ways in which people of same-sex orientation have their rights and their loving and caring relationships recognised and cherished in a culture of difference, while respecting the uniqueness of the male-female relationship. I know that the harshness with which the Irish Church treated gay and lesbian people in the past – and in some cases still today – may make it hard for LGBT people to accept that I am sincere in what I am proposing. Marriage is not simply about a wedding ceremony or about two people being in love with each other. We are all children of a male and a female and this must have relevance to our understanding of the way children should be nurtured and educated.
>
> (Diarmuid Martin, Archbishop of Dublin, quoted in MacDonald, 2015d)

While Diarmuid Martin is prepared to understand LGBT relationships as deserving of 'rights', he seeks to carve out a special place for marriage based in male and female procreative relationships. 'Cherishing difference' but arguing for the 'uniqueness' of heterosexual marriage is not paradoxical nor does it equate with treating gay and lesbian people harshly. Rather simplistically equating marriage, sexual procreation and social nurturing and education, Martin suggests that Church abuse of LGBTQ people is a thing of the past, and that the Irish Church remains central to decisions regarding families, children and education despite the exclusion of 'lesbian and gay people'. In this way, he reconciles a faith that asks for love with a Church that has power and has abused this power in relation to 'lesbian and gay people'.

Vote Yes: priests supporting same sex marriage

Contrary to the official position of the Irish Church, some Catholic clergy broke ranks and spoke in favour of same sex marriage. Supportive clergy were much less prevalent than oppositional ones, and our search of the mainstream Irish press found 16 reported statements. Priests who argued in favour of same sex marriage took a variety of positions, including coming out as gay themselves, as Martin Dolan did. Those supporting the Yes vote did so from a position that overtly recognised the weakness of the Church, while seeking to support those who had been hurt by the Church.

Similar to the paradoxical positions taken by those who argued for a No vote, some priests who supported a Yes vote did not necessarily approve of same sex relationships:

> It is not what I would see as the ideal, in fact I would disagree with it but I am willing to allow those that believe to live out their lives. [. . .] It would be sinful for me [to judge same sex marriage as sinful] but to use that lovely phrase of Pope Francis, 'Who am I to judge?' I might disagree with them and I wouldn't be able to participate in such a ceremony I admit that, but at the same time I am willing to accept the opinion of those who have that view. [. . .] I am not in any way calling for a yes or no vote, I was simply asked how I would vote myself. All I am saying is that if a yes vote is carried or if the no vote is carried it won't affect me in the slightest, I will still be a believer and provocateur of catholic values and catholic marriage. Maybe I am a 'fuddy duddy' on this one but I am a believer in marriage for life, in heterosexual marriage between a man and a woman. I see that as Catholic marriage, that is the one I believe in but I am willing in civil law, and I am not changing Church law in any way – in civil law, the state is a secular reality and the state legislates for all its citizens including those who have different views on marriage.
>
> (Iggy O'Donovan, Priest, quoted in Hayes, 2015)

Iggy O'Donovan's position nicely illustrates an interesting paradox. O'Donovan does not believe in same sex marriage, would not conduct such a ceremony and continues to regard Catholic marriage as between a man and a woman. Nevertheless, as a personal matter he would vote in favour of same sex marriage. He supports this position by referencing the Pope's statement that it is not his place to 'judge' gays and lesbians. He also supports the separation of state/Church, framing the Irish constitutional referendum as pertaining to 'civil marriage' which is, in effect, 'about giving statutory recognition and protection, irrespective of sex, to the relationships of all people who publicly want such recognition by the State, nothing more, nothing less' (Brian Ó Fearraigh, Curate in Gaoth Dobhair, Co Donegal, quoted in Harkin, 2015b).

As we noted earlier, traditional Catholic orthodoxy holds that anything less than heterosexual marriage is detrimental to children and to society as a whole. Nevertheless, some priests did question whether same sex marriage was actually detrimental to society:

> I ask if it [same sex marriage] is in the interests of society, and in this instance I think it is and that is why I will be voting Yes. There are so many different

> types of families. From the nuclear ones with a mam and a dad and children to single parents of children from one father and single parents of children from different fathers – as well as same-sex couples. I believe in relationships and family and marriage in all those different types of situations. In every community there are same-sex couples, and as a priest you get a sense of how people live, and there is nothing like staying with a same-sex couple and their families to make you change any preconceptions you might have had about them. Quite a few male suicides are rooted in the struggle over sexuality, and anything we can do to de-stigmatise the old thinking and the old prejudices about sexuality is welcome. My worry is if the referendum is defeated, what message will it send to people who are struggling with their sexuality?
>
> (Gerry O'Connor, Dublin Priest, quoted in Feehan, 2015)

Gerry O'Connor pushes back against the homogenous portrayal of parents as necessarily only male and female and instead recognises the validity of distinctive family forms. His quote humanises same sex couples in ways that challenge 'old thinking and old prejudices'. These old ways are potentially damaging and hurtful, and a No vote, as O'Connor suggests, may affect LGBTQ people in adverse and potentially life threatening ways.

Tony Flannery extends this argument by suggesting that denying same sex marriage is 'morally wrong' and by drawing on Pope Francis to suggest that a Yes vote is 'the Christian thing' to do:

> Pope Francis has brought us back to some of the very basic teachings of Jesus. He constantly tells us that love, compassion and mercy are fundamental Christian attitudes. If this country rejects the proposal put before us in this referendum, I fear that gay people will hear it as a further rejection, another example of society telling them they are lesser human beings.
>
> [. . .]
>
> Because of the struggle they have experienced, first in coming to terms with themselves, and then with the negative attitudes in society, they have developed particularly sensitive antennae to rejection of any sort. For me, the really Christian thing is to give them a strong and clear message that they are loved and accepted just as they are, and that they deserve to be treated with the same dignity as the rest of us.
>
> (Tony Flannery, Priest, suspended from Church, 2015)

One of the key strategies of the Yes campaign was to personalise the Yes vote, that is, to make the vote about real people whose lives would be severely affected by a No vote. This strategy included encouraging many prominent Irish celebrities to come out, and by asking LGBTQ people to come out to their families and friends. By coming out, LGBTQ people could appeal directly to family and friends to support a Yes vote, thereby engaging the straight public citizen at a very personal level (Mulhall, 2015). This campaign tactic was seen as central to the success of the Yes vote, as it moved the conversation away from abstract ideas of family and morality towards personal narratives and the lives that could be enhanced.

Besides trying to be conciliatory and 'nice', the Church also recognised the potential damage it might suffer in opposing same sex marriage. Brendan Hoban for the Association of Catholic Priests argued that 'haranguing' for a No vote would have negative implications not only for 'gay people' but also for the Church:

> Individually or collectively has that triumvirate – the bishops, Iona and the hard-line fundamentalists – any idea of the damage they're doing to the Church they profess to serve with such devotion? For the Catholic Church, it can be argued that the result of the referendum on same-sex marriage will matter less than the fall-out afterwards. A positive result for 'Catholic' forces (the defeat of the referendum) could do huge damage to the Irish Catholic Church. In every Catholic congregation, for instance, there are gay people and straight people who have gay members of their family and straight people who have gay friends. And haranguing them into voting No in the referendum, regardless of the substance of the arguments offered, will have the effect of driving more and more of them out of the church and out of the Church.
>
> (Brendan Hoban, Association of Catholic Priests, 2015)

Rather than just fearing further rejection with a No vote, some priests turned this around to speak of the positive messages a Yes vote would send:

> I believe it's the right thing to do now. It's time that gay people had the same rights as everyone else. The Church has its own rules for marriage within it . . . but this is something different entirely. The Church cannot lay down its rules for everyone. The Church has made statements saying that they respect all people, gay or straight. This is a way for them to show that this is true – that somebody can come and ask that their love be blessed. Many people in the gay community feel that the Church is against them, and this would be a way to show that this isn't so. . . . We are taught that God is love.
>
> (Pádraig Standún, parish priest of Carna in Connemara, quoted in Anon, 2015)

The idea that 'God is Love' could be demonstrated through the Church's acceptance and respect. This was in contrast to not only the damage that might be done to the Church by a No vote, but also to the controversies that have weakened the Church:

> I would be very slow to bring a crowd onto a field where we ourselves are vulnerable. In view of our recent history, our street credibility in these areas is not very high.
>
> (Iggy O'Donovan, Priest, quoted in MacDonald, 2015e)

Bringing a crowd to the Church's field may not have been desired, but these debates highlighted the Church's tenuous and uncertain place in a modern Ireland. Moreover, as these two sections have shown there were significant divergences in the Catholic Church, with clergy who pushed for a Yes vote directly contradicting those who sought a No vote. This speaks to pluralist positionings of Roman

Catholic clergy in Ireland and questions any easy linkage of religious freedom with rejecting constitutional change.

Conclusion

There can be little doubt that in many ways Ireland remains a 'Catholic country' created through a historically central and cultural (if not doctrinal) Catholicism. The position of the Church has changed significantly over the past century, with modernising forces acting alongside the secularisation of the state, public life and citizenship. Most recently, sex abuse scandals have limited the moral ground upon which the Church traditionally claimed authority. The loss of the working class vote in the referendum was another blow to the Church, where social conservatism and adherence to Catholicism is expected (see Mulhall, 2015).

Whilst officially the Catholic Church in Ireland opposed any amendments to the Constitution that made civil marriage available to couples outside of the binary of male/female, this chapter demonstrates that, as with most research that explores LGBT and religious relationships, such a view only partially captures the complexities at the intersections of sexualities and religions. The presumption that Christianity and its manifestation through the Roman Catholic Church in Ireland is necessarily and uniformly exclusionary, is problematic. The Catholic Church, composed of diverse individuals and priests, reflected multiple and contested views in the national newspapers. This failure to present a thoroughly unified rhetoric was eventually blamed for the referendum 'loss'. However, as the chapter has shown, only some see this as a loss for the Church. In addition to dissenting priests, during the referendum debates parishioners were reported as walking out of Mass during sermons which opposed gay marriage and criticised sportspeople that supported a Yes vote.

Considering sexualities, spiritualties and nationalism, this chapter highlights the precarious ways in which 'official' state and national religions are operationalised in relation to sexual and gendered difference. It emphasises the contestation over the control of religious identities, doctrines and practices in relation to sexual equalities, challenging a coherent stasis regarding spiritualities and sexualities, even in a national context such as 'Catholic Ireland'. In doing so, it has shown the tensions regarding inclusion of lesbian, gay, bi and trans people that continue to be a source of conflict within and beyond the Roman Catholic Church. Further examining the detailed encounters between spiritual and religious discourses and sexual/gender identities will develop the understandings of these complexities, opening spaces to new possibilities beyond the dichotomous presumption that pits religious freedom against sexual/gendered liberations, and thus sees religions generally, and Christianity specifically, as exclusionary and hostile to LGBT people.

Notes

1 LGBTQ is used here as it is one of the accepted acronyms that represent a variety of sexual and gender differences from heteronormativities. Heteronormativity is the normalisation of heterosexuality within normative man/woman, male/female understandings of

gender. In this chapter, we use LGBTQ to describe the populations affected by these discussions, and we use other terms and sets of wording to highlight how LGBTQ people are addressed in these debates.

2 King James Bible, Leviticus 18:22, Standard English Version 'You shall not lie with a male as with a woman; it is an abomination'.

3 The chapter uses a digital archival data collection method. It uses online material from the three mainstream Irish newspapers, namely the *Irish Times*, *Irish Independent* and *Irish Examiner*. The aim of the data collection was to capture public pronouncements from the Catholic clergy and laity. We used a date range of 1 January 2014–22 May 2015. This date range covered the entire referendum campaign and avoided post referendum proclamations, analyses and revisions. Practically, the data collection began with the *Irish Times*. We used an online archive search covering key terms such as 'marriage referendum', 'same sex marriage', 'Catholic Church'. Following this, new articles were added from the *Irish Independent*, this allowed for data to be corroborated by multiple sources. It also meant that a broad range of material was gathered, but that duplications were not counted so as not to inflate the coverage of Roman Catholic clergy in the mainstream press. When searching the *Irish Examiner* no new material was found, indicating saturation. The data was coded and then analysed for the purposes of this chapter. This followed a for/against categorisation. Key arguments were identified and these were explored to develop the thinking for this chapter.

4 Parish priests work at the local level and if there is more than one there will be a hierarchy here. Overseeing regional districts, or dioceses, are Bishops, and there are 26 dioceses covering the Republic of Ireland and Northern Ireland. These are contained within four provinces each led by an Archbishop.

References

Andersen, K. 2010. Irish secularization and religious identities: evidence of an emerging new Catholic habitus. *Social Compass*, 57 (1), 15–39.

Andersson, J., Vanderbeck, R.M., Valentine, G., Ward, K. and Sadgrove, J. 2011. New York encounters: religion, sexuality, and the city. *Environment and Planning A*, 43 (3), 618–633.

Anon. 2015. Connemara parish priest supports Yes vote in marriage poll. *The Irish Times*. Available from www.irishtimes.com/news/politics/connemara-parish-priest-supports-yes-vote-in-marriage-poll-1.2208664 [Accessed 14 March 2017].

Baklinski, T. 2015. Irish bishop: Catholics can back gay 'marriage' in good conscience. *LifeSiteNews*. Available from www.lifesitenews.com/news/irish-bishop-catholics-can-back-gay-marriage-in-good-conscience [Accessed 13 March 2017].

Böhm, B., Zollner, H., Fegert, J.M. and Liebhardt, H. 2014. Child sexual abuse in the context of the Roman Catholic Church: a review of literature from 1981–2013. *Journal of Child Sexual Abuse*, 23 (6), 635–656.

Breen, M.J. and Reynolds, C. 2011. The rise of secularism and the decline of religiosity in Ireland: the pattern of religious change in Europe. *The International Journal of Religion and Spirituality in Society*, 1 (2), 195–212.

Browne, K., Munt, S.R. and Yip, A.K.T. 2010. *Queer spiritual spaces: sexuality and sacred places*. Burlington, VT: Ashgate Publishing.

Browne, K. and Nash, C.J. 2013. Special issue: new sexual and gendered landscapes. *Geoforum*, 49, 203–205.

Browne, K. and Nash, C.J. 2014. Resisting LGBT rights where 'we have won': Canada and Great Britain. *Journal of Human Rights*, 13 (3), 322–336.

Commission to Inquire into Child Abuse. 2009. *Final report of the commission to inquire into child abuse*. Available from www.childabusecommission.ie/rpt/pdfs/ [Accessed 14 March 2017].

Crowe, C. 2008. The ferns report: vindicating the abused child. *Éire-Ireland*, 43 (1), 50–73.

Donnelly, S. and Inglis, T. 2010. The media and the Catholic Church in Ireland: reporting clerical child sex abuse. *Journal of Contemporary Religion*, 25 (1), 1–19.

Drennan, M. 2015. Letter on marriage and the family. *Irish Catholic Bishops Conference*. Available from www.catholicbishops.ie/2015/05/16/bishop-martin-drennan-letter-marriage-family/ [Accessed 13 March 2017].

Dunne, E.A. 2004. Clerical child sex abuse: the response of the Roman Catholic Church. *Journal of Community and Applied Social Psychology*, 14 (6), 490–494.

Elections Ireland. 2015. *Referendum 2015*. Available from http://electionsireland.org/results/referendum/refresult.cfm?ref=201534R [Accessed 14 March 2017].

Feehan, C. 2015. Dublin priest will vote Yes in referendum because there is 'more than one type of family'. *The Irish Herald*. Available from www.herald.ie/news/dublin-priest-will-vote-yes-in-referendum-because-there-is-more-than-one-type-of-family-31217078.html [Accessed 14 March 2017].

Flannery, T. 2015. God's love is not diminished by one's sexual orientation. *The Irish Independent*. Available from www.independent.ie/opinion/comment/gods-love-is-not-diminished-by-ones-sexual-orientation-31191224.html [Accessed 14 March 2017].

Garrett, P.M. 2013. A "catastrophic, inept, self-serving" church? Re-examining three reports on child abuse in the Republic of Ireland. *Journal of Progressive Human Services*, 24 (1), 43–65.

Girvin, B. 1996. Church, state and the Irish Constitution: the secularisation of Irish politics? *Parliamentary Affairs*, 49 (4), 599–615.

Harkin, G. 2015a. Voting No to same-sex marriage is not homophobic, says bishops. *The Irish Independent*. Available from www.independent.ie/irish-news/politics/voting-no-to-samesex-marriage-is-not-homophobic-say-bishops-31192852.html [Accessed 14 March 2017].

Harkin, G. 2015b. Priest goes against his bishop to back 'Yes' campaign. *The Irish Independent*. Available from www.independent.ie/irish-news/priest-goes-against-his-bishop-to-back-yes-campaign-31208250.html [Accessed 14 March 2017].

Hayes, K. 2015. Other priests will vote yes in referendum, says Fr Iggy. *Irish Examiner*. Available from www.irishexaminer.com/ireland/other-priests-will-vote-yes-in-referendum-says-fr-iggy-317613.html [Accessed 14 March 2017].

Hoban, B. 2015. Winning battles, losing the war. *Association of Catholic Priests*. Available from www.associationofcatholicpriests.ie/2015/02/winning-battles-losing-the-war/ [Accessed 24 July 2016].

Hogan, L. 2011. Clerical and religious child abuse: Ireland and beyond. *Theological Studies*, 72 (1), 170–186.

Hunt, S. 2016. *Contemporary Christianity and LGBT sexualities*. London: Routledge.

Hyland, P. 2012. Number of Catholics at record high, despite lowest percentage ever – CSO. *The Journal*. Available from www.thejournal.ie/regious-statistics-census-2011-640180-Oct2012/ [Accessed 14 March 2017].

Inglis, T. 2007. Catholic identity in contemporary Ireland: belief and belonging to tradition. *Journal of Contemporary Religion*, 22 (2), 205–220.

Irish Census. 2011. *Irish census 2011*. Available from http://faithsurvey.co.uk/irish-census.html [Accessed 27 July 2016].

Johnson, P. and Vanderbeck, R. 2014. *Law, Religion and Homosexuality*. London: Routledge.
MacCormaic, R. 2015. Same-sex marriage referendum countdown begins. *The Irish Times*. Available from www.irishtimes.com/news/politics/same-sex-marriage-referendum-countdown-begins-1.2073120 [Accessed 14 March 2017].
MacDonald, S. 2015a. Archbishop Martin hits out at 'obnoxious jibes' at gay community from 'No' camp. *The Irish Independent*. Available from www.independent.ie/irish-news/archbishop-martin-hits-out-at-obnoxious-jibes-at-gay-community-from-no-camp-31081097.html [Accessed 14 March 2017].
MacDonald, S. 2015b. Three bishops address marriage referendum. *Catholic Ireland*. Available from www.catholicireland.net/bishops-issue-statements-marriage-referendum/ [Accessed 14 March 2017].
MacDonald, S. 2015c. Bishop: gay couples with children 'are not parents'. *The Irish Independent*. Available from www.independent.ie/irish-news/bishop-gay-couples-with-children-are-not-parents-31053862.html [Accessed 14 March 2017].
MacDonald, S. 2015d. Archbishop Diarmuid Martin to vote 'No' in same-sex marriage referendum. *The Irish Independent*. Available from www.independent.ie/irish-news/news/archbishop-diarmuid-martin-to-vote-no-in-samesex-marriage-referendum-31202074.html [Accessed 14 March 2017].
MacDonald, S. 2015e. Fr Iggy breaks ranks to support 'yes' vote. *The Irish Independent*. Available from www.independent.ie/irish-news/news/fr-iggy-breaks-ranks-to-support-yes-vote-31053860.html [Accessed 14 March 2017].
McGarry, P. 2015. 'No obstacle' to gays marrying, just not each other, says bishop. *The Irish Times*. Available from www.irishtimes.com/news/social-affairs/religion-and-beliefs/no-obstacle-to-gays-marrying-just-not-each-other-says-bishop-1.2115808 [Accessed 14 March 2017].
McLoone-Richards, C. 2012. Say nothing! How pathology within Catholicism created and sustained the institutional abuse of children in 20th century Ireland. *Child Abuse Review*, 21 (6), 394–404.
Mulhall, A. 2015. The republic of love: On the complex achievement of the same sex marriage referendum in Ireland. *Bully Bloggers*. Available from https://bullybloggers.wordpress.com/2015/06/20/the-republic-of-love/ [Accessed 24 July 2016].
Murphy, F.D., Buckley, H. and Joyce, L. 2005. *The ferns report*. Available from www.lenus.ie/hse/bitstream/10147/560434/2/thefernsreportoctober2005.pdf [Accessed 14 March 2017].
Murphy, J.Y., Mangan, I. and O'Neill, H. 2009. *Report of the Commission of Investigation into the Catholic Archdiocese of Dublin*. Available from www.justice.ie/en/JELR/Pages/PB09000504 [Accessed 14 March 2017].
Nash, C.J. and Browne, K. 2015. Best for society? Transnational opposition to sexual and gender equalities in Canada and Great Britain. *Gender Place and Culture*, 22 (4), 561–577.
O'Brien, T. and McGarry, P. 2015. Bishops urge their congregations to vote in referendums. *The Irish Times*. Available from www.irishtimes.com/news/ireland/irish-news/bishops-urge-their-congregations-to-vote-in-referendums-1.2215487 [Accessed 14 March 2017].
O'Leary, J. 2015. Yes to marriage equality in Ireland – an inspiration for battles to come. *rs21*. Available from https://rs21.org.uk/2015/05/23/yes-to-marriage-equality-in-ireland-an-inspiration-for-battles-to-come/ [Accessed 24 July 2016].
Pilgrim, D. 2011. The child abuse crisis in the Catholic Church: international, national and personal policy aspects. *Policy and Politics*, 39 (3), 309–324.

Rodgers, B. 1995. The Radical Faerie Movement: a queer spirit pathway. *Social Alternatives*, 14 (4), 34–37.

Savage, R.J. and Smith, J.M. 2003. Sexual abuse and the Irish church: Crisis and responses. *The church in the 21st century: from crisis to renewal*, Occasional paper #8.

Silvera, A. 2015. Marriage is not equality: thoughts on #MarRef from a worried radical queer. *Feminist Ire*. Available from https://feministire.com/2015/05/21/marriage-is-not-equality-thoughts-on-marref-from-a-worried-radical-queer/ [Accessed 24 July 2016].

Terry, K.J. 2015. Child sexual abuse within the Catholic Church: a review of global perspectives. *International Journal of Comparative and Applied Criminal Justice*, 39 (2), 139–154.

Towey, N. and Duncan, P. 2015. Senior clergy figures speak in favour of No vote in referendum. *The Irish Times*. Available from www.irishtimes.com/news/social-affairs/senior-clergy-figures-speak-in-favour-of-no-vote-in-referendum-1.2206825 [Accessed 14 March 2017].

Vanderbeck, R.M., Andersson, J., Valentine, G., Sadgrove, J. and Ward, K. 2011. Sexuality, activism, and witness in the Anglican Communion: The 2008 Lambeth conference of Anglican Bishops. *Annals of the Association of American Geographers*, 101 (3), 670–689.

Yip, A.K.T. 2008. Researching lesbian, gay, and bisexual Christians and Muslims: Some thematic reflections. *Sociological Research Online*, 13 (1).

4 Building sacred modernity

Buddhism, secularism and a geography of 'religion' in southern Sri Lanka

Tariq Jazeel

1

In 1979, Sri Lanka's most famous tropical modern architect, Geoffrey Bawa, was commissioned by the United National Party (UNP) government to design and build a new parliamentary complex in a site just 10 km from Colombo. In a post-colony that had long since turned its back on its post-independent commitment to the multi-ethnic accommodation of Sinhalese, Tamils, Muslims and Burghers, the geographical conception of this new parliamentary complex was very much in keeping with Sinhala-Buddhist nationalist intent to fashion a society in which Buddhism informed the polity.[1] Kotte, the location earmarked for the parliamentary complex, was chosen because it was a historic Sinhalese metropolitan centre from which a former King, Parakramabahu VI, was reputed to have fought invading South Indian forces in the mid-fifteenth century in an attempt to re-establish Sinhala-Buddhist rule over the whole island.

Although revisions to the Sri Lankan constitution in 1972 and 1978 respectively were notable for the ways that they, first, accorded Buddhism the foremost place amongst Sri Lanka's other religions (Hinduism, Christianity and Islam), and second, offered it special protection in the national polity (Bartholomeusz 1999, p. 185), the country still to this day professes a notional secularism through its commitment to parliamentary democracy and political modernity. Indeed, that abstract commitment to political modernity has been essential for the state to be able to pronounce itself a mature institution firmly under the control of human, not religious, will. As the secularization thesis clearly holds, 'in order for a society to be modern it has to be secular and for it to be secular it has to relegate religion to nonpolitical spaces because that arrangement is essential to modern society' (Asad 2003, p. 182). Despite the machinations of Sinhala-Buddhist nationalism in mid-1970s Sri Lanka, secularism still continues to perform a valuable operation for post-independent Sri Lanka insofar as its geographical excision of 'religion' from the engine rooms of political decision-making was precisely what produced the state *as* a mature and modern political institution.

In this sense then, the choice of Geoffrey Bawa to design the new parliamentary complex was not incidental. Bawa was a modernist, and as such he held a deep commitment to the idea of 'art for art's sake'. His work can be situated within the global circuits of international modernism and landscape design, specifically their tropical variants (Robson 2002, p. 238; Jazeel 2013a; Jones 2011). Having qualified as an architect from London's *Architectural Association* (AA) in 1957, only thereafter did Bawa return to Sri Lanka to practice professionally. Architecturally, the clean lines and sharp edges of many of his early buildings betray his European training and a range of western influences, including art nouveau, international modernism and brutalism in particular. If these modernist architectural sensibilities were to remain integral to his work, Bawa's story – and the story of Sri Lankan tropical modernism more generally – can also be understood through his attempts to adapt to the tropical materialities and demands of a South Asian environmental context (see Jazeel 2013a). Bawa's training at the AA coincided with the establishment in 1953 of Otto Koenigsberger's newly conceived Department of Tropical Studies (Pieris 2007a, p. 64), where he learnt the latest European theories regarding how modernism practised in the tropics might express regional and national particularities, providing 'authentic' reactions to European and North American Functionalism (Goad and Pieris 2005; Pieris 2007a, 2007b pps.1–16). As a result, what has become known as Sri Lanka's own iteration of the regional modern was gradually consolidated (see Robson 2007).

Bawa's design for the parliament complex (Figure 4.1) was, as we might expect then, a striking and sprawling monument to the post-independent nation-state; one

Figure 4.1 The Parliament Complex, Sri Jaywardenapura Kotte, Sri Lanka: architect, Geoffrey Bawa.

Source: Tariq Jazeel.

resolutely modernist by design, and thus befitting of a mature and notionally secular post-colonial nation-state. It consists of a series of interconnected pavilions comprising one main structure surrounded by five satellite buildings, all of which are separated by a series of walkways and piazzas. The pavilion structures are set in the midst of an artificial lake, and as Bawa's chief architectural commentator David Robson (2002, p. 150) has written, 'everything below the roof has been designed in an abstract Modernist mode with a simple elegance'. The debating chamber was planned as a symmetrical rectangle based on the Westminster model, containing galleries for MPs and public viewing spaces rendering transparent to public scrutiny the national political process. Characteristically though, the complex references diverse architectural times and spaces, and has been described as a cosmopolitan and internationalist edifice gesturing variously toward Mogul Lake palaces, South Indian temples and Chinese palaces (ibid., p. 148). As Lawrence Vale (1992, p. 194; also see Perera 2013) has written, 'Bawa's capitol complex stands squarely between the abstract universalism of high modernism and literal localism'. Indeed, it is the abstract universalism of these architectural referents that enables a reading of Bawa's capitol complex as a suitable monument to a post-colonial nation-state committed to political modernity and free from the vagaries of religious interference.

2

There is, however, far more to Bawa's parliamentary complex. Just as the complex's architectural modernism signifies the kind of secularization key to political modernity gestured to above, it simultaneously instantiates what, after Raymond Williams (1977), I refer to as Sinhala-Buddhist 'structures of feeling' that are neither 'religious' *nor* 'secular' (in the Enlightenment sense of those terms). These are what I refer to here, and elsewhere in much more depth (see Jazeel 2013a), as sacred modernity: structures of feeling in everyday life and in modernity wherein Buddhist metaphysics and historical resonances are made palpably and affectively present for and by the subject. As I suggest, sacred modernity is a concept-metaphor that betrays the existence of Buddhism not as 'religion' per se, but moreover as a problem of difference for scholars attuned to 'religion's' colonial history in South Asia. That is to say, to stress that Bawa's parliament complex instantiates Buddhist structures of feeling is not to suggest that Buddh*ism* is present in this space. It is to provincialize our understandings of what the sacred is positioned to name in the Sri Lankan context.

To be clear, my point here is not that conceiving of Bawa's parliament complex as a straightforward concretization of the secularization thesis is in any sense wrong per se, but rather that doing so mistakenly implies that if the space is secular, it cannot at one and the same time be sacred. In other words, if the sacred and secular exist in a binary relation to one another then spatially the secular must necessarily exist outside the sacred, outside religion that is to say. However, to reason as such is to gloss the colonial continuities of self-certain analytical understandings that portend 'religion' to be a universal and stable Enlightenment category (Asad 2003, p. 35). As such, part of the work of this chapter is to stress the postcolonial imperative for critical and introspective engagements with 'religion' as a concept in South Asian contexts, for 'religion' is itself a knowledge domain with its own colonial histories (see Suthren Hirst and Zavos 2011, pps. 16–20). To this extent, sacred modernity bears some methodological similarity to the ways that this volume mobilizes spirituality as shorthand for the everyday and practical instantiations of religion conceived as an abstraction. However, it also marks an important difference insofar as my argument is that sacred modernity in the Sri Lankan context should be understood on its own terms, not through extant categorical nouns like religion, or spirituality.

The essentialization of the sacred as an external power emerged as European encounters with the non-European world began to deploy 'religion' as a universal category through which the West could identify and map different variations on the things the concept was thought to name (Asad 2003, p. 35). In other words, 'religion' as a concept, and one which implies a rigid sacred/secular binary, has since the colonial era (the nineteenth century in particular) been part of an Orientalist gaze that has effectively disciplined and organized certain elements of South Asian culture and society that were not familiar to the European gaze (Suthren Hirst and Zavos 2011, pps. 18–19). As the anthropologist David Scott (1999, pps. 53–69) has demonstrated, Buddhism was not simply 'discovered' to exist in place in colonial Ceylon. Its emergence *as* a formal 'religion' in nineteenth-century colonial Ceylon had everything to do with a 'comparative science of religion' driven by Orientalist scholars whose obsession was to identify, classify and interpret the existence of 'other religions' extant in the world. By 'other religions' we must emphasize that world religion scholars at the time were operating with a normative, that is to say Enlightenment, conception of 'religion' in which secularism was already implicated (Abeysekera 2002, p. 40). 'Religions' came to be – explicitly at first, then tacitly – understood as textualized systems of doctrines-scriptures-beliefs for which the operation of Christianity provided a template of recognition. Once other 'religions' were identified by these hallmarks (doctrines-scriptures-beliefs), their truth statuses could be investigated, compared (implicitly and explicitly against Christianity) and disputed. As Scott (1999, p. 58) puts it: 'the emergence of the modern concept of "religion" and its plural, "the religions", occurred pari passu with the emergence of the comparative science of religion. Each was, so to speak, the condition of the other's possibility'.

What this reveals reaches beyond just the history of organized religious Buddhist orthodoxy in Sri Lanka. (The emergence of a politicized, majoritarian 'religious' community in late nineteenth and early twentieth-century Ceylon has been characterized as the rise of 'protestant Buddhism' (see Obeysekera 1970; Perera 2002) precisely because its organized institutional structures were derived from the forms of colonial Protestantism at large in the colony). In terms of a history of concepts it also reveals 'religion's' contemporary force as an 'authoritative categor[y] through which the histories of the colonial and postcolonial worlds have been constituted as so many variations on a common and presupposed theme' (Scott 1999, p. 54). In this sense, one of the fundamental problems of the straightforward post-secular thesis for any engagement with Sri Lanka is that it leaves the very taxonomic category of 'religion' in place, thus dissimulating the *different* ways that Buddhist structures of feeling produce space from the inside out in the Sri Lankan context. The post-secular implies spaces that some-time, or somewhere, were once secular and are now 'religious'. The postcolonial challenge in South Asia is to think Buddhism beyond the coordinates of the concept 'religion'. This is the challenge of what, in a similar context, the Sri Lankan anthropologist Pradeep Jeganathan (2004, p. 197) has referred to as the simple elaboration of an unravelling, a slow, uncertain immersion into what has become the ordinary. And it is to those ordinary Sinhala-Buddhist resonances of Bawa's parliament building despite its secularism that I turn now; resonances that exist both iconographically and affectively in ways not reducible to the sacred/secular binary that inheres in 'religion' as a concept-metaphor.

3

Colonial Ceylon's first parliament building was located in the centre of Colombo. It was completed in 1929, and built in the Anglo-Palladian style by Austin Woodeson, chief architect at the time of the Public Works Department (Robson 2002, p. 146). It was in many senses a concretization of colonial legislative power; an elaborate colonial edifice deliberately located in the centre of a city whose preeminence within the colony emerged because of its importance to the plantation economy (see Pererra 1998). In this context, the very decision to relocate Ceylon's administrative capitol from Colombo, the colonial city, to Kotte, a site so resonant historiographically in the Sinhala chronicles, was itself as symbolic as it was practical. It signified a conscious anti-colonial attempt to step outside colonial time, and into the pre-colonial temporality of an island that nationalists thought to be Sinhala and Buddhist historically, and by nature (see Jazeel 2013a, forthcoming). Nonetheless, this was not in itself Geoffrey Bawa's decision given the site was selected well before he was commissioned.

A closer look at the parliament building itself, however (Figure 4.1), reveals more clearly the forms of neither religious, nor entirely secular, sacred modernity that Bawa has built at Kotte. Despite the complex's abstract universalism, it contains a litany of quite deliberate references to non-metropolitan times and spaces, all of which consciously look away from the (colonial) city, instead referencing the (pre-colonial) Sinhala village, an agrarian landscape geography and the former interior kingdom of Kandy. For example, the main building's double pitched roof is a direct reference to the distinct roof style characteristic of Kandyan architecture. The four pillared pavilions surrounding the main building and horizontal concrete pillars that adorn the four sides of the main structure also recall audience or assembly halls across Kandyan towns and villages which historically have provided shelter and rest to travellers and Buddhist pilgrims alike. The complex itself is built on reclaimed land set amidst an artificial lake, and Bawa deliberately created an extensive network of stepped, ornamental terracing across the grounds, making strong visual connections to Sri Lanka's two millennia of tank (reservoir) building and the agrarian paddy cultivation on which the prosperity of pre-colonial Sinhala kingdoms was built (Jazeel 2013a, p. 119). Not usually prone to narrativizations of his own work, Bawa himself once remarked that the whole look of the complex is meant to reflect 'the visual formalities of the old Sinhalese buildings' (Bawa, quoted in Robson 2002, p. 148).

For Bawa then, the parliament complex was meant to extend out into the pre-colonial geography, and indeed temporality, of a nation-state that was retroactively being fashioned as Sinhala and Buddhist all the way back. He also remarked how:

> We have a marvelous tradition of building in this country that has got lost. It got lost because people followed *outside influences* over their own good instincts. They never built right 'through' the landscape. I just wanted to fit [Parliament] into the site, so I opened it into blocks. You must 'run' with site; after all, you don't want to push nature out with the building.
>
> (ibid., my emphasis)

Neither Buddhism nor the Sinhala ethnos are mobilized explicitly or directly here, but his words resonate with popular nationalist refrains of the time concerning the 'outside influences' on an interior and native kernel that is implicitly framed as Sinhala, and just as implicitly thereby Buddhist (even though Bawa himself was not Sinhala-Buddhist). Buddhism then is mobilized not as a 'religion', or religious influence here, but instead as an ornamental facet of the broader effort to historiographically realign the nation-state *in* and *with* its own native modernity. This is the 'literal localism' to which Lawrence Vale (1992, p. 194) refers (quoted above) when he stresses that 'Bawa's capitol complex stands squarely between the abstract universalism of high modernism and literal localism'. If Bawa considered his work to be beyond the divisive politics of ethnicity ('art for art's sake' that is to say), being a modernist he deemed an integral part of his craft to be the recuperation of an appropriate and authentic architectural, artistic and ultimately spatial language for the expression of the nation-state's historical identity. A rooted Sinhala ethnos intractably linked to a historical narrative of Buddhist practice (not 'religion') was part of this anti-colonial modernity. That is part of this space's sacred modernity.

But Geoffrey Bawa's architectural production of this kind of sacred modernity was not just instantiated iconographically. The fluidity and transparency of his architecture was equally if not more important in his attempts at making palpably present these post- and anti-colonial temporalities and environmental aesthetics of the nation-state. Historically, tropical modern architecture across the continents has characteristically blurred the boundaries between inside and outside space (see Goad and Pieris 2005). In large part this has been a stylistic innovation born from the historical necessity to build well-ventilated structures through which light, air and breeze can flow with maximal ease in challenging environmental contexts (see Chang 2016). And in the case of Sri Lankan tropical modernism, tropical architectural innovations in the service of thermal comfort must also be positioned in a historical-political context where expensive imported air conditioning units were increasingly scarce. In Bawa's architecture, these seamless transitions between inside and outside were common, and beyond their techno-political origins they have come to epitomize the types of fluid spatial experience typical of Sri Lankan tropical modernism. He typically employed verandahs, internal courtyards, terraces, folding doors or columns in place of walls, and open hallways as transition spaces and techniques for softening the stark divisions of inside-outside, natural-cultural, public-private (see Figure 4.3). And just as typically, though these architectural devices were in reality drawn from a range of historical influences (Muslim, Hindu, Mughul architecture), they often came to be narrativized as historically Sinhala architectural traditions, often by Bawa's commentators more than himself.

Although the parliament complex is not the best example of his experiments in opening structures out (security requirements limited his capacity to do this at the Kotte site), it is conceived and realized with much of Bawa's characteristic attention to the drama and fluidity of spatial experience. As much as it was a concrete edifice, for Bawa the parliament building was a spatial event extending to the outside and back again. As Nihal Perera (2013, p. 87) writes of the complex:

> . . . the rooms are open to terraces and outside lakes. There are strong thresholds in the Parliament House, not least due to security. Yet, the people who enter walk through covered and artificially lit corridors to arrive at rooms in gardens and offices opening to terraces reminiscent of paddy-fields which are again replicated on the site below, thus creating continuity.

It is not just these smooth transitional features that create a sense of continuity in the parliament complex, Bawa's use of water also aimed at the production of fluid space. His use of reflecting pools and water-retaining structures opened the building's internal spaces out, but also served to link those structures with the wider spatiality of the complex whilst facilitating temporal continuities with places celebrated in popular accounts of anti-colonial Sinhala historiography. As one Sri Lankan archaeologist put it, Bawa's considered use of water 'reflects the ancient traditions of Anuradhapura, Polannaruwa, and Sigiriya' (Senake Bandaranayaka, quoted in Perera 2013, p. 88).

All of these architectural devices aimed, as I have stressed, at producing particular kinds of spatial experience for the user of these built spaces, and elsewhere I have written in depth on how Bawa's architecture, as well as the architecture of other Sri Lankan tropical modernists, has been experienced, lived in, consumed (see Jazeel 2013a, 2013b). In her work on Brazilian artistic tropical modernism, Nancy Leys Stepan (2001, p. 230) suggests how similar artistic managements of tropical nature in mid-twentieth century Brazil aimed at fashioning an appropriately Brazilian disposition to the natural world against a history of European tropical vision. Similarly, if Bawa's tropical modern architecture aimed at creating the experiential illusion that there is little between nature and social space, he did so as a way of expressing something of an 'appropriately Sri Lankan' disposition to the natural world. The effect of building with and into a site like this was, for Bawa, the production of built space that affectively was felt to emerge from the surrounding tropical environmental context, and equally, as we have seen,

Figure 4.2 The transparent and fluid spatiality of the Guest House at Lunuganga, Bentota, Sri Lanka: architect, Geoffrey Bawa.

Source: Tariq Jazeel.

Figure 4.3 Columns and terrace leading to outside space at the back of the Main House, Lunuganga, Bentota, Sri Lanka: architect, Geoffrey Bawa.

Source: Tariq Jazeel.

from a particular historical milieu that was being written as ethnically Sinhala and aesthetically Buddhist. In other words, he was intent on building spatial experience rather than visually prominent structures, and he intended his work to be experienced as *ordinary* components of landscapes not easily divisible into their human and non-human components.

4

The idiom of the ordinary spatial experiences Bawa attempted to create through his work are a crucial component of sacred modernity, and to elaborate on the idiomatic register of Bawa's landscape experience I move now from the architect's parliament complex to his rambling estate, Lunuganga, on the south-west coast of Sri Lanka. Lunuganga was an old, disused rubber estate fringed by a lake. Bawa bought it in 1948. He chose to keep and renovate the main house on the estate's northern hill, and gradually over the next half century he opened up the landscapes and vistas around it with slow and steady precision, imagination and purpose. He experimented by building forms, shapes and structures across the estate, but always in ways attuned to what he perceived as the genus of this place. The garden and estate evolved in texture and dimension, and today its open spaces, terraces and ornamental paddy fields are liberally sprinkled with statues, pavilions and walls, all of which form part of the estate's careful choreography. But as David Robson (2002, p. 240) has written, '[t]oday the garden seems so natural, so established, that it is hard to appreciate just how much effort has gone into its creation'.

Elsewhere, I have written in more depth on Lunuganga (Jazeel 2007, 2013a, 2013b), and it is not my intention here to elaborate on the estate itself. However, Bawa's treatment of the estate stands as an important testimony to the idiomatic configuration of the ordinary spatial experience that Geoffrey Bawa attempted to instantiate through his built space. In other words, his authorship of the estate speaks of the kind of sacred modernity that mobilizes Buddhist structures of feeling as an historical, aesthetic and ornamental component of places that are at once resolutely modernist and thus secular.

Compositionally, Lunuganga is characterized not only by the ways that outside space blends with inside space within its boundaries (Figures 4.2 and 4.3). As at the parliament complex, it also extends out and into the environment beyond the estate itself, into the landscape and nation-state beyond, so to speak. In a glossy, illustrated coffee-table book on the estate, Bawa is himself quoted as saying that 'Lunuganga from the start was to be an extension of the surroundings – *a garden within a garden*' (in Bawa, Bon and Sansoni 1990, p. 11, my emphasis). It is in this context that we should read Bawa's work to ornamentally, and scopically, draw into Lunuganga a view of the gleaming dome of the Katakuliya temple, a Buddhist dagaba, positioned on a hill some distance beyond the estate itself. Indeed, this carefully choreographed long view to the south that 'ended with the temple' (ibid., p. 13), was Bawa's favourite from the estate. Rumour has it that Bawa even paid the monks at the temple to keep the temple's dome white and clean enough such that it was always visible from his vantage point on the estate. If Lunuganga then was to be a 'garden within a garden', it is precisely this kind of work that evidences Bawa's desire that at tropical modernism's core was a naturalization of Sinhala tradition and Buddhist structures of feeling. Bawa's single minded work, typical of modernism, to make the temple central within his landscape composition, such that as he also wrote, it 'now looks as if it had been there since the beginning of time' (ibid.), leaves us under no illusion that the idiom of the larger garden – the garden of the post-independent nation-state, so to speak – is Buddhist ornamentally and historically, if not religiously.

But as I have been suggesting, the idiom of the ordinary in this tropical modern architecture reveals itself not just visually and ornamentally, but also affectively or aesthetically. And here, the coffee-table book, entitled simply *Lunuganga*, is once again useful. The book, published in 1990, is a hardback montage of black and white photographs taken at Lunuganga. The montage is accompanied by a short English language essay on the estate, as well as some of Bawa's sketches and plans of the estate. The book's price tag is discerningly high, and the combination of text and image as well as the book's high production values suitably convey the aesthetic qualities of Lunuganga. All in all, it is a fitting tribute to the special meaning this haven held for Bawa and his closest friends and collaborators.

Precisely because of this, it is also a text that betrays the Buddhist structures of feeling I have been suggesting are key to tropical modernism's sacred modernity. The book's short epilogue is a first person narrative reflection on the estate written by Bawa. In its very last line, he defers to the reaction of a visiting lorry driver who took the opportunity to walk around the estate during a delivery. Bawa (in Bawa, Bon and Sansoni 1990, p. 219) describes the encounter thus:

> . . . when his bricks were being unloaded – [the lorry driver] said to me "මේක නං හරි සීදේවි තැනක්" (but this is a very blessed place).

The significance of this passage is twofold. First, it is in the fact that Bawa chooses to leave the final endorsement of his garden to a working-class Sinhalese lorry driver (we know he is Sinhalese from the Sinhala script). It suggests something of his own desire that, despite his work's quite evident class exclusions, the broader Sri Lankan public might embrace his modernist vision for an appropriately national form of landscape architecture. In other words, the lorry driver's endorsement of the estate is an allegory of its acceptance by the Sri Lankan folk in ways that speak directly to Bawa's lifelong desire to develop a suitably national modern architecture equipped to bring the post-colony into modernity on its own terms.

Second, however, and not at all unconnected to this, the significance of this passage is in the simple desire to reprint the lorry driver's compliment in the language in which it was uttered, Sinhala, in what is an English language publication. For Bawa, the richness of the lorry driver's compliment inheres in its linguistic and cultural idiom in a national context where language politics have a troubled anti-colonial nationalist history. In 1956, the Sinhala-Buddhist nationalist SLFP (Sri Lanka Freedom Party) government passed a Sinhala Only Language Act. In doing so they replaced English with Sinhala (the language of the majority Sinhalese) as the post-colony's official language, at once marginalizing Tamil speaking minorities which included Tamil and Muslim communities. This, however, is but a historical backdrop, and my intention is not to equate Bawa's decision to relay this compliment in Sinhala with the divisive politics of linguistic nationalism in Sri Lanka. As I have stressed, Bawa was always keen publically to distance himself and his work from national politics. Rather, it is the comment's apparent untranslatablity that interests me here, and the precise ways that such untranslatability might be activated as what Emily Apter (2013, pps. 1–27) refers to as a theoretical fulcrum for techniques of reading for difference. In this case, the untranslatability of the Sinhala expression offers a way of comprehending the idiom of the very ordinary, yet radically different, structure of feeling that Bawa means to equate with his architecture; its sacred modernity so to speak. By retaining the Sinhala script, Bawa suggests that the literal English translation cannot capture the essence of the compliment. In other words, he conveys the sense that the English language cannot capture the essence of this place; an essence on which the lorry driver seems to have put his finger. But this is a brief passage caught between untranslatability and translation for the simple fact that it *is* translated for the English language reader. And crucial within this context is the significance of the English language word 'blessed' offered in brackets *as* translation, because it is a word used frequently to refer to the Lord Buddha's enlightened metaphysical state. It is a word that describes an affective state of oneness. In other words, what the lorry driver names is a residual structure of feeling in the spatial present that is quintessentially *Buddhist*, yet at the same time un-nameable in the English language as Buddh*ism* for all the 'religious' connotations this precipitates. As an affect, this is not in any way non-representational, but it is not reducible to any affective resonance that the English language can adequately name; the translation is precisely what transports the language beyond its own limits (Spivak 2008, p. 189). This is Sri Lankan tropical modernism's sacred modernity.

5

It is my argument that this very same sacred modernity, with its characteristic Buddhist structure of feeling, is key to the production of tropical modern architectural space more generally, and equally thus at Bawa's parliament complex. It is pivotal to my argument that we recognize this Buddhist structure of feeling as not 'religious' in the Enlightenment sense of the term. As I have suggested, 'religion' names a self-contained historically European concept with its own objective reality identical to itself the world over. Insofar as the sacred/secular binary is inherent in Enlightenment conceptions of 'religion', then spatially 'religion' implies a secular outside some*where*. On the one hand then, the parliament complex *is* a materially secular institutional space, and it is its very modernism that performatively produces it as a secular space; a secularism on which the proper functioning of political modernity in Sri Lanka depends. On the other hand, however, when we conceive of Buddhism not as a 'religion' per se, but instead as an historical and metaphysical register, Bawa's parliament complex is at one and the same time a space replete with Buddhist structures of feeling produced ornamentally, architecturally and affectively. In this way, sacred modernity is not a politically benign formulation. It serves a dual purpose: first, to give the lie to the secularism inherent to, and essential for, political modernity, and second, to spatially produce the post-colony in modernity as historically, essentially and metaphysically Buddhist and Sinhala all the way back. This is precisely what makes it impossible for Tamil, Muslim and other non-Sinhala-Buddhist others to be anything but guests in a national polity spatially produced as such.

Note

1 In this chapter, I use the terms 'post-colony' and 'post-colonial' to refer to the Ceylon/Sri Lanka's status after formal decolonization and thus the time period after colonialism. I use the term 'postcolonial' on the other hand to name methodological and theoretical approaches attentive to the ideological presence of colonialism in the present and attempts to transcend those colonial remains.

References

Abeysekara, A., 2002, *Colors of the robe: religion, identity and difference*, University of South Carolina Press: Columbia, SC

Apter, E., 2013, *Against world literature: on the politics of untranslatability*, Verso: London and New York

Asad, T., 2003, *Formations of the secular: Christianity, Islam and modernity*, Stanford University Press: Stanford, CA

Bartholomeusz, T., 1999, 'First among equals: Buddhism and the Sri Lankan state', in I. Harris [Ed.], *Buddhism and politics in twentieth century Asia*, Continuum: London and New York, pps. 173–193

Bawa, G., C. Bon and D. Sansoni, 1990, *Lunuganga*, Times Editions: Singapore

Chang, J.-H., 2016, *A genealogy of tropical architecture: colonial networks, nature and technoscience*, Routledge: London and New York

Goad, P. and A. Pieris [Eds.], 2005, *New directions in tropical Asian architecture*, Periplus: Singapore

Jazeel, T., 2007, 'Bawa and beyond: reading Sri Lanka's tropical modern architecture', *South Asia Journal for Culture* 1, pps. 7–26

Jazeel, T., 2013a, *Sacred modernity: nature, environment, and the postcolonial geographies of Sri Lankan nationhood*, Liverpool University Press: Liverpool

Jazeel, T., 2013b, 'Dissimulated landscapes: postcolonial method and the politics of landscape in southern Sri Lanka', *Environment and Planning d: Society and Space* 31 (1), pps. 61–79

Jazeel, T., forthcoming, 'Urban theory with an outside', *Environment and Planning D: Society and Space*

Jeganathan, P., 2004, 'Discovery: anthropology, nationalist thought, Thamotharapillai Shanaathan and an uncertain descent into the ordinary', in N. L. Whitehead [Ed.], *Violence*, School of American Research Press: Santa Fe, New Mexico, pps. 185–202

Jones, R., 2011, 'Memory, modernity and history: the landscapes of Geoffrey Bawa in Sri Lanka, 1948–98', *Contemporary South Asia* 19 (1), pps. 9–24

Leys Stepan, N., 2001, *Picturing tropical nature*, Reaktion: London

Obeyesekere, G., 1970, 'Religious symbolism and political change in Ceylon', *Modern Ceylon Studies* 1

Perera, N., 2002, 'Indigenizing the colonial city: late 19th century Colombo and its landscape', *Urban Studies* 39 (9), pps. 1703–1721

Perera, N., 2013, 'Critical vernacularism: multiple roots, cascades of thought, and the local production of architecture', in N. Perera and W.-S. Tang [Eds.], *Transforming Asian Cities: intellectual impasse, Asianizing space and emerging translocalities*, Routledge: London and New York, pps. 78–93

Pererra, N., 1998, *Society and space: colonialism, nationalism and postcolonial identity in Sri Lanka*, Westview Press: Boulder, CO

Pieris, A., 2007a, *Imagining modernity: the architecture of Valentine Gunasekera*, Stamford Lake (Pvt) Ltd. and Social Scientist's Association: Colombo, Sri Lanka

Pieris, A., 2007b, 'The trouser under the cloth: personal space in colonial modern Ceylon', in Scriver and Prakash [Eds.], *Colonial modernities: building, dwelling and architecture in British India and Ceylon*, Routledge: London and New York, pps. 199–218

Robson, D., 2002, *Geoffrey Bawa: the complete works*, Thames and Hudson: London

Robson, D., 2007, *Beyond Bawa: modern masterworks of monsoon Asia*, Thames and Hudson: London

Scott, D., 1999, *Refashioning futures: criticism after postcoloniality*, Princeton University Press: Princeton, NJ

Spivak, G., 2008, *Other Asias*, Blackwell Publishing: Oxford

Suthren Hirst, J. and J. Zavos, 2011, *Religious traditions in modern South Asia*, Routledge: London and New York

Vale, L., 1992, *Architecture, power and national identity*, Yale University Press: New Haven, Conecticut

Williams, R., 1977, *Marxism and literature*, Oxford University Press: Oxford

5 'I renounce the World, the Flesh, and the Devil'

Pilgrimage, transformation, and liminality at St Patrick's Purgatory, Ireland

Richard Scriven

Each of us, in turn, each of us kneels and says three Our Fathers, three Hail Marys, and one Apostles' Creed at St Brigid's Cross – a cross marked on the exterior of the basilica – before standing with our backs to the cross, with arms fully outstretched, and say three times aloud 'I renounce the World, the Flesh, and the Devil'. This embodied prayer captures an essence of Lough Derg pilgrimage. We, as pilgrims, intentionally separate ourselves from the everyday world to pursue a temporary life of prayer and personal contemplation. Within this space, the pilgrim's journey has physical practices interlinked with metaphysical layers of spirituality and emotionality. It is the voluntary entering into a transitionary social and spiritual state with the intention of achieving a form of renewal or rejuvenation. This potential for spiritual or personal transformation marks pilgrimage out as a distinct form of journey.

St Patrick's Purgatory, or Lough Derg as it is more popularly known, is a Roman Catholic pilgrimage site in northwest Ireland. Pilgrims spend three days on a lake-island where they withdraw from the rest of the world and complete a set of requirements to focus on the more meaningful and spiritual dimensions of life. It is a centuries' old practice of prayer, fasting, going barefoot, and keeping vigil. Magan (2014) describes it as involving 'three days of fasting and prayers, while standing on sharpened rocks. It's not for everyone, but there must be a reason why people return each year'. This account conveys the distinct nature of St Patrick's Purgatory as a pilgrimage that offers meaningful encounters to thousands of people annually. The requirements of the pilgrimage combine to facilitate liminal experiences through which pilgrims can reflect on their beliefs, their lives, and themselves. Within this space, pilgrims find new meanings and reach fresh insights (Maddrell and Scriven 2016). My study occurs within such a space.

This chapter on the transformative aspects of Lough Derg is based on an auto-ethnographic field study comprising of participation in the pilgrimage and interviews with pilgrims both at the site and afterwards. In addition, I examined historical and published accounts of the island. This aligns with recent research which foregrounds direct engagements with the experiences of pilgrimage (Frey 1998; Maddrell 2013; Maddrell and della Dora 2013; Michalowski and Dubisch 2001). I blend these strands to explore how Lough Derg can further understandings of the emergence of liminality and the enabling of personal and spiritual

Figure 5.1 Pilgrims at St Brigid's Cross. One pilgrim stands outstretched reciting the prayer, while others kneel in prayer before standing themselves.

Source: Richard Scriven.

renewal. In general the research participants are described as 'pilgrims' to reflect the character of Lough Derg, which adopts a loose definition incorporating all who come to the island in search of something beyond themselves (Lough Derg 2016).

Pilgrimage is a dynamic phenomenon which has witnessed a considerable revival in recent decades (Alliance of Religions and Conservation 2012; Jansen 2012). Moreover, definitions have broadened beyond a religious focus to encompass cultural, nationalistic, and personal journeys (Coleman and Eade 2004; Gale, Maddrell and Terry 2016). In this shifting context, the examination of the characteristics of pilgrimage become more relevant. It is broadly understood as involving a journey to a specific place for religious-spiritual, and/or cultural-emotional, reasons. The outer physical journey enables a corresponding inner metaphysical one (Hyndman-Rizk 2012; Schmidt and Jordan 2013). Within this framework the 'inner intellectual, emotional and spiritual journey' is often seen as being the most significant component (Maddrell 2011, p. 16). Ritualistic practices facilitate reflective and transformative experiences through which pilgrims can (re)consider their spiritual identity and place in the world (Osterrieth 1997; Turner and Turner 1978). This transformative capacity speaks to the distinct nature of pilgrimage as a form of meaning journey. It is the effective 'destination', rather than a shrine or a significant place. My focus on this aspect highlights an essence of pilgrimage, which will help progress understandings of the activity's contemporary spiritual and socio-cultural role.

In this chapter, I consider how Lough Derg is instilled with a pronounced transformative potential through a focus on embodied performances and liminality. Conceptual understandings of renewal within pilgrimage are enlivened in accounts of the island as research participants reveal sincere feelings concerning the reaffirmation of faith or an appreciation for the important things in life. Different forms of holistic individual renewal unfold within the micro-geographies of the pilgrimage. As Lough Derg is a Roman Catholic site with the accompanying structures and connotations, the religious-spiritual aspect is foregrounded; however, there are multiple layers to this process as the emotional and social are equally present. The role of the pilgrim enables the emergence and temporary nurturing of the religious-spiritual, which is generally understated or neglected in everyday lives. Accompanying these shades of awareness is a more general affective recognition for the actual priorities in life. It is a reaffirmation of both traditional and contemporary interpretations of the site and of pilgrimage. The transformative capacity is visceral and authentic in these settings, making a palpable impact on the pilgrims.

My discussion opens with an exploration of pilgrimage as a journey of transformative capacity, informed by both the tropes of the practice and interpretations from the field of pilgrimage studies. I draw from research increasingly focused on the 'embodied-emotional-spiritual-social-spatial relations' (Gale, Maddrell and Terry 2016, p. 2), which intervenes within the journey space through performance and engagement with pilgrims (Dubisch 1995; Frey 1998; Coleman and Eade 2004; Maddrell 2013; Maddrell and della Dora 2013; Michalowski and Dubisch 2001). In the following section, a description of Lough Derg and the different components of the three-day tradition establish the distinct character of this space. Embodied spatial practices forge this reflective space, in which the pilgrims, as active agents, co-generate metaphysical journeys in conjunction with the physical and social context. Next, the focus falls on the emergence of liminality through the structural arrangements of Lough Derg and the resultant pilgrim experiences. A genuine separation from the normative is affectively registered as participants occupy a contemplative state. This enables significant encounters which are explored in the following section. Participants offer accounts of how emotional and spiritual rejuvenation is nurtured on the island and how they return to the world in a refreshed state. The chapter concludes by outlining the significance of the transformative dimension of pilgrimage.

Pilgrimages: journeys of transformation

Pilgrimages are journeys. Most obviously, they are journeys to shrines, sacred places, and ritualised locations across faiths, cultures, and traditions. Interlaced with these outer physical journeys are inner personal journeys of emotional, spirituality, and personal growth (Maddrell 2013; Rountree 2006). Pilgrimage's unique role exists in the merging of these aspects. It presents an appealing and structured means of undertaking a meaningful journey. Pilgrimages continue to be larger religious-spiritual practices, with five million Muslims making the Hajj annually, approximately twenty million Catholic pilgrims going to Guadalupe in Mexico, and twenty-eight million Hindu pilgrims travelling to the River Ganges. More recently, the concept of pilgrimage is being recognised as incorporating a range of cultural, nationalistic, and personal journeys, such as visits to war graves, Elvis's Graceland, and ancestral homelands (Campo 1998; Coleman and Eade 2004).

Pilgrimage can be seen 'as a ritual of transformation of the self' (Gemzöe 2012, p. 42). It offers a distinct means of moving beyond normativity by 'looking for an experience outside the margins of material interest and the simplistic pursuit of gain' (Oviedo, Courcier and Farias 2014, p. 441). Osterrieth (1997, p. 27) explains it in anthropological terms as a ritualised quest that 'stems from an individual decision and aims at personal transformation'. This involves a separation from home and all associated social normativity to enter a marginal or liminal state, through a journey or ritualised approach to a specific site. This enables spiritually or personally meaningful encounters, before the pilgrim returns as with a new identity having been spiritually and/or emotionally revived. By consciously breaking with everyday life, undertaking a journey, and 'ritually' participating in the pilgrimage site, participants engage in a process of renewal through which they can establish a new identity or sense of self. This replicates broader understandings of rites of passage in which the ritual subject undergoes separation, transition, and, then, incorporation (Gennep 1960). In this process, the old self of the pilgrim 'dies' and a new self is (re)born; a self which returns to the everyday emotionally or spiritually transformed (O'Giolláin 2005). In religious pilgrimages, it involves the movement from the secular world to sacred spaces where the divine is more easily encountered through more immersive spiritual experiences (Gesler 1996). Moreover, in a Christian context, the trope of transformation is theologically central (Maddrell 2011). Traditional medieval pilgrimages focused on penitential exercises as the means for spiritual progression, whereas modern pilgrimages emphasise renewal through prayer, reflection, and developing a personal relationship with God.

The transformative potential of pilgrimage is understood to be facilitated by entering a liminal state. The concept of liminality, established as one of the main theoretical tools in the study of pilgrimage by Victor and Edith Turner (1978), describes how pilgrims can experience a marginal state as a ritual subject between two identities or social positions. In this anti-structural position they become ambiguous, existing between definite social states. Pilgrimage 'provides a carefully structured, highly valued route to a liminal world where the ideal is felt to be real, where the tainted social persona may be cleansed and renewed' (Turner and Turner 1978, p. 30). There is a disengagement from normative rules, routines, and responsibilities which facilitates personal and spiritual transformation (Osterrieth 1997). It involves 'a fundamental ritual pattern of transformation by means of a spatial, temporal, and psychological transition' (Bell 1997, p. 248). Although the idea of liminality has gained widespread use more recently, within this framework, it has a distinct theoretical function. Critiques of the concept have highlighted how pilgrimage is never entirely separated from its social and political contexts (Coleman and Eade 2004). Indeed, there have been numerous studies which have focused on pilgrimage as a site of contestation or a political activity (Digance 2003; Galbraith 2000; Pazos 2012). Refinements, however, now appreciate liminality as an ontological state that is shaped by the features of the pilgrim journey (Slavin 2003). Pilgrimage studies tend to employ it in terms of a balance between its social commentary and conceptual meaning.

Pilgrimages are appreciated as an interlacing of an outer physical journey with an inner spiritual or emotional one. The meanings participants bring 'imbu[e] the actions, objects and spaces with considerable significance for individuals and whole communities' (Scriven 2014, p. 252). Traditionally this has taken the form of religious-spiritual beliefs in the form of travelling to shrines and sites of miraculous events, whereas trips motivated by secular convictions or cultural involvement are equally appreciated in contemporary discussions. These meanings are explored and revitalised in an inner journey which is facilitated by the outer physical journey. In many cases, especially for believers, the 'inner intellectual, emotional and spiritual journey' is seen as being more significant than the 'demands and challenges of the outer physical journey' (Maddrell 2011, p. 16). The meanings that are brought to pilgrimage generate and forge the significance of the journey. The practices and physical exertions are the embodied expression of these beliefs. Together they combine to produce not only substantial encounters, but also transformative experiences.

Lough Derg

Lough Derg offers a space of spiritual retreat centring on practices that have been inherited from at least the early modern period, including sets of prayers, fasting, going barefoot, and keeping an all-night vigil. It reaches towards a medieval past, while remaining firmly located in the present. While the island is continually modernised and developed, it retains a character that appeals to thousands of pilgrims who are drawn by the capacity of this place to facilitate journeys of personal and spiritual reflection and transformation.

The origins and history of Lough Derg reinforce its role as an exceptional place. It is believed that in the fifth century St Patrick spent the religious season of Lent – the six weeks preceding Easter – on retreat in a cave on the lake island, during which he received a vision of the afterlife. These miraculous events marked the island out as a sacred space, or a thin place, where the boundary between the natural world and the spiritual realm was permeable. The earliest written records date from the twelfth century and associate the site with the Roman Catholic doctrine of purgatory (a transitory state during which souls are cleansed before entering heaven) (Flynn 1986). By imitating St Patrick's asceticism, pilgrims believed that they could spiritually purify themselves and achieve salvation by enduring an earthly purgatory (Cunningham and Gillespie 2004). It gained relative prominence in medieval Christendom as a site of pilgrimage.

Since 1780 Lough Derg has been administered by the Roman Catholic Diocese of Clogher and has been developed with the addition of dormitories, services, and St Patrick's Basilica (Flynn 1986). The site is headed by a diocesan priest, called the Prior, and is staffed by both lay and religious, including a pastoral team of other priests and counsellors. Over 10,000 pilgrims undertake the three-day pilgrimage annually. This number has fluctuated over the past century, from 8,000 in 1921 to 34,645 in 1952 (Duffy 1980).

Structurally, the different aspects of the pilgrimage align to produce a liminal location where normativity is voluntarily suspended. Within these conditions participants become more open to transformative encounters. The pilgrimage begins at midnight with a seventy-two-hour fast consisting of one meal a day, of dry bread or toast, oatcakes, and tea or coffee, without milk. Water can be consumed freely throughout and soft drinks are allowed on the third day (when people are travelling home). Fasting – a penitential activity which was practiced in the early medieval Celtic Church (Wooding 2003) – is appreciated as a sacrifice of earthly desires that enables a focus on spiritual concerns.

On the morning of the first day, pilgrims get a boat across to the island. This physical withdrawal from the world is reinforced by turning phones/devices off, severing a constant connectivity. On the island, shoes are removed and the barefooted state begins. Through the bodily register of feet meeting the surfaces of the island, the aesthetics of a medieval pilgrimage are felt and lived, adding to the sense of timelessness. While on the island, nine Prayer Stations are performed. These are a pattern of prayers involving the repeated reciting of specific prayers – Our Father, Hail Mary, and the Apostles' Creed – while walking around and kneeling at different features. Through these prayer states and numerous religious ceremonies pilgrims enter a 'liturgical life' (della Dora 2012, p. 969). That night the twenty-four-hour Vigil begins with participants staying up all night performing four prayer stations and keeping each other's spirits up. This is often seen as the very heart of the pilgrimage.

The following day is marked by personal reflection and religious ceremonies, including mass and confessions. The Vigil ends with night prayers that evening. The final morning begins with mass, followed by the final prayer station. Many people pick up religious items and souvenirs, which are blessed during mass, to give to family, friends, and neighbours. Pilgrims return to the shore filled with a renewed spirit and begin their transition back into ordinary life. However, the fast continues until midnight, extending the pilgrimage experience beyond the shrine and disrupting clear sacred-profane boundaries.

Lough Derg is a combination of these features, as body and meaning, and performance and place meet in the enactments. Liminalities emerge in these interactions, facilitating spiritual, more-than representational, and numinous experiences. The structures of the pilgrimage are reinforced by leaflets given to each pilgrim and by the staff who advise. By entering into this framework, pilgrims are freed from everyday concerns, enabling a concentration on prayer and contemplation.

Lough Derg is a removal from the world. Similar to other pilgrimage centres, it is a place where people seek experiences that offer release from the limitations of daily life (Osterrieth 1997). It is the embracing of a liminality that seems to disrupt modern sensibilities in search of something beyond the everyday and observable.

Lough Derg can be located within this wider context. Although it is clearly Roman Catholic in nature, it is a pilgrimage destination that attracts a wide variety of people. Devoted members of the denomination walk barefoot alongside those with only a loose affiliation to Catholicism and those who define themselves as spiritual, rather than religious. Lough Derg's website (2016) emphasises that it 'welcomes those from all religious practices and backgrounds and regular Church attendance is not a pre-requisite of completing the pilgrimage'. Moreover, the information and promotional literature present that pilgrimage in broad terms as a place of prayer, reflection, and searching for meaning. While being inherently theistic it avoids the denominational character or overt doctrine of other pilgrimage sites.

Liminality

Lough Derg has a pronounced liminal capacity. In many ways it aligns with the conceptual ideals outlined by Turner and Turner (1978). Not only is there a clear break from the everyday, but we also inhabit a temporary disconnected world which exists almost parallel to the quotidian. A further sense of otherness, outside of the din of modern living, is generated through the physicality of the island in a remote valley (Ivakhiv 2003). This simple watery barrier separates us from the world. On crossing to the island, a distinct departure is enacted, while sitting looking towards the mainland reinforces our separation on practical and affective registers. In addition, the requirement for pilgrims not to use phones/devices is an equally significant means of withdrawing from contemporary society.

The physical, symbolic, and felt conditions of the pilgrimage align to facilitate these liminal experiences (Figure 5.2). Our voluntary involvement generates the liminal conditions we encounter. We become pilgrims. We become of this liminality. In talking with Eleanor, who was on the second day of her pilgrimage, we discussed this sense of detachment from the world. We are seated near the lake shore looking towards the entrance and buildings on the mainland. She explains how:

> It's just a different world . . . there could be anything happening beyond those pillars there [*points to the main entrance] and we won't know.

She touches on this intentional isolation that we have entered into. Although we can see across to the mainland, which stands in for the rest of the world, we can feel our separation. In considering this distance, it becomes a soothing chasm, a buffer between us and the world. There is a further liberation in this realisation as we settle into our detached role which invites time for personal reflection and prayers.

Figure 5.2 Lough Derg statue of St Patrick the Pilgrim, with the island in the background, the lake waters separating it from the rest of the world.

Source: Richard Scriven.

The virtual disconnection of leaving your phone behind or switched off is increasingly felt as being a significant aspect of the pilgrimage. It is frequently mentioned by pilgrims as being one of the most welcome components of the pilgrimage. Contemporary existence, no matter how much it is criticised or bemoaned, involves a tethering to our mobile devices. Even when on holidays or annual leave, we have a persistent feeling of an obligation to check emails or update social media. However, the requirement of the pilgrimage, combined with the character of the place, provides a palpable relief. The island is emancipatory. Another pilgrim, Ann, mentions the importance of this aspect for her:

> We're all so busy. Everybody is on. I find at work with email, and then I have an iPhone, so you never get off-line. . . . I think it's good that things, you know, that we can cut off and just get back to basics, maybe listen to the silence for a while.

Crucially, this intentional disengagement, as Ann alludes to, is not only about separating ourselves from the world, but also using this condition to create a space for contemplation. The default setting of turning to our mobile devices is disrupted and we are gladly forced to sit and think, to reflect or chat.

These themes are woven into the pilgrim experience through the performances and structures of the pilgrimage. Fasting and bare feet combine with the features of the island named after prominent ascetic saints creating an affective liminal landscape. The aforementioned St Brigid's Cross on the side of the basilica brings these aspects into sharp relief as each of us, during our prayer stations, renounces the world in an assertion of spiritual separation from earthly concerns. John, who had been to Lough Derg several times, emphasised the role of St Brigid's Cross in his pilgrimages:

> You say, 'I renounce the world, the flesh, and the devil'. Em, and I think I had to say, I said it out loud the first time, but it's a very, very unusual thing to do. And, it's almost like you are renouncing your physical body, but, em, the entire world.

In describing the importance of this point, John illustrates how the larger spiritual significance is materialised and verbalised in this act. The liminality which is facilitated by the structures of the shrine becomes personalised as each individual stands at the cross, holding out their arms, as they make their declaration three times. It is a self-conscious action as we are all called on to not only perform this prayer, but also to reflect on what it means. We kneel praying in preparation, aware of our fellow pilgrims standing up to renounce, before each of us must ourselves make this prayer aloud. It is an individual and collective testament. A clear purpose is added to the seclusion from the world as each person takes on an ownership of their pilgrim journey and we as a group share this commitment. Moreover, this point is reinforced throughout the three days, as the St Brigid's Cross prayer and action is repeated during each of the nine stations. St Brigid's Cross then becomes a touchstone for the whole three days, encapsulating the processes that enable the emergence of liminality. Here John presents an account which aligns with the sentiments of other pilgrims, but for him it is tied to the renouncing at that Cross. This is a defining point on his spiritual journey where he is conscious of his separation from the everyday world and all it entails.

Through structural, symbolic, and emotional modalities, Lough Derg manifests a distinct form of liminality. The physical characteristics of a lake-island blend with the requirements of the three-day pilgrimage and the commitment of each individual to produce an almost textbook example of the liminal experience. We are separated from the world in a way unlike other settings. We genuinely leave behind deadlines, appointments, and to-do lists. The island becomes a liberating space as we are both allowed to and allow ourselves to leave all of it behind. Inhabiting this space enables us to consider other parts of our lives as the deeper stiller waters begin to surface. We move further on our journey.

Transformation and renewal

Processes of transformation and renewal emerge from active engagements with the meaningful aspects of life that are enabled through the pilgrimage. The shedding of everyday concerns is not 'disengagement from the challenges of one's life, but rather a journey toward the transformative possibility that the journey itself contains' (Schmidt and Jordan 2013, p. 67). It is valued as a distinct opportunity to reflect on the more important parts of ourselves and our lives. In this liminal space, deeper, more profound sentiments, ideas, and feelings are allowed to surface (Slavin 2003; Turner and Turner 1978). Lough Derg provides the time, space, and mind-set for these purposeful reflections. These considerations are manifest in pilgrims performing the prayer stations, praying privately, sitting quietly overlooking the lake, chatting with fellow pilgrims, or in conversation with the staff. Journeys of transformation and renewal are nurtured in this reflective space.

While the pilgrimage has a broader Christian spirituality appeal, it remains a Roman Catholic site of devotion where religious pilgrims travel annually to prayer for special intentions and to take time out to communicate with God. For one such pilgrim, Kathleen, the prayerful and peaceful aspects stood out for her. These three days were a special way of relating to God and developing her faith:

> It's so peaceful, you kind of get an inner calm when you come here, you know? . . . You get time to connect to God, the prayers now and the singing, it's just lovely like, you know? I mean at home you go to Mass, you're probably rushing home to make the dinner or something. . . . Whereas at least here, you can slow down, you know? There's time to slow down. It's peaceful and tranquil. . . . You have time for God and time for yourself.

Kathleen's pilgrimage takes what is generally classified as a traditional form relating to prayer and developing personal faith. Even though she is someone who practices her Catholicism regularly, she is aware of how weekly mass fits in as another component of ordinary life. Worship and prayer are regularised and scheduled in a manner, which although necessary, can erode their purpose. At Lough Derg, where there is a deliberately slower pace, she has time to pray and to participate in the liturgies in a more rewarding and resonant way. Her connection with God is experienced in an impactful manner through the pilgrimage. This serves to strengthen her faith in very real ways. It is a spiritual revival.

Comparably Martin locates his time at St Patrick's Purgatory as being an important feature of his spiritual life. He returns to the pilgrimage regularly as it offers him a means of revitalising his faith:

> It's very much a nourishing point for my faith. It was such a positive experience. You know, I can find it hard to be prayerful in my life, every day or every week, you know, or throughout the day. Whereas something like that I find such a profound spiritual experience. It's nice to have this. . . . I suppose, refreshment or nourishing point, you know, going back to daily life with, having had this experience.

There is a very clear sense from Martin's words of the significance he attaches to his pilgrimage. Not only is it a special event which offers him the time and space for reflection, it is an intensely significant spiritual encounter. The everyday world is not conducive to reaching a prayerful state with little time available to truly be still and pray. A spiritual immanence is facilitated by the conditions of the Lough Derg as Martin settles into the rhythms and sensibilities of pilgrimage. The ethereality of faith, which often remains beyond his grasp in normative circumstances, becomes a felt and lived experience on this journey.

The experiences of both Kathleen and Martin correspond to one of the broader rationales for pilgrimages, as they let 'believers act out religious tenets in concrete ways' (Gesler and Pierce 2000, 228). Moreover, it is appreciated how faith and spirituality can be reaffirmed in such visceral and intense moments (Beckstead 2010). These journeys then take on an important role as a form of religious transformation that reinvigorates their faith and themselves personally.

The morning of the third day is frequently mentioned as a highpoint of the pilgrimage. An earned satisfaction pervades, one drawn from enduring the difficulties of the rites. The aggregation of the journey results in a celebratory atmosphere, as individually and collectively the pilgrims begin to return to the world renewed (Osterrieth 1997). We have completed our prayer stations, put back on our shoes, and are getting ready to leave the island. As we depart, there is an intense sense of completion and renewal. Having endured the challenges of the pilgrimage, we now return to the world spiritually and emotionally refreshed. Gráinne, who had completed the pilgrimage several times, explains her experiences of that last morning:

> And, we all just remember that epic feeling we have on day three, and we kind of forget how awful it's been on like day two or the vigil night, and it's only when you come back again that you realise how that feels, but somehow it's like when you revised for your exams all you remember are the results, you don't remember that revision period.

This 'epic feeling' is the completion of the pilgrimage. She appreciates how the hardships of the three days generate the sense of revitalisation on departing the island. The distinct challenges of Lough Derg led to this crescendo. It is only because of the trials involved that the achievement is so purposeful. Moreover, Gráinne's regular participation in the pilgrimage illustrates the value she places on these experiences and its reviving nature.

The impact of Lough Derg can last well beyond the immediacy of three days as people carry the insights and feelings with them into their ordinary lives. For religious believers it reinforces their faith, while for those with a more spiritual or agnostic outlook it can be equally personally reviving. Grace, who I talked with after her pilgrimage, described having a meaningful experience at Lough Derg which stuck with her:

> I don't particularly have a wonderful belief in the Catholic Church or in faith, or that kind of thing; but I kind of thought: 'I'll go and see'. . . . I felt wonderful! Still feel wonderful after it. You feel a lot lighter. Mmm, I don't know

> what it is, why it works the way it does; but, it does seem to make you feel a bit lighter afterwards. . . . It clears away all the material stuff, all the rubbish of your daily life and you're just back to basics, aren't you? You just can think about what's happening at the minute. All the other stuff doesn't [matter].

Sensibilities of liminality and transformation come through in Grace's account of her pilgrimage. Her journey was not based in a religious faith but nonetheless drew from the exceptional nature of Lough Derg as a transitional and reflective space.

A core characteristic of pilgrimage is the desire to search for a 'mystical or magico-religious experience' through which pilgrims 'experience something out of the ordinary that marks a transition from the mundane secular world of their everyday existence to a special and sacred state' (Collins-Kreiner 2010, p. 442). This feature is distinctly evident on Lough Derg as pilgrims withdraw from the world and dwell in a spiritual context which facilitates meaningful experiences. These spiritual and personal reflections help people take stock of aspects of their lives, re-consider issues, and strengthen their faith. In different ways the research participants articulated a sense of renewal which they carried with them back to the world.

Conclusion

In this chapter, I have articulated the transformative dimension of the Lough Derg as a form of personal and spiritual rejuvenation. Pilgrims pursue metaphysical journeys through the embodied practices of the island, as medieval aesthetics, a retreat from the world, and personal motivations intermix. They are afforded the space and disposition to reflect on themselves, their lives, and their spiritualties. These opportunities are valued as being a rarity amongst the demands of everyday living that allow the pilgrims to be still and truly contemplate. Within this process, corporeal and affective resonances present an avenue towards genuinely meaningful encounters. These considerations allow for a re-appreciation for the role of the transformative within pilgrimage journeys.

I emphasise how a contemporary pilgrimage is manifest as an active process of change. By engaging in a temporary performance through a ritualised journey, participants can induce personal change in the form of spiritual or emotional progression. There is a distinct agency involved with each person intentionally entering into the space and enacting the embodied practices. This draws attention to the continuing cultural relevance of the pilgrim as a social role which is being adopted for religious-spiritual reasons, alongside more secular motivations. I build on recent trajectories in pilgrimage studies which are intervening in the spaces and experiences of the journey by considering how the transformative dimension is manifested through embodied spatial practices on an individual scale. This unravels how the concept of the pilgrimage journey is encountered in the realities of a Western Christian site, revealing how many of the tropes are present but are experienced very personally.

In foregrounding the transformative dimension of pilgrimage, I have examined how purpose and practice interweave to facilitate personal rejuvenation. While transformation is considered to be a significant component of pilgrimage and one of the features that distinguishes it from other religious/spiritual and cultural activities, it needs to be appreciated on the scales in which it occurs. Embellished religious accounts and popular concepts can tend to emphasise the expressive or spectacular components of pilgrimage; however, a more nuanced consideration understands the experiences to often be quiet, personal, and subtle. The encounters presented by the research participants in this chapter are of this order. It is a subdued, yet resonant, form of renewal that is manifest in spiritual and emotional registers. These feelings are slowly arrived at through the tranquility and rhythms of located ritual practices. This highlights the significance of investigating both the process and modalities of transformation within pilgrimage.

These features prompt further questions about the nature of liminality and how we conceive of it in relation to pilgrimage. While the original concept of the liminal as a social state has been disrupted and developed, its role in processual, practiced, and embodied terms needs to be more fully explored and articulated. There are rich ontological and practical aspects of the concept within pilgrimages; in particular, how it facilitates pilgrims experiencing transformative encounters. This also highlights the need for research to occur in the midst of pilgrimages as these liminal spaces are being forged by the participants on their journeys of belief, searching, and contemplation. While such interventions need to be practiced in a conscientious and thoughtful manner, they can yield rich insights into these momentary and ephemeral worlds.

The prayer of 'I renounce the World, the Flesh, and the Devil' is repeated continually at St Brigid's Cross throughout the pilgrimage. Pilgrims of varying religious beliefs and spiritual dispositions each make this prayer on their Lough Derg journeys. It encapsulates a commitment to the spirit of this pilgrimage, as each person rejects the normative, at least temporarily, and embraces the transitionary status of the pilgrim. It is in such acts and intentions that the liminalities and potential transformations of Lough Derg are nurtured and enabled.

References

Alliance of Religions and Conservation, 2012. *Pilgrim numbers*. ARC (Alliance of Religions and Conservation). Retrieved May 15, 2016, from www.arcworld.org/projects.asp?projectID=500

Beckstead, Z., 2010. Commentary: Liminality in Acculturation and Pilgrimage: When Movement Becomes Meaningful. *Culture Psychology* 16, 383–393. doi:10.1177/1354067X10371142

Bell, C.M., 1997. *Ritual: Perspectives and Dimensions*. New York, Oxford University Press.

Campo, J.E., 1998. American Pilgrimage Landscapes. *The Annals of the American Academy of Political and Social Science* 558, 40–56. doi:10.1177/0002716298558001005

Coleman, S., and Eade, J., 2004. Introduction: Reframing Pilgrimage, in: Coleman, S., and Eade, J. (Eds.), *Reframing Pilgrimage: Cultures in Motion*. London, Routledge, pp. 1–15.

Collins-Kreiner, N., 2010. Researching Pilgrimage: Continuity and Transformations. *Annals of Tourism Research* 37, 440–456. doi:10.1016/j.annals.2009.10.016

Cunningham, B., and Gillespie, R., 2004. The Lough Derg Pilgrimage in the Age of the Counter-Reformation. *Éire-Ireland* 39, 167–179.

della Dora, V., 2012. Setting and Blurring Boundaries: Pilgrims, Tourists, and Landscape in Mount Athos and Meteora. *Annals of Tourism Research* 39, 951–974. doi:10.1016/j.annals.2011.11.013

Digance J., 2003. Pilgrimage at Contested Sites. *Annals of Tourism Research* 30, 143–159. doi:10.1016/S0160-7383(02)00028-2

Dubisch, J., 1995. *In a Different Place: Pilgrimage, Gender and Politics at a Greek Island Shrine*. Princeton, NJ, Princeton University Press.

Duffy, J., 1980. *Lough Derg Guide*. Dublin, Irish Messenger Publications.

Flynn, L.J., 1986. *Lough Derg: St Patrick's Purgatory*. Dublin, Eason.

Frey, N.L., 1998. *Pilgrim Stories: On and Off the Road to Santiago, Journeys Along an Ancient Way in Modern Spain*. Berkeley, CA, University of California Press.

Galbraith, M., 2000. On the Road to Częstochowa: Rhetoric and Experience on a Polish Pilgrimage. *Anthropological Quarterly* 73, 61–73. doi:10.2307/3317187

Gale, T., Maddrell, A., and Terry, A., 2016. Introducing Sacred Mobilities: Journeys of Belief and Belonging, in: Maddrell, A., Terry, A., and Gale, T. (Eds.), *Sacred Mobilities: Journeys of Belief and Belonging*. London, Routledge, pp. 1–17.

Gemzöe, L., 2012. Big, Strong and Happy: Reimagining Femininity on the Way to Compostela, in: Jansen, W., and Notermans, C. (Eds.), *Gender, Nation and Religion in European Pilgrimage*. Aldershot, UK, Ashgate Publishing, pp. 37–54.

Gennep, A. van, 1960. *The Rites of Passage*. Chicago, University of Chicago Press.

Gesler, W., 1996. Lourdes: Healing in a Place of Pilgrimage. *Health and Place* 2, 95–105. doi:10.1016/1353-8292(96)00004-4

Gesler, W.M., and Pierce, M., 2000. Hindu Varanasi. *Geographical Review* 90, 222–237. doi:10.1111/j.1931-0846.2000.tb00332.x

Herrero, N., 2008. Reaching Land's End: New Social Practices in the Pilgrimage to Santiago de Compostela. *International Journal of Iberian Studies* 21(2), 131–149.

Hyndman-Rizk, N., 2012. Introduction: Pilgrimage and the Search for Meaning in Late Modernity, in: Hyndman-Rizk, N. (Ed.), *Pilgrimage in the Age of Globalisation: Constructions of the Sacred and Secular in Late Modernity*. Newcastle upon Tyne, Cambridge Scholars Publishing, pp. xviii–xxiii.

Ivakhiv, A., 2003. Nature and Self in New Age Pilgrimage. *Culture and Religion* 4, 93–118. doi:10.1080/01438300302812

Jansen, W., 2012. Old Routes, New Journeys: Reshaping Gender, Nation and Religion in European Pilgrimage, in: Jansen, W., and Notermans, C. (Eds.), *Gender, Nation and Religion in European Pilgrimage*. Aldershot, UK, Ashgate Publishing, pp. 1–18.

Lough Derg, 2016. Lough Derg Three Day Pilgrimage FAQ. *Lough Derg*. Retrieved June 10, 2016, from www.loughderg.org/faqs

Maddrell, A., 2011. "Praying the Keeills": Rhythm, Meaning and Experience on Pilgrimage Journeys in the Isle of Man. *Landabréfið – Journal of the Association of Icelandic Geographers* 25, 15–29.

Maddrell, A., 2013. Moving and Being Moved: More-Than-Walking and Talking on Pilgrimage Walks in the Manx Landscape. *Culture and Religion* 14, 63–77. doi:10.1080/14755610.2012.756409

Maddrell, A., and della Dora, V., 2013. Crossing Surfaces in Search of the Holy: Landscape and Liminality in Contemporary Christian Pilgrimage. *Environment and Planning A* 45, 1105–1126. doi:10.1068/a45148

Maddrell, A., and Scriven R., 2016. Celtic Pilgrimage, Past and Present: From Historical Geography to Contemporary Embodied Practices. *Social and Cultural Geography*, 17(2), 300–321. doi:10.1080/14649365.2015.1066840

Magan, M., 2014. Is This the Toughest Pilgrimage in the World? *The Guardian*. Retrieved August 20, 2014, from www.theguardian.com/world/shortcuts/2014/aug/15/-sp-toughest-pilgrimage-st-patrick-purgatory

Michalowski, R., and Dubisch, J., 2001. *Run for the Wall: Remembering Vietnam on a Motorcycle Pilgrimage*, New Brunswick, NJ, Rutgers University Press.

Ó Giolláin, D., 2005. Folk Culture, in: Cleary, J., and Connolly, C. (Eds.), *The Cambridge Companion to Modern Irish Culture*. Cambridge, Cambridge University Press, pp. 225–244.

Osterrieth, A., 1997. Pilgrimage, Travel and Existential Quest, in: Stoddard, R., and Morinis, E.A. (Eds.), *Sacred Places, Sacred Spaces: The Geography of Pilgrimages*. Baton Rouge, Louisiana State University, pp. 24–39.

Oviedo, L., de Courcier, S., and Farias, M., 2014. Rise of Pilgrims on the "Camino" to Santiago: Sign of Change or Religious Revival? *Review of Religious Research* 56, 433–442.

Pazos, A.M., 2012. Introduction, in: Pazos, A.M. (Ed.), *Pilgrims and Politics: Rediscovering the Power of the Pilgrimage*. Aldershot, UK, Ashgate Publishing, pp. 1–8.

Reader, I., 2007. Pilgrimage Growth in the Modern World: Meanings and Implications. *Religion* 37, 210–229. doi:10.1016/j.religion.2007.06.009

Rountree, K., 2006. Performing the Divine: Neo-Pagan Pilgrimages and Embodiment at Sacred Sites. *Body and Society* 12, 95–115. doi:10.1177/1357034X06070886

Schmidt, W.S. and Jordan, M.R., 2013. *The Spiritual Horizon of Psychotherapy*, London, Routledge.

Scriven, R., 2014. Geographies of Pilgrimage: Meaningful Movements and Embodied Mobilities. *Geography Compass* 8, 249–261. doi:10.1111/gec3.12124

Slavin, S., 2003. Walking as Spiritual Practice: The Pilgrimage to Santiago de Compostela. *Body and Society* 9, 1–18. doi:10.1177/1357034X030093001

Taylor, L.J., 2007. Centre and Edge: Pilgrimage and the Moral Geography of the US/Mexico Border. *Mobilities* 2, 383–393. doi:10.1080/17450100701597400

Turner, V.W., and Turner, E., 1978. *Image and Pilgrimage in Christian Culture: Anthropological Perspectives*. New York, Columbia University Press.

Wooding, J.M., 2003. Fasting, Flesh and the Body in the St Brendan Dossier, in: Cartwright, J. (Ed.), *Celtic Hagiography and the Saints' Cult*. Cardiff, University of Wales Press, pp. 161–176.

6 Ministers on the move

Vocation and migration in the British Methodist Church

Lia D. Shimada

Introduction

In the popular imagination, images of 'the Christian minister' tend toward the static: a black-robed, white-collared figure (usually male) presiding behind a pulpit or, perhaps, at the edge of a grave. Pulpit and grave: These, it would appear, are the iconic spaces for ministry. Television shows like the BBC's *The Vicar of Dibley* and *Rev* may place their ministers in a wider context – village chapel for the former; inner-city parish for the latter – but the storylines nonetheless unfurl in these specific locations, their fictional geographies bounded by the practicalities of 60-minute storytelling.

Yet there is another way to think about a Christian minister: as a walking, breathing geography experiment; as an exercise in religion-as-practice.

In 2010, mere months after finishing my doctorate, I accepted a three-year post to implement the national diversity strategy for the British Methodist Church. On spec, this was a marvelous job for a newly minted cultural geographer – and not just because 'mapping' appeared as a designated task in my workplan. I also came to this job as a professional mediator, well-versed in the art of dealing with conflict. Little did I know that both sets of skills would be pressed into service, in equal measure, time and again. From knowing next to nothing about British Methodism, I swiftly learned a new language, a new organisational culture, and a new category of person called 'clergy'. In the process, I dealt day in and day out with ordained ministers: some lovely, some infuriating, all of them unavoidably human with a vocation to the divine. It didn't take long to realise that my generic perceptions of ministerial location (pulpit, parish, chapel, graveside, etc.) were merely wayfaring marks in the larger scope of British Methodist ministry.

Quite simply, I learned that ministers are constantly on the move. Mobility is ingrained in Methodist DNA, with modern-day clergy treading in the restless wake of their long-ago founder, John Wesley. From one appointment to the next, over the course of his or her ministry, a British Methodist minister[1] may live and serve in drastically different geographical contexts, from the Channel Islands in the south to Shetland in the north, from Wales and the Isle of Man to the coastal fringes of East Anglia. Migration, however, is anything but a story of Britain's interior. Increasingly, the currents of globalisation bring ministers from distant parts of the planet back 'home' to serve the British 'Mother Church'. Migration, for Methodists, expresses history in the present tense. The ministers who migrate to Britain – from Australia to Zimbabwe and everywhere in between – bring with them the versions of Methodism carried to their shores by well-meaning missionaries of centuries past.

Migration, for Methodists, is above all a story of change. Not surprisingly, diverse streams of migration give rise to countless tensions: between local and global, between rural and urban, between theory and practice, between tradition and innovation. Migration reveals rifts in expectations and experience and, above all, in theology. At its best, migration is a source of spiritual renewal for congregations and for the ministers themselves. At its worst, it can be a recipe for raging conflict.

By and large, academic geographers have left Methodism to the historians. With its founding narrative (more below) firmly couched in the visually alluring eighteenth century, bolstered by thousands of archived sermons and hymns, the Methodist Church seems to have a natural home in the historical sciences. Yet Methodism should appeal to geographers, too. The denomination is inherently geographical, scaling between the local and the global in dynamic, ever-shifting ways. As such, it makes a fine case study for exploring the spatial politics of spirituality, as viewed through the prism of Methodist ministry.

In this chapter, I explore the relationship between ministry, movement and the complex (and potentially contested) spiritual spaces through which ministers move. I take as my starting point an understanding of 'ministry' as religion-in-practice, with a startling variety of vocational forms for both laypeople and for those who are ordained. For the purposes of this article, I focus on the ministry of the latter. In the story of British Methodism, two broad types of movement fascinate me: **itinerancy** and **reverse mission**. Both have profound implications for geographies and theologies of ministry; both are steeped in longstanding historical and theological traditions. They now unfold in contemporary times – to the tune of changing social contexts and conflicts, with human beings at their heart.

A brief history of the British Methodist Church[2]

> Our servant came up and said, 'Sir, there is no travelling today. Such a quantity of snow has fallen in the night that the roads are quite filled up.' I told him, 'At least we can walk twenty miles a day, with our horses in our hands.' So in the name of God we set out. The northeast wind was piercing as a sword and had driven the snow into such uneven heaps that the main road was impassable. However, we kept on, afoot or on horseback, till we came to the White Lion at Grantham.
>
> John Wesley's journal entry, 18 February 1747

John and Charles Wesley were born in Epworth, Lincolnshire to a vicar father and a formidable mother, who not only bore 19 children but lived to tell the tale. In between home-schooling her 10 surviving children, she also managed to organise

Sunday afternoon meetings in her kitchen for as many as 200 people at a time. The home-schooling bore fruit; John and Charles eventually made their way to Oxford University. There, they both became more serious about Christianity, developing a spirituality that combined inward faith with outward commitment to serving those in need. They formed a small group, nicknamed 'The Holy Club', with a handful of other like-minded students. Together, they would go into town and to the local prison to perform their good deeds. Their fellow Oxonians mocked them with insults that, by today's standards, sound positively benign: 'Bible Moths', 'Enthusiasts', 'Supererogationists' . . . and 'Methodists'. Thus began the remarkable movement that would carry the Wesley name across continents, oceans and centuries, founding a global phenomenon that would one day be known as the Methodist Church.

Charles may have been the musical prodigy (he composed between 6000 and 9000 hymns of variable quality, depending on how you count the output),[3] but John was the real organising genius, with a keen eye for expanding into new markets for the soul. Although they followed their father into the ordained priesthood, John and Charles soon chafed against Establishment Anglicanism. They were alive to the discontent of the working classes, feeling keenly their sense of exclusion from local parish churches. In 1739, at the age of 36, John reluctantly preached his first open-air sermon. He never looked back. So-called 'field preaching' – at pitheads of mines, on village greens, wherever anyone would gather to listen – became a key feature of the Methodist religious revival. In his wake, across the countryside and in industrialising towns, John's fiery sermons spurred his followers to band together in new communities of faith and fellowship.

Eighteenth-century Britain was a place of excess and enthusiasm, over which reigned the Enlightenment and enormous social inequalities. Into this mix thundered John and Charles Wesley, with their heart-stirring preaching, gusty hymn-singing, spirit-infused assemblies and boundless energy. Like any oral movement, the Methodist revival struck an emotional chord that resonated with the times, rippling outward with each encounter. Unlike those straight-laced Calvinists, with their conviction in God's absolute sovereignty and their gloomy doctrine of predestination, the Wesley brothers believed fervently that no one – absolutely no one – was beyond salvation. To John, Charles and their ever-expanding community of followers, religion was meaningless if it did not combine faith with good works: caring for widows, orphans and the poor; prison reform; education. A tireless campaigner to the end, John's last known letter urged the abolition of slavery. The Wesleys' message fell on grateful ears and hearts. This was a movement in which women and the working classes found voice and status they would have otherwise been denied. Increasing numbers of preachers were trained – women as well as men – to spread the Wesleyan message. By the time John died, 72,000 people belonged to Methodist societies in Britain alone. The real number of adherents, on both sides of the Atlantic, was far higher.

As an ordained Anglican priest, John never set out to create a new denomination; his mission as he saw it was to revive the Church of England from its fusty, tired elitism. However, by the dawn of the next century, 'the people called Methodists' subscribed to a version of faith distinctly different from their parent body. In 1795, the Methodist Church separated formally from the Church of England: no longer Anglican, but a Christian denomination in its own right.

John Wesley's restless mobility remains imprinted upon modern-day Methodism; contemporary ecclesiastical structures clearly reflect their founding history. Local Methodist *churches* are congregations based on the original Methodist 'societies' that met initially within the Church of England. The *circuit* is the standard administrative and missional unit – normally a group of churches served by a team of ministers. Here is where new initiatives and changes in the pattern of church life unfold, where closures of chapels are debated, where new expressions of church emerge, and to which a minister is appointed. The *district* comprises a collection of circuits, much as an Anglican diocese gathers its parishes under its wings. Unlike an Anglican map of Britain, however, Methodist boundaries are not concrete but notional – a faint trace of history in the contemporary landscape. A circuit, after all, can also be imagined as the distance a wandering preacher might feasibly cover on horse. From the outset, Methodism moved through interlinking spaces of spirituality, and continues to do so today. Church, Circuit, District: These are the building blocks of Methodist geography, transposed on the older, Anglican map of Christian Britain.

Arching over all is the *Connexion.* The old-fashioned spelling is a direct legacy from the Wesleys' time. Today, the British Methodist Church continues to adopt a connexional (as opposed to a congregational) structure of spiritual governance. In other words, the whole Church acts and decides together in this large, connected community; no local congregation is independent. The Connexion provides the spiritual geography for all British Methodists, regardless of the extent to which they choose – or not – to acknowledge it. To be a member in one, local Methodist congregation translates as full membership of the British Methodist Church, across the entirety of the Connexion. And for those individuals who discern a vocation to ordained ministry, the Connexion becomes part of the warp and weft of their identities. When Methodist ministers are ordained, they are 'received into Full Connexion.' In doing so, they enter into a lifelong, covenant relationship with the Methodist Church, through which their adventures in ministry now unfold.

Itinerant ministry

Over the course of his remarkably long life (1703–1791), John Wesley travelled over 250,000 miles and preached over 40,000 sermons. Today, ordinary Methodists would hardly expect their ministers to demonstrate competency in horseriding, and no one would bother to count the number of sermons delivered. However, like their historical forefathers, ministers are expected to move where Church and God direct them. To be ordained is to belong to an ordered group of people with a common discipline. For ordained Methodist clergy, this 'common discipline' translates, in part, to undertaking a lifetime of movement – to 'exercise a ministry of visitation to particular groups of disciples and particular situations in the wider world' (The Methodist Church in Britain, 2002: 458). Contemporary clergy tread in the footsteps of the earliest itinerant Wesleyan preachers, sent forth to be 'extraordinary messengers' to help people discern the needs of the Kingdom (*ibid*: 459).

To understand how distinct is this calling, we can compare it to, say, the monastic Benedictine discipline of stability. In the early Christian and medieval tradition, Benedictine monks dedicated their lives to God within the four walls of a cell, within the closed brotherhood of a monastery. Here, through exercising the discipline of stillness, a Benedictine monk could heed his vocation, rooted in a single place, where one's spirituality could deepen and thus flourish (see De Waal, 1999). In contrast, Methodism is a movement that flings itself across a far wider geography, with ordained ministers as the vanguard.

Intriguingly, for all that itinerancy is a defining characteristic of Methodist ministry, it leaves only faint traces in the written record. In the process of conducting research for this article, I consulted several books dedicated to Methodist theology (for example, Marsh *et al*., 2004; Luscombe and Shreeve, 2002; Langford, 1998). With each, I flipped to the index in search of 'itinerant' or some variation thereof; not once did I find an entry. Official publications from Methodist headquarters in London also proved to be elusive. In part, this is due to the peculiar shape of the Methodist 'Conference.' The Conference is the governing body, comprising representatives from across the Connexion, which meets annually to confer, debate and ultimately to agree policy for the British Methodist Church. Through a lineage that stretches back to John Wesley's day, Methodist theology has evolved over time. In large part, the articulation of Methodist theology can be traced through lengthy written reports that are submitted to the annual Conference for discussion. Every year, Conference receives hundreds of pages of these documents, after which they are added to Methodism's hefty archive. In this way, the collected reports to Conference form a sort of canon for British Methodist ecclesiological and theological thought and practice.

For a member of the general public, however, the most readily accessible reference to itinerancy is a document for people who are considering a vocation to the ordained ministry. Embedded on page 33 of 40 is this warning:

> The covenant relationship with those in ministry means that the Church will place you in a circuit and whilst the Church makes every effort to support ministers and their families, no-one should think that the relocation to a new circuit and home is an easy formality without challenge.
>
> (The Methodist Church in Britain, 2015: 33)

Every November, there is a national gathering for the leaders, known as 'Chairs', of each of the 31 districts within the British Methodist Church. Each Chair comes armed with the names of ministers seeking a new appointment, and the names of circuits seeking a new minister. Each minister and circuit will have written a lengthy profile, highlighting their interests, passions and hopes. Over five days, in a delicate and complex process, the Chairs confer and pray together, matching ministers and circuits across the Connexion. As they do so, the spatial breadth of British Methodism becomes concentrated in one, specific location. According to Stephen,[4] a former Chair with long experience of the stationing process: 'In that room, there is knowledge of every single chapel and minister in (British) Methodism.' Within this room, during these intensive days of wrestling with people, places and paperwork, the Chairs will seek to discern, together, God's will for the spiritual geography of Methodist Britain.

Once matched, the minister (and family, if relevant) will then arrange to visit the circuit; each will hope that the other lives up to the hyperbole of the profile. If the visit is deemed successful by both sides, the minister will then be 'stationed' to the circuit. Over the next few months, the minister, plus his or her family, will prepare to leave their current home and then move – sometimes hundreds of miles – to a new community. These decisions are not taken lightly, as evidenced by the prayer, time and consideration devoted over many months, by many individuals. Taken at face value, however, the Methodist stationing process appears to be a strange hybrid of internet-, speed-, and blind-dating, conducted in the giddying hope of an arranged marriage.

Today, a standard appointment lasts five years, with the option of extending if both the minister and the circuit agree. (Depending on your point of view, this is either a drastic improvement or a shocking step backwards from John Wesley's day, when a standard appointment lasted merely one year.) In theory, however, a minister could be moved at any time. Rarely will a minister stay longer than seven to ten years in any one appointment. Not infrequently, the stationing process goes awry; sooner or later, minister and circuit – or both – may realise that this is anything but a match made in Heaven. In some cases, the minister and circuit will stumble on, for better or for worse. In other cases, the match will end with the minister curtailing his or her appointment. Sometimes curtailment is a healthy, healing process, but usually it is painful and fraught for all involved. Itinerancy can be reviving and exciting, yes, but it is also deeply – and for some, dangerously – precarious.

So, what does this look like in practice? Let us consider the life of Alexander, now a retired minister with many years of active service behind him.

Born and bred in the moors of Northern England, as a young man Alexander travelled south, to the flatlands of Cambridge University, where he trained as a Methodist minister. Newly married and newly ordained, he embarked on a ministry marked by myriad bends in the road. In the beginning, Alexander and his wife, Beth, moved to Manchester, where they lived on a council estate in one of the most deprived wards in England. This was inner-city ministry based firmly in and of the community. Alexander relished the ecumenical nature of this appointment, working alongside like-minded colleagues from the Church of England and other denominations. Eventually, and now with young children in tow, they relocated to a rural market town for Alexander's next appointment. This was the base from which he served a half-dozen tiny chapels, strung along the folds of the dales – the smallest of which boasted six members on paper, of which only three attended regularly. Here, Alexander and Beth's children grew and thrived, while

Beth drove daily to the regional city where she flourished in her own career. This was, in many ways, an idyllic appointment, but shadows lurked on the horizon from the beginning. Alexander would learn that as an itinerant minister, he threatened the status quo of small chapels in communities which rarely welcomed new blood into their midst. These were congregations that, in Alexander's words, 'did not want to be moved' from their traditional, rooted ways of doing and being. The growing tensions boiled to the surface when the lay members of his circuit voted – by the narrowest of margins – not to extend Alexander's appointment. He and his family – including two teenagers, the youngest in a vulnerable position as she entered her exam year – now had a matter of months to move out of the manse and wrap their minds around a new place to call home.

Toward the end of his vocational career in the British Methodist Church, Alexander took one final appointment, this time in the heart of London. For a few years, he reveled in the many opportunities the capital offered to exercise his passion for social justice. Before long, however, a combination of exhaustion, toxic congregational conflict and ill health led him to take early retirement. In a fitting tribute to Methodism's energetic history, a minister's time in service is known as 'years of travel'; when a minister approaches retirement, he or she asks for 'permission to sit down.' Seven years after he 'sat down', I asked Alexander if he had any reflections he wanted to share about itinerant ministry. His reply: 'I'm glad it no longer has anything to do with me.'

Theologies of itinerancy

In the absence of a glossy, readily accessible 'official' theology from the powers-that-be at Methodist headquarters, I asked a handful of ministers to articulate their own understanding of the itinerant nature of ministry. Alexander, whom we met above, took a pragmatic stance:

> I think we invent things, and if we are that way inclined, we think of a theology to justify it.

To others, however, itinerancy is a crucial strand of their spiritual life and work. Wilson is a 40-something-year-old minister currently based in London, where he is serving his third circuit appointment. Without hesitation, he summarised his theology as 'Pilgrimage':

> So, it's Abraham, you know. In [the biblical letter to the] Hebrews, they talk about a pilgrim passing through . . . a stranger and alien in the land. So we (Methodist ministers) come along, we pitch our tent for a while, alongside people, and then we move on.

In this invocation, Wilson reaches not just for John Wesley but much further back in time, into the biblical foundations of Abrahamic faith itself. Itinerant ministry, as articulated here, is a spiritual geography shaped around a call to nomadic ways of being and of relating: 'we pitch our tent for a while, alongside people.' Wilson's theology can be read as strongly geographical, in the way it speaks to the spatial, directional energy that is harnessed in developing one's spiritual faith: pilgrimage *through* a point in space or time; pilgrimage *toward* closer union with God; pilgrimage as journey itself.

For Terri (30-something-year-old minister, now serving her second appointment in a market town within London's commuter belt), itinerant ministry is an expression of social justice. Like Alexander, Terri trained to be a minister in Cambridge. Her first appointment took her to a Northern seaside city characterised by a slumping economy, high levels of deprivation and a sizeable population of refugees and migrants seeking asylum in Britain.

> There is something about equality. I like itinerancy because everyone is served regardless of their ability to pay or the attractiveness of the area. With [Methodist] stationing, you don't get a choice. Everybody is served. The whole country is covered.

She compared the Methodist system favourably to its Anglican counterpart. During her time in this appointment, Terri worked ecumenically with vicars in the Church of England, who struggled to attract colleagues to serve the parish due its widespread perception as an unattractive city. Terri expressed concern that the Methodist Church could follow suit:

> Now, the reality is that's changing, given that we (the British Methodist Church) are getting so short of ministers now . . . which means there is an element of choice. There is a buyer's market, so to speak, for ministers. Which is very sad, actually, because we are losing the theology of itinerancy.

Terri's use of the word 'choice' is significant, as it signals to her the end of the current framework in which – in theory, at least – the Methodist Connexion and the minister discern, together, the will of God in the mission of the Church.

At its best, Methodism's vision of 'corporate discernment' pairs ministers and circuits for the greater good, each bringing out the best in the other, and in doing so galvanising congregations toward growth and constructive change. Like Terri, Alexander compared the Methodist practice of itinerancy favourably to the Anglican approach to parish appointments, which he described as a 'freehold' leading to 'a sort of staleness.' Terri acknowledged openly that she would not have chosen, of her own accord, to live and work in the economically depressed, physically unattractive city to which she was stationed. Yet once there, buoyed by the belief that she was following her vocation and that she had been sent for a reason, Terri flourished in her role. In the process, she developed an impressive set of skills and knowledge base for working with refugees:

> I didn't have any qualifications, but I developed the skills necessary because they were needed. I did my best.

Without the new experiences offered through itinerancy, Terri suspects that she would have 'pigeonholed' herself in one area. Without the discipline of itinerancy, and the world-opening opportunities that emerged, Terri would have flung herself into youth work and carried on doing it throughout her ministry. Itinerancy took Terri far from her comfort zone – geographic *and* spiritual – and opened new vocational horizons she may not have heeded otherwise.

If the introduction of a new minister holds the potential to rejuvenate a congregation, so too does the unsettling period of transition that marks the end of a minister's appointment in a circuit. For Wilson, leavetaking is a process that, whether individual or communal in scale, calls on the minister to enact a symbolic, representative role:

> I think that an important part of what we do is leaving. It's a bit like a pastoral visit. They've been visited by 'the Church', but then we leave, and they breathe a sigh of relief when we go. Not because it's been a bad experience, but because we take stuff when we go. And I think that's the same for leaving a circuit as well.

In Wilson's description are echoes of the sacrament of confession, with the minister removing 'stuff' that may have hampered the spiritual life of the individual, or (more broadly) the congregation. Along similar lines, Terri finds itinerancy 'very useful, in that you can actually tackle conflict [in a congregation], knowing you're going.' Read this way, itinerancy is nothing less than a catalyst for transformation.

The future of itinerant ministry?

In John Wesley's day, 'the travelling preacher' was a perfect creation for its context and its time. However, the world has changed enormously, in ways which John, Charles and their early followers would never have imagined. As the contemporary Church grapples with ageing congregations, chapel closures and declining financial resources in an increasingly secular society, the question must be asked: What will – and should – happen to itinerant ministry?

In parallel with the declining numbers of Methodist members, the Methodist Church is also experiencing a decline in the number of vocations – of people coming forward to offer for the ministry. As the pool of active ministers dwindles and ages, itinerancy as currently practiced may require a drastic overhaul. Moreover, advances in technology have opened new channels through which the Methodist Church can speak of and to the world. These days, anyone with access to a laptop and an internet connection can go online and stream a sermon. How does itinerant ministry respond – or not – to these new frontiers?

Contemporary British society no longer reflects the social landscape in which itinerant ministry once took root and thrived. Reflecting from the far side of active ministry, Alexander (whose trajectory we traced in a previous section) recognises that the world in which he grew up is far removed from the world he now inhabits.

In the world I was born into, it really wasn't that difficult for a minister to up and move. Increasingly, ministers have spouses – male and female spouses – with careers, and that makes it difficult.

Wilson put it more bluntly:

It's a system that grew up in an age when the ministers were men and the wives were housewives.

Alexander and his wife, Beth, managed to combine Alexander's vocation to the Methodist ministry – with its requisite commitment to itinerancy – with Beth's own career trajectory. When I interviewed Alexander for this article, he expressed, at first, a remarkably sanguine approach to the human dimension of itinerancy: 'Most people in the modern world do a certain amount of moving. Some people are more inclined than others to move about in their lives.' Nonetheless, there was no denying the toll taken on his family, and the disruptions and dislocations they endured.

Other ministers whom I interviewed for this article (some of whom chose not to be quoted directly) were more vocal about the negative – even cataclysmic – effects of their itinerant vocation on their personal lives and relationships. Terri spent more than a decade living on the opposite side of the country to her partner and parents, while she served her first appointment in the northern, seaside town to which she was stationed. She spoke to me, at length, of the financial, time and above all emotional costs of itinerant ministry. The repercussions continue today: 'I've missed out on every member of my family's wedding, baptism, etc. I'm just not invited anymore.'

Time will reveal whether the British Methodist Church can sustain its insistence on itinerant ministers. The apparent absence of a theology of itinerancy which is widely understood, shared and embraced across the Connexion amongst ordained clergy and laypeople alike, may make itinerant ministry an increasingly difficult practice to maintain. Already, anecdotal evidence suggests that the prospect of itinerancy may deter potential candidates from offering for the ministry – at least in its present form. Yet running alongside the Church's narrative of decline is a fascinating new story of migration. The historical tides of itinerancy and mission are now reversing, bringing new Methodists 'home' to Britain.

Reverse mission

> I look upon all the world as my parish; thus far I mean, that, in whatever part of it I am, I judge it meet, right, and my bounden duty to declare unto all that are willing to hear, the glad tidings of salvation.
>
> John Wesley's journal entry, 11 June 1739

Even during his lifetime, John Wesley refused to let the limitations of horse travel halt the spread of his message. Early in their ministry, John and Charles crossed the Atlantic in an attempt to convert the Native Americans of Georgia. Closer to home, as the Methodist Revival gained pace, they made several trips across the Irish Sea. As Birtwhistle (1983: 1) observes: 'The very nature of the Methodist Revival made it impossible that its energies could be confined to one small country.' In its heyday, Methodism made inroads on six continents, flourishing in several places far removed – geographically and culturally – from Britain. Facilitated by the currents of British imperialism, Methodist missionaries – like 'clever parasites' (Hempton, 2005: 19) – carried Wesleyan theologies across the globe.

In the twentieth century, as empires collapsed and imperialism became associated less with progress and more with oppression, so the Christian missionary movement that had developed (at least pragmatically) off the back of imperialism required radical rethinking. Indeed, as missionaries themselves became involved in nationalist movements in the countries in which they served, so their sending Churches were forced to grapple with a new generation of missional thinking (see Hempton, 2005; Koss, 1975). As the British Empire evolved into the vast Commonwealth, the Methodist Church gradually loosened its oversight and governance on the international stage. Once upon a time, and not so long ago, 'overseas districts' answered to the British Church. Over several decades, these districts became autonomous Methodist Churches (or 'Conferences') in their own right. The last of these once-subsidiary districts evolved into full independence as 'The Methodist Church, The Gambia' in 2008. In place of its former, imperially-marked missionary approach, today the British Methodist Church sees itself (at least in theory) as one partner of many in the global Methodist network.

In Britain, the Methodist Church may be a denomination in decline, yet Methodism continues to thrive in many parts of Asia, Africa and the Caribbean. Global currents bring migrating Methodists – the fruits of long-ago missionary labour – to Britain, where in many places they are reviving the ailing 'Mother Church.' This trend, which is by no means limited to Methodism, has given rise to the phenomenon of 'reverse mission.' Ojo (2007: 380) offers this definition:

> The sending of missionaries to Europe and North America by churches and Christians from the non-Western world, particularly Africa, Asia and Latin America, which were at the receiving end of Catholic and Protestant missions as mission fields from the sixteenth century to the late twentieth century.

These 'reverse missionaries' migrate for a variety of reasons – not least out of a desire to evangelise the 'dark continent of Europe' (Catto, 2012) and its post-secular populations (see also Catto, 2013). By identifying and naming this trend, sociologists of religion have performed a valuable service. The next question, though, may be one for the geographically-inclined: What does 'reverse mission' look like on the ground, in three dimensions? How do these global itinerants express themselves on the local stage? What are the broader implications of 'reverse mission' for Methodist itinerancy in Britain?

During the years I spent working as a geographer/mediator-cum-church bureaucrat, I frequently found myself on the frontline of sharp questions. Tensions were emerging in the gap between the British Church's aspirations toward multicultural inclusion and the realities of the present. Nowhere was this more apparent than with appointments involving international ministers, who hail from across the Methodist diaspora to serve the people of Britain.

A common example: In a quiet, rural village – in a tiny chapel in which John Wesley himself may have preached – an ordained minister from Sierra Leone (or Singapore, or Antigua) may struggle to serve a community long accustomed to a certain type of white British minister. For ears only used to hearing sermons delivered by ministers trained in Bristol, Birmingham or Cambridge, what challenges does an unfamiliar accent (say, Korean) and a different set of cultural references present from the pulpit? For the ministers themselves, the pitfalls are legion, with the annual Harvest Sunday service a depressingly predictable stumbling block. Local congregations can harbour fierce expectations for this service – often forgetting that the agricultural calendar and culture so familiar to them in Derbyshire may seem positively alien to a Brasilian.

What has long been considered 'the norm' is now in flux, as new voices, new ways of worship and, crucially, new theologies reshape British Methodist ministry. Over two centuries, British Methodism navigated its identity as a mainline Christian denomination that could encompass a reasonably wide but nonetheless recognisable theological spectrum. Today, the denomination is witnessing a theological sea change, as ministers from the diaspora bring diverse strands of Methodism back to Britain. For those early missionaries, who went forth with a broadly unified Methodist message, the contemporary, post-colonial world would be unrecognisable – not least in the variety of Methodist theologies and practices now present in British ministry.

As this article goes to print, I will be deep in a study of ordained ministers from the Methodist Church of Southern Africa who are currently serving appointments in Britain. Through a case study of 'Ubuntu', this project will explore the broader dynamics of migration, ministry and theologies of leadership, identity and place. The concept of Ubuntu originated from the southern region of Africa; its high-profile populisers include none other than the Archbishop Desmond Tutu. In broad terms, Ubuntu can be summarised as: 'I am because we are.' This cohort of Southern African ministers (hailing from six countries: South Africa, Namibia, Botswana, Mozambique, Lesotho and Swaziland) are stationed to circuits across the length and breadth of Britain. They are male and female, black and white, able-bodied and not, with vastly different life experiences between them. What these ministers have in common is a familiar reference point in Ubuntu. How are their theologies of Ubuntu shaping their practice of ministry in Britain? How is their understanding of Ubuntu shifting as they settle into British culture and community? For the people and the congregations they serve, how does the presence of a minister 'from afar' reshape their own Christian spirituality?

From South African ministers preaching from the pulpit, to Fijian soldiers serving on British army bases, to second- and third-generation Ghanaians worshipping in London or Leicester, the congregations of the British Methodist Church have never been more multicultural. These patterns of migration, and the encounters they produce, are changing the cultural practices of Methodism in Britain. In doing so, they are re-shaping the denomination – theologically, culturally, geographically – in dramatic ways.

Conclusion

For the Wesley brothers, the Methodist movement may have begun as a noun – just one corner within the familiar Anglican Church in which they were raised and to which John and Charles were ordained. From its inception, however, the Methodist movement was also a verb, full of restless energy to spread its good news to all who would listen. Over the course of three centuries (and counting), Methodism moved across the face of Britain and beyond, carrying its distinctive theologies and spiritual practices across the globe. This was the work of itinerant ministry.

At its heart, 'itinerant' can be defined as 'one who travels from place to place' (Collins Dictionary). Through itinerant ministry, the scales of geography condense, collide and expand, as the memories and the experiences of diverse places accumulate and thicken. In sparking a movement, John Wesley and his followers created fresh spiritual terrain, connecting 'the people called Methodists' to one another and to God in new scales of belonging, engagement, accountability and worship. Weaving through all was – and is – the vocation to itinerancy. At his or her ordination service, a minister is 'received into Full Connexion' and thus becomes a public representative of the Methodist Church – a living embodiment of connexional spirituality, carried into the local congregations and circuits he or she is then sent forth to serve. Ministers may be trained in particular locations, but each individual is shaped long before and afterwards by unique cultural forces and quirks of geography. Distances become condensed, while simultaneously horizons widen: a congregation is altered by its minister, while the minister is broadened by each place to which he or she is stationed. And then, it's time to move again, carrying the accretion of all these places and spiritual experiences into the next appointment. In Methodist ministry, it can be very difficult to decipher where 'geography' ends and 'spirituality' begins.

The time is indeed ripe for multi-faceted interrogations of 'the geographies of spirituality' and 'the spaces of spirituality.' What, though, might be unearthed if the phrasing were flipped? How might 'a spirituality of geography' open new ways of thinking about geography – about spaces, places, scales and their attendant politics? As fascinating as it may be to apply a geographical lens to practices of spirituality (of which vocation and ministry are prime examples), what more may be gained by considering questions of geography through the lens of religion and spirituality? As Methodist itinerancy illustrates – and particularly in the form of 'reverse missionaries' – the politics and practices of one's spiritual vocation are profoundly shaped by the geographical context in which a person is formed. Itinerant ministry is forever shaping, and being shaped by, the spiritual spaces, places and scales through which a minister moves. Places are palimpsests: This is a well-known trope of geography. So too are the ministers of Methodism, as they carry 300 years of spiritual heritage into, and through, the twenty-first century. Through the human figure of the Methodist minister, spirituality and geography remain constantly on the move.

Notes

1 The British Methodist Church has two orders of ministry: Presbyters (whose vocation can be loosely summarised as 'Word and Sacrament') and Deacons ('Service'). For the purposes of this article, I refer to members of both orders as 'ministers.'
2 Books on Methodist history are plentiful and widely available. Classics include Baker (1970), Davies (1976), Davies *et al.* (1983) and Southey (1890). See also Turner (2005) and The Methodist Church in Britain website (www.methodist.org.uk/who-we-are/history).
3 Outside the Christian tradition, where he is celebrated for penning such classics as *Christ the Lord is Risen Today* and *Love Divine*, Charles Wesley is probably best known for the lyrics to the Christmas carol *Hark! The Herald Angels Sing.*
4 All names, and some locations, have been changed.

References

Baker, F. (1970) *John Wesley and the Church of England.* London: Epworth Press.
Birtwhistle, N. A. (1983) Methodist missions. In R. Davies, A. R. George and G. Rupp (Eds.), *A History of the Methodist Church in Britain*, Volume 3. London: Epworth Press, pp. 1–116.
Catto, R. (2012) Reverse mission: From the Global South to mainline churches. In D. Goodhew (Ed.), *Church Growth in Britain: 1980 to the Present.* Farnham, Surrey: Ashgate, pp. 91–103.
Catto, R. (2013) Accurate diagnosis: Exploring convergence and divergence in non-Western missionary and sociological master narratives of Christian decline in Western Europe. *Transformation*, 30(1), 31–45.
Collins Dictionary. Retrieved from www.collinsdictionary.com/dictionary/english/itinerant.
Davies, R. E. (1976) *Methodism.* London: Epworth Press.
Davies, R., George, A. R., and Rupp, G. (Eds.). (1983) *A History of the Methodist Church in Britain.* London: Epworth Press.
De Waal, E. (1999) *Seeking God: The Way of St Benedict.* Norwich: Canterbury Press.
Hempton, D. (2005) *Methodism: Empire of the Spirit.* New Haven: Yale University Press.
Koss, S. (1975) Wesleyan and empire. *The Historical Journal*, 18(1), 105–118.
Langford, T. A. (1998) *Methodist Theology.* Peterborough: Epworth Press.
Luscombe, P., and Shreeve, E. (Eds.). (2002) *What Is a Minister?* Peterborough: Epworth Press.
Marsh, C., Beck, B., Shier-Jones, A., and Wareing, H. (Eds.). (2004). *Unmasking Methodist Theology.* London: Continuum.
The Methodist Church in Britain. (2002) *Releasing ministers for ministry.* Retrieved from www.methodist.org.uk/downloads/conf-releasing/ministers-for-ministry-2002.pdf
The Methodist Church in Britain. (2015) *Called to ordained ministry?* Retrieved from www.methodist.org.uk/media/1765496/called-to-ordained-ministry-0715.pdf.
The Methodist Church in Britain. Retrieved from www.methodist.org.uk/who-we-are/history.

Oden, T. C. (2008). *Doctrinal Standards in the Wesleyan Tradition*. Nashville: Abingdon Press.

Ojo, M. (2007) Reverse mission. In J. J. Bonk (Ed.), *Encyclopedia of Mission and Missionaries*. New York: Routledge, pp. 380–382.

Southey, R. (1890) *The Life of Wesley and the Rise and Progress of Methodism*. London: George Bell and Sons.

Turner, J. M. (2005) *Wesleyan Methodism*. London: Epworth Press.

Wesley, J. (1951) Journal entry for 18 February 1747. In P. Livingstone Parker (Ed.), *Journal*. Chicago: Moody Press.

Section 2
The spiritual production of space

Steve Pile

The phrase 'the production of space' is closely associated with the work of Henri Lefebvre (1974). The phrase contains a hidden 'social': that is, Lefebvre's intention is to critique the social production of space. For him, the social order produces space. That is, social processes such as globalisation, commodification, neoliberalism, financialisation all produce spaces in ways that support the exercise of the power relations through which they operate. Thus, space is a social product of the power relations inherent in the social processes that produce space. This, Lefebvre wryly observes, might seem a tad circular (1974, page 36). However, Lefebvre argues that the production of space is never complete, never that there are no fractures in the relationship between space and the social order. As importantly, once it has been produced, space then becomes productive of social relations. Thus, socially produced space can act back on the social order, creating unexpected and surprising outcomes.

Indeed, Lefebvre recognised the significance of religious and spiritual ideas in the production of space (e.g. 1974, page 40–41). We can see this both in his analysis of the Judeao-Christian thought that underlies the western production of urban space, but also through his discussion of the myths, symbols and language that produce the lived experience of the body (as a space). Both the body and the city, in Lefebvre's view, are produced spaces that are constitutive of everyday life. As such, they warrant critical analysis, for they are a product of power relations, and therefore are political. Thus, we can say that Lefebvre supports the idea that space is produced spiritually and that the production of spiritual space would actively constitute and reconstitute the social order (though few have taken up this idea). Yet, following Lefebvre, we must remember that spiritual space can also act back on the social and the spatial in unexpected and surprising ways. In this section, the chapters explore those unexpected and surprising ways in the context of the spiritual production of social and political spaces.

Highways and byways

As Claire Dwyer observes, there has been a shift in geography away from dedicated religious sites towards wider studies of the relationship between expressions of religion and everyday life. This has created an increasingly wider frame of reference for the study of religion and space. However, the increasing concern with everyday life has begun to alter and expand both the ways everyday spaces are understood, and also how religious practices, beliefs and ideas might intersect with the social production of space. And, drawing out questions of spirituality (in and beyond religion) only adds to this. In her chapter, Dwyer explores a very mundane space: Highway 99. In part, her focus is historical, exploring the siting of religious buildings along the road. The road affords the opportunity to build churches, but what happens once those religious buildings are viewed along a highway? The proximity of religious buildings has several effects: not only does it produce a spiritual landscape curiously laminated to the prosaic experience of the commute and of the car, it also renders that landscape diverse and, moreover, open to many Gods. This is paradoxical: on the one hand, it might suggest that no particular experience of expression of the divine is dominant, yet on the other hand it democratizes the divine, making its experiences and expressions more legitimate. In this way, the spiritual production of suburban space maps on to the more familiar question of the cosmopolitanism and heterogeneity of the city. Yet, under spiritual production, the space of the highway becomes as metaphysical as it is physical: not simply a mundane connection between Place A and Place B, but a road to prayers and miracles, to goddesses and God, and significantly to extraordinary experiences.

The unexpected diversity of religion and spirituality in particular places was captured well by Paul Heelas and Linda Woodhead's foundational study of Kendal's spiritual landscape (Heelas and Woodhead, 2005). This unexpected diversity found expression both in the range of spiritualities they discovered, and also in the surprising places where that diversity could be spotted. Karin Tusting and Linda Woodhead return to Kendal, asking what has been learned and what has changed since the original fieldwork back in 2001–2002. On the one hand, the core thesis of the original study has held. They identified two trends: one towards the decline of 'official religion' and the other towards the proliferation of alternative spiritualities. If anything, they argue, they underestimated these trends. However, they identify the rise of 'no religion' as particularly significant. Indeed, there is evidence that the UK is now a majority 'no religion' country (according to the annual social survey in 2017). The category of 'no religion' has become increasingly normalised, yet its content – that is, what people mean by saying they are 'no religion' – is highly diverse. Indeed, it is a reflection of the proliferation of different ways that people choose to be spiritual. Surprisingly, then, Kendal may look increasingly unChristian, but this does not mean it is less spiritual. Importantly, the spaces of spirituality are proliferating and democratising, rather than being localised in the church.

This accords with Jennifer Lea's research on yoga, which reveals how mainstream yoga and yogic practices have now become. At the outset of their chapter, Jennifer Lea, Chris Philo and Louise Cadman observe that the spaces of spirituality have proliferated in British cities, not just in the creation of centres for different religions and different kinds of spiritual practices, but also in the weaving of spiritual practices into ordinary spaces, such as the workplace and the home. Even so, they argue, witnessing this proliferation can overlook the small stuff of spirituality in the production and experience of everyday life. To get at the small stuff of spirituality, they explore the idea of stillness. In their hands, the seeming boundary between spiritual life and modern life is rendered open and unsustainable. Their study reveals how people utilise spiritually derived conceptual frameworks and practical techniques to manage the stresses and strains of everyday life. Significantly, they can do so without even realising this is what they are doing. Alongside the increasingly visible proliferation of spiritual ideas that Tusting and Woodhead note, then, are the quieter and almost invisible uses of spirituality in everyday life.

By witnessing the proliferation of religion and spirituality through the landscape and through everyday life, three things begin to happen. First, the seeming secularity of modernity begins to unravel. Second, religion itself begins to turn upside down, becoming less associated with its singular expression in a faith than with its multiplicity and its connection to the worlds that lie beyond it (physically and metaphysically). Third, the question of the political begins to become more present as the spiritual construction of everyday life becomes more in evidence.

A world turned upside down

By creating a space where the dominant social order could be temporarily suspended, by installing their own Kings and Queens and their own (im)moral order, Peter Stallybrass and Allon White (1985) argue that the mediaeval Carnival turned the world upside down. For this reason, they were banned or continued only in highly proscribed and regulated forms. Even so, the impulse to suspend the social and moral order so as to see the world from underneath, however temporarily, remains. Religion, of course, can be seen as a constitutive or significant part of the social order that needs to be transgressed or overturned. Yet, religion itself can be used to negotiate (and survive), if not overturn exactly, the dominant social order (which, as we know, can be immoral and irreligious).

Turning the analysis of religion on its head, Elizabeth Olson, Peter Hopkins and Giselle Vincett explore the role of sacrilege in the production of youthful spiritualities. Curiously, sacrilege reveals the different forms through which religious thinking can enter young people's lives. It shows that spirituality can be as easily constructed out of an engagement with witchcraft, the occult, Satanism and vampirism as out of the more authorised versions of Christianity. Significantly, Olson, Hopkins and Vincett use the paradoxical phrase 'connection across difference' to describe one instance of the use of spirituality in everyday social encounters. Thus, sacrilege and blasphemy should not necessarily be seen in opposition to religious life, but as a way of negotiating it and of producing and inhabiting a spiritual life. Indeed, shifting between alternative positions is a way for people to define exactly where they want to be, spiritually. This shifting of positions is not simply about navigating a map of fixed spiritual positions, but about creating new spiritual maps; new spiritual maps that better account for, and help diagnose or negotiate, people's place in the world. What Olson, Hopkins and Vincett emphasise is the paradox that resistance to religion – through sacrilege and blasphemy – actually becomes part and parcel of spirituality.

Meanwhile, religion can act as a way to negotiate and resist the dominant social order. Olivia Sheringham and Annabelle Wilkins show how migrants from Brazil and Vietnam draw upon transnational religious connections to create a home for themselves in London. Significantly, religion affords migrants the possibility of creating a sense of belonging in an otherwise strange place. It does not do so exclusively through its official or institutional forms, but also through informal and everyday practices. This is doubly paradoxical: a religion on the margins of the social order, experienced through its liminal everyday forms, provides a sense of belonging in an alien social order. It can achieve this precisely because it operates in in-between spaces between the migrant experience and the place they find themselves in. As importantly, the lived experience of religion and spirituality is capable of connecting people back to a sense of place and home far away. In this way, religion not so much turns the world on its head as enables it to be shaped and moulded across distance, across borders, from underneath the social order. In this sense, spirituality produces space in ways that can make it stretch and fold, create home and belonging, as well as enable the difficulties of everyday migrant life to be recognised, acknowledged and negotiated.

Everyday life, indeed, can be not just challenging, it can be harmful; it can be deadly. Sat in the West, it can be easy to assume that the world is secularising and that there is no place (literally) for religion and spirituality. Yet, Kim Beecheno's work starts with the religious conversion of women in low-income, high-crime areas of Latin America. Her study of São Paula shows how women use religious conversion to negotiate and challenge violence, and especially domestic violence. This reveals a paradox: often religious conversion requires women to submit to a highly patriarchal and conservative form of religion, yet this submission enables women to side with those patriarchal and conservative power relations and turn them against forms of violence, in the home and on the street. As Beecheno shows, religious conversion does not guarantee that violence against women is stopped; indeed, it can just as easily get worse. However, what this shows is how religion can be swerved or inverted to produce effects opposite to those we might expect. Paradoxically, in the moment of submission, religion can be empowering.

Moving away from religion in its institutional and official forms shows that the spiritual production of space can act as a mode of critique of the social order and of the production of everyday life; it can also be a resource for negotiating everyday life, both its harms and enchantments; and, it is also generative of new spaces, new forms of spirituality and new ways of inhabiting the world. In this, it is political and transformative, as we show in the final section.

References

Heelas, P. and Woodhead, L. with Seel, B., Szerszynski, B. and Tusting, K., 2005, *The Spiritual Revolution: Why Religion Is Giving Way to Spirituality*. Oxford: Basil Blackwell.

Lefebvre, H., 1974 [1991], *The Production of Space*. Oxford: Basil Blackwell.

Stallybrass, J. and White, A., 1985, *The Politics and Poetics of Everyday Life*. London: Methuen.

7 Suburban miracles

Encountering the divine off Highway 99

Claire Dwyer

Introduction

Along a three kilometre stretch of Number 5 Road, the road which marks the eastern edge of the suburb of Richmond in Vancouver running parallel to the interstate Highway 99, is a remarkable cluster of more than thirty diverse religious buildings, including temples, churches, mosques and schools known locally as *Highway to Heaven* (see Figure 7.1).[1] This agglomeration of religious buildings is the product of a local planning designation which provided space for religious buildings on land reserved for agricultural production in a creative attempt to offset municipal responsibilities for land management and control suburban sprawl. The scale and diversity of religious construction was unanticipated, however, creating an 'accidental landscape of religious diversity' (Dwyer *et al.* 2016: 5) in this suburban edge-city of Vancouver. *Highway to Heaven* is celebrated locally through a secular lens of multiculturalism, framing its diverse faith communities in a language of cultural rather than religious diversity. This is a framing not necessarily rejected by the faithful themselves who may strategically use markers of cultural diversity to access favourable planning outcomes, preferring that religious practice remains largely invisible in the public political domain. Yet fieldwork with the diverse faith communities of Number 5 Road[2] suggests another reading of *Highway to Heaven* as a distinctive edge-city geography of spirituality shaped by religious belief and practice and animated through a 'performative presencing' (Dewsbury and Cloke 2009: 696) of the sacred. These performative practices are evident not only in the materialisation of religious buildings and their animation as sacred, but also in accounts of the felt presence of the divine on Number 5 Road. This chapter explores Number 5 Road, as a distinctive landscape of spirituality manifest in the seemingly mundane, ordinary and functional spaces of edge-city North American suburbia.

Geographies of religion have shifted from focusing on dedicated religious sites to the 'unofficially sacred' (Kong 2001; Woods 2013) and to wider explorations of the everyday geographies of spirituality (Bartolini *et al.* 2017; MacKian 2012). Yet conventional worship space remains important, not least because of the symbolic power of such buildings, particularly those of minority ethnic faiths and migrants which are often contested through the normative values of planning legislation.

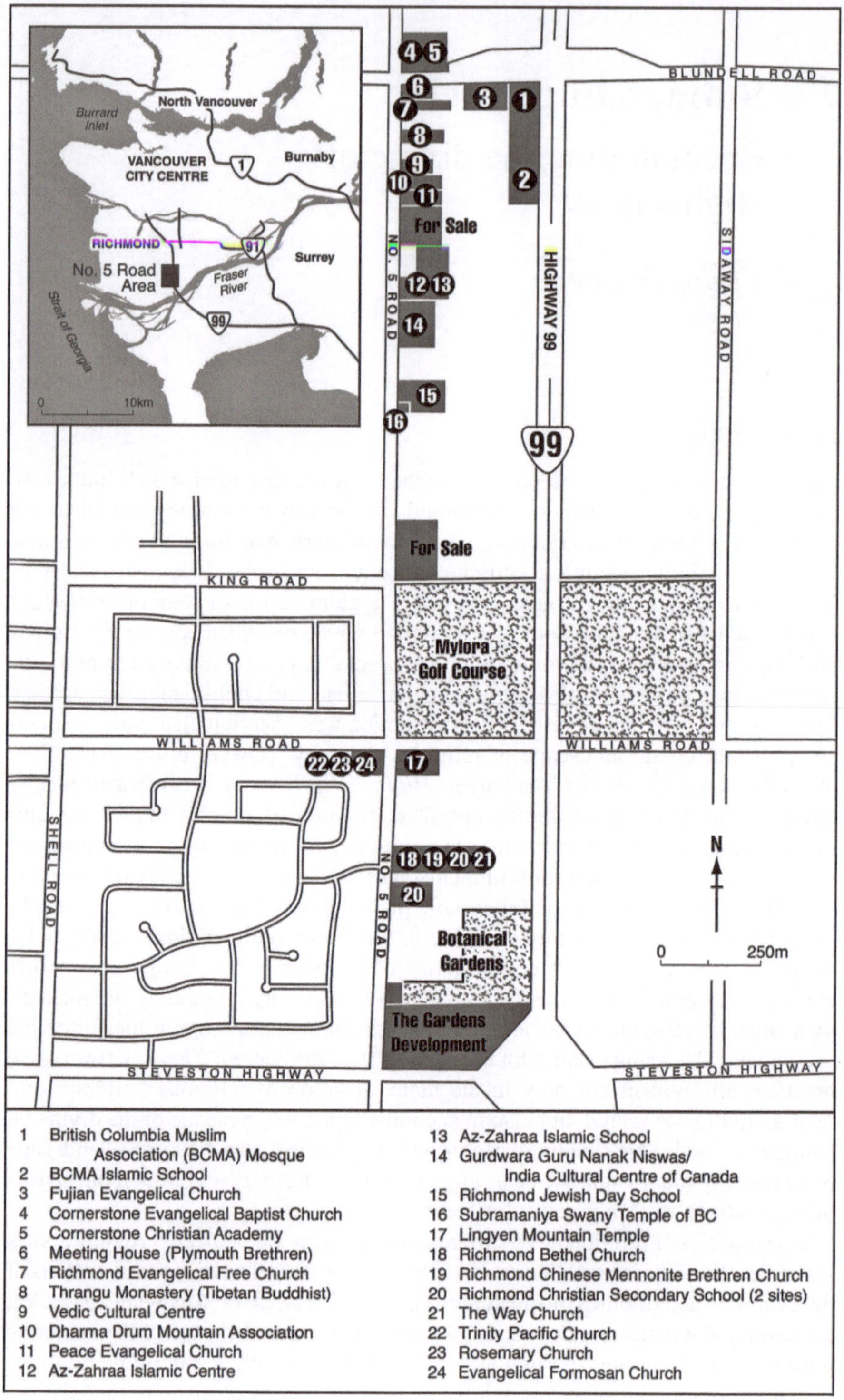

Figure 7.1 Map of Number 5 Road. Reproduced with kind permission of Miles Irving, UCL Department of Geography.

Nonetheless most studies, including some of my own, discuss 'religious landscapes' (Peach and Gale 2003) focusing on the institutional and architectural dimensions of new religious buildings rather than engaging more directly with the narratives and sensibilities of believers who use these buildings. In thinking about Number 5 Road as a spiritual landscape, I have drawn from Dewsbury and Cloke's characterisation of spiritual landscapes as 'co-constituting sets of relations of bodily existence, felt practice and faith in things that are immanent, but not yet manifest' (2009: 696). For Dewsbury and Cloke (2009: ibid.) spiritual landscapes are 'not just about religion, but open out spaces that can be inhabited or dwelt, in different spiritual registers'. Dewsbury and Cloke (2009: ibid.) suggest that a 'performative presencing of some sense of spirit' characterises the spiritual. This term resonates with Julian Holloway's understanding of embodied spiritual practice and the performative realisation of 'space-time as infused with the divine that the faithful enact and continually re-presence' (Holloway 2011: 399; Holloway 2003). These understandings of embodied spiritual practice are echoed in recent explorations of the 'ephemeral and affective geographies that produce and are produced by embodied practices of prayer and worship' (Williams 2016) and 'embodied religiosity' (Olson *et al.* 2013). In this chapter I use the idea of 'performative presencing' in an exploration of both how the new worship spaces along Number 5 Road become sacred places and how the road itself becomes a spiritual landscape.

In the first part of the chapter I focus on how the different faith communities are engaged in creating meaningful sacred space in the suburban edge-city. The diverse religious architectures of Number 5 Road are explored not simply as material manifestations of religious identities but also as animated sacred spaces. Responding to Lily Kong's focus on '*how* place is sacrilized' (Kong 2001: 213, emphasis added) requires a recognition that 'sacred space needs to be understood not as a static thing, not as a disembodied set of practices of discourse, but as an assemblage, always made or remade' (Della Dora 2016: 23). Starting by outlining different approaches to the engineering of affective sacred space by different faith communities, I move to focusing on the role of practice in rendering these spaces spiritually active. The second part of the chapter explores the wider geographies of Number 5 Road, suggesting ways in which spatialities and temporalities are unsettled by the faithful in the making of a distinctive 'spiritual' landscape through the performative presencing of the divine in the ordinary everyday spaces of the edge-city. Finally, I reflect on the implications of taking seriously the extraordinary and the miraculous in the midst of the mundane and the ordinary. The chapter begins with a brief contextualising account of the history of the presence of diverse faith communities on Number 5 Road.

Highway to Heaven: the creation of a multicultural religious landscape

Part of metro Vancouver, the city of Richmond is located to the South of downtown Vancouver and adjacent to the Fraser River and international airport. Established as an agricultural municipality in the 1860s, although also the site of a salmon canning industry which recruited Japanese and Chinese labour migrants, the city has grown rapidly since the 1990s and has an estimated population of 213,891.[3] Richmond is one of the most ethnically diverse cities in Canada, with 70.4% of the population defined as a 'visible minority', the majority identifying as 'Chinese' (47%) while 8% identify as 'South Asian'.[4] The substantial population of those with Chinese ethnicity reflects transnational migration circuits from Hong Kong, China and Taiwan particularly since 1997 (Ley 2010), and the emergence of Asian-themed shopping malls and restaurants in Richmond has prompted its description as an 'ethnoburb' (Li 2009; Pottie-Sherman and Hiebert 2015). Many of the more recent religious buildings along Number 5 Road were established by Chinese immigrants, including three Buddhist temples and four Chinese-language Christian Churches. However, religious institutions along Number 5 Road include a Sikh Gurdwara, two Hindu temples and two mosques as well as Jewish, Muslim and Christian schools.

The earliest buildings on Number 5 Road are churches established in the 1970s. The oldest non-Christian institution was the British Columbia Muslim Association (BCMA) Mosque which bought land in 1976 although their mosque did not receive planning permission until 1982. In the 1980s facilities were established by Sikh and Hindu communities prior to the completion of purpose built temples in the early 1990s.[5] In 1990 Richmond Council created a new land use category which designated part of the eastern side of Number 5 Road as an area zoned for 'Assembly Use', a category that specifically included religious institutions and religious schools. This new planning designation required any religious communities building on the road to maintain a proportion of their land in agricultural use to satisfy the requirements of British Columbia's Agricultural Land Reserve policy. The planning policy effectively created Number 5 Road's distinctive religious landscape with faith communities choosing to relocate there because of the favourable planning environment, even if the agricultural stipulations were less attractive (see Dwyer *et al.* 2016).

Since the 1990s faith communities which have re-located to Number 5 Road include:

- a mosque, Az-Zahraa Islamic Centre and school which opened in 2002
- the Fujian Evangelical Church, a church started by Canadians of Filipino-Chinese descent
- several churches founded by Hong Kong Chinese migrants, including the Richmond Evangelical Free Church (Figure 7.2), Peace Evangelical Church and the Richmond Chinese Mennonite Brethren Church
- The Lingyen Mountain Buddhist temple (Figure 7.3), a Pure Land Taiwanese origin Buddhist temple in traditional style opened in 1999
- Dharma Drum Buddhist temple opened in 2006 (Figure 7.4)
- the Thrangu Tibetan monastery, another traditional style building opened in 2010
- Smaller communities include the South Indian Subramaniya Swamy temple and the Plymouth Brethren Meeting Hall.

Figure 7.2 Richmond, Evangelical Free Church, Number 5 Road.

Source: Claire Dwyer.

Figure 7.3 Lingyen Mountain Temple.

Source: Claire Dwyer.

Figure 7.4 Dharma Drum Mountain Temple.
Source: Claire Dwyer.

The result is a diversification of the edge-city where an agricultural landscape of blueberry farms is now punctuated with the gleaming spires and minarets of a gurdwara, a mosque and the ornate roofs of Chinese temples (Figure 7.5). Although largely celebrated as evidence of an emblematic Canadian multiculturalism, there is local opposition to the increasing scale of some new buildings (see Dwyer *et al.* 2016). Number 5 Road features in local official tourist literature as an interesting site of 'cultural' diversity, and 'cultural tours' of the neighbourhood are arranged several times a year by the local museum. Some limited attempts have been made to develop inter-faith initiatives recognising the particular opportunities provided by proximity (see Dwyer *et al.* 2013a; Agrawal 2015). It is this proximity of different faith communities and institutions which I want to engage more explicitly in this chapter by raising the possibility that Number 5 Road might also be narrated as a distinctive spiritual landscape.

Building the sacred: animating suburban architectures

Visitors to Number 5 Road are most likely to encounter its religious landscape from a car window as they drive along a four-lane highway. Drawing on our typology for framing suburban religious geographies (Dwyer *et al.* 2013b), Number 5

Figure 7.5 Number 5 Road showing India Cultural Centre (Gurdwara) [on right] and Az-Zaharra Mosque [on left].

Source: Claire Dwyer.

Road is a distinctive 'edge-city faith' landscape of drive-in churches and temples with large car parks for its faith-commuters. If work on suburban megachurches (Wilford 2012) has emphasised the ways in which such new religious formations have reworked the mobility and transience which characterises the edge-city, for transnational faith communities the suburban fringe offers expansive sites for new purpose-built worship spaces which may act as regional centres (Shah *et al.* 2012). For the faith communities on Number 5 Road the primary reason for their location is because of the 'assembly district' zoning. The advantages of this edge-city location, such as space for more expansive facilities and parking and proximity to the highway, are off-set by the challenges of a location outside the main city which is hard to reach by public transport and affords little immediate connection with the local neighbourhood.

Number 5 Road's unusual planning designation, which has mitigated some of the planning difficulties facing communities elsewhere, has produced a religious landscape made up entirely of relatively new purpose-built facilities, all built by fundraising within faith communities. Against the characterisation of this as a touristic site, even a 'Disney-land'[6] of religious diversity, we sought to explore how faith communities engaged with this edge-city landscape to produce a meaningful, perhaps a 'sacred' space for religious practice. Research with the diverse faith communities along Number 5 Road suggested three broad architectural approaches: a functional approach to a purpose-built facility, the reproduction of a traditional form of religious architecture in this new suburban context and a new form of architectural innovation which might engage explicitly with the local context. The production of a meaningful religious place may be architectural in the self-conscious engineering of affective sacred spaces (Gilbert *et al.* 2016) but also reveals the salience of religious *practices* in rendering a new space of worship spiritually active.

The churches along Number 5 Road blend in effortlessly to an edge-city suburban vernacular architecture of office parks or large houses. They are brick or concrete built with very little ornamentation, usually by local architects and builders who also build residential and commercial buildings, and identified by commercial-like signs from the highway. For their communities, this is intentional. As the pastor at the Richmond Evangelical Free Church explains: '[Our church is] pretty plain, but the primary thought behind it was to have a place that was functional . . . to create a space that could be used for multiple purposes'. Similarly the pastor at Peace Evangelical Church describes his church as 'it's very functional. We are not the kind of church that pursue our look. Because the true definition of the church is the people'. These functional buildings are sometimes described in contrast to the more elaborate buildings of their neighbours (the Thrangu Tibetan Monastery is immediately adjacent to the Richmond Evangelical Free Church) and a reflection of careful use of resources: 'we need to be financially responsible'. For these evangelical faith communities the narration of their building as a functional space reflects their theology – that it is the faith community, the people, which constitutes their Christian witness.[7] As the pastor at the Richmond Mennonite Brethren Church reiterates: 'Jesus is in the building, Jesus is not the building'. Location on Number 5 Road is a pragmatic choice, facilitating space for large churches with purpose-built auditoriums for Sunday gatherings of up to 500 people and large parking spaces since everyone arrives by car (see Figure 7.2).

If these churches seem to mirror the functional landscape of the edge-city, these buildings are none the less different from their secular counterparts in their divine intention. The pastor of Peace Evangelical Church describes a large donation from a practitioner in Hong Kong which allowed completion of the church as 'miraculous', emphasising that the church was spiritually ordained. Similarly, other pastors described their foundation as being led by the Holy Spirit. Our interviews also revealed some theological reflections on the ways in which buildings might have some role in shaping worship practice. The pastor of the Richmond Evangelical Free Church reflected: 'there is something to be said about worship in a place that is awe-filled. There's certainly a dialogue of the younger evangelicals now about the importance of art and trying to recover that'. Nonetheless for the Christian communities on Number 5 Road their juxtaposition alongside more ornate places of worship from other faith traditions produces a reiteration of a theological approach to understanding places of worship as functional spaces for groups to gather. Participation in services was to understand the powerful affective dimensions of collective worship using music, bible study and spontaneous unscripted prayer to create a temporal presencing of the divine within largely plain and functional spaces. The spirit moved within these ordinary spaces.

Perhaps more surprisingly, the three older diasporic faith communities on Number 5 Road, the BCMA mosque, the Vedic Cultural Centre and the Indian Cultural Centre/Gurdwara Guru Nanak Niwas also framed their buildings through a discourse which emphasised functionality over distinctive religious architectures. The chair of the BCMA explains that when their mosque was built in 1982, 'money was so scare they were just trying to get an ordinary building'. Subsequently the association network has built more modern and experimental mosques in Vancouver's outer suburbs which respond to local vernacular architectures or environmental issues. However an emphasis remains, like the evangelical Christians, upon a building as simply a space for Muslims to gather.

The Vedic Cultural Centre is a two-storey pink concrete building which houses a sanctuary for Hindu deities on the second floor and rooms below for meetings, eating and yoga as well as accommodation for resident priests. With the exception of some Hindu signage on the front of the building there is little to distinguish the building as a site of Hindu worship. One of the founders of the Centre, which opened in 1998, describes an explicit intention to create a building which 'did not look like a temple'. Instead, she explains, 'We wanted to call it multicultural so that everybody feels comfortable coming here. We wanted it to look like a cultural centre', emphasising the guidance of their spiritual leader Swami Chakradhari. The religious geographies of this space are evident, however, once inside, with the first floor of the temple designated a 'sacred space' where being in the presence of the deities requires removing shoes and only those properly initiated within practice can be involved in worship or food preparation. The Vedic Cultural Centre was founded by Fijian Indian migrants although it now attracts a diverse range of worshippers of Indian heritage. The success of the temple in attracting devotees is understood by its founders as evidence of its efficacy as a sacred site of Hindu worship so that 'the temple is like a god itself'. The Vedic Cultural Centre thus emerges as a powerfully active site of religious worship where cumulative practice and devotion have animated and strengthened the sacredness of the space. The naming and external architectural style of the Vedic Cultural Centre reflects the founders' desire to engage wider publics, but this is not a 'secular' framing of religious identity and practice, and the temple within the building is understood by its worshippers as sacred space through the efficacy of the deities within.

At the Gurdwara Guru Nanak Niwas, also named the Indian Cultural Centre of Canada, a somewhat similar positioning was evident although for different reasons. Opened as a new purpose-built gurdwara in 1993, the Number 5 Road gurdwara is narrated by its founders as a direct successor to two earlier Sikh temples in Vancouver; the first Sikh temple in Vancouver, a wooden structure which opened on Second Avenue, central Vancouver in 1909 and the Ross Street gurdwara which was designed by celebrated Canadian architect Arthur Erikson and opened in the 1970s. Tracing their ancestry to the pioneer generation of Sikh migrants who came to work in the lumber industry in British Columbia in the

1880s, many of those who founded the Number 5 Road gurdwara had also been involved in establishing the Ross Street gurdwara. However conflict over religious practices between these second and third generation Indo-Canadians and more recent Indian migrants centred on the wearing of turbans and beards and the use of tables and chairs in the langhar (eating) halls (see Nayar 2010) had prompted a split within that gurdwara and the quest for a new site. Establishing the Number 5 Road gurdwara as a private organization named the 'Indian Cultural Centre' safeguarded its members against attempts to change its religious norms and practices by new migrants who were described as not understanding 'Canadian values, [they] did not understand multiculturalism'. The building itself was designed pragmatically to hold large numbers for gatherings like weddings or holy days, and there were insufficient resources at this third temple for more innovative architecture.

The relatively plain concrete structure, distinguished only by a large exterior painting of the Guru Nanak and a gold dome, was also described as a sensible architectural response to the Canadian environment. Their chairmen compared their own modern, air-conditioned building with another gurdwara in Richmond (Guru Nanaksar Gurskih Gurwara on Westminister Highway) built in a traditional Punjabi style with an elaborate painted façade, and now deteriorating due to its unsuitability for the climate. The gurdwara on Number 5 Road is thus a religious building which communicates a particular understanding of a distinctively *Canadian* Sikh identity for its members. Inside the building, like most gurdwaras, it is the presence of Sikh priests who come on short term visas from India as religious workers (as they do at the Vedic Cultural Centre) which ensures an authentic religious experience to a large and diverse membership for whom the gurdwara is a communal as well as a religious space. Thus the 'functional' approach to building religious buildings on Number 5 Road might be understood somewhat differently for some non-Christian faith communities. The 'edge-city' opportunities allowing large assembly room space and parking remain important, but plain or utilitarian building styles mark an explicit choice to blend in with a suburban, Canadian vernacular articulated through a language of multiculturalism. Nonetheless, it is through practice (as discussed in greater detail below) that such spaces become spiritually animated.

In contrast to the assimilationist aesthetic which characterizes the Vedic Cultural Centre and the gurdwara, two of the Buddhist temples on Number 5 Road are characterised by their reproduction of a traditional form of religious architecture – the Lingyen Mountain Temple, a pure land Buddhist Temple (opened in 1999) and the Thrangu Tibetan Monastery which opened in 2010. Lingyen was built as a sister temple to the Ling Yen Shan monastery in Taiwan founded by Master Miao Lien in 1984 and is home to a resident community of mainly Taiwanese monks and nuns. Built in traditional Chinese-style (see Figure 7.3), the temple was built by Canadian architects to ensure it met local building and environmental regulations. For the temple's predominantly Chinese worshippers, who are from Taiwan, Hong Kong and mainland China, the temple represents both a familiar space within which religious practices are maintained and a spiritual anchor for more recent migrants – daily visitors offer prayers and light incense sticks as they share their petitions. Resident monks, most Taiwanese, offer regular retreats and instruction. Lingyen's efficacy as a religious space, explained by the success of the petitions of the faithful, has produced controversy as the numbers of visitors have increased and the temple's successive attempts to build a much larger traditional style temple have been rejected (see Dwyer *et al.* 2016; Dwyer 2017b).

Thrangu Tibetan Monastery is the newest building on Number 5 Road and the largest. The building is described by its Canadian architects and builders as a 'full emulation' of a temple in Tibet, a reproduction which was only possible through the creative use of steel, glass reinforced concrete and fibre-glass to recreate the appearance of 'authenticity' (Dwyer 2017a). The ambitious building was funded by wealthy Hong Kong Chinese donors and was built under the instruction of the resident monastic community. This instruction ensured that the spiritual capacities of the building were foundational in its construction with different stages of construction marked by key rituals which involved both blessing the foundations but also embedding sacred offerings into the building. Days prior to the opening ceremony the monks related that a 'miracle' had occurred in the temple – that several of the small 'medicine buddhas' (a devotional object of healing) housed in cases along the sides of the main shrine hall had been inexplicably moved.[8] The miraculous movement served to effectively sanctify the new building as a site of worship and pilgrimage, to authenticate a new Tibetan Monastery described by its founders as the first 'traditional-style Tibetan Monastery in North America'.[9] At both Lingyen Mountain Temple and Thrangu Tibetan Monastery the authority of such spaces as religious is conferred through their traditional architectural style which carefully emulates a foundational temple elsewhere. Interestingly, in the case of the Thrangu Monastery the founders were able to gain a relaxation in local planning regulations concerning the scale of new buildings in the Assembly District precisely by invoking the *religious authenticity* of their building style. These 'traditional- style' buildings are those which have also attracted most external attention on Number 5 Road given their 'exotic' appearance – both favourably as new sites of cultural tourism and more controversially in relation to local residents who have opposed the scale of suburban change (see Dwyer *et al.* 2016).

Despite their narration as 'authentic' reproductions, it could be argued that both Lingyen and Thrangu are somewhat hybrid spaces since the replication of a traditional style requires a creative engagement by their builders in response to the challenges of the local environmental and planning context (see Dwyer, 2017b; Shah *et al.* 2012). However, this is not part of the narrative by which these religious spaces are understood by their followers. In contrast, a final two examples of the architectural making of religious space on Number 5 Road are more suggestive of a narrative of architectural innovation. The Az-Zahraa Mosque which opened in 2002 is an imposing structure built in a recognisable Islamic style with minarets and a green painted dome. The Shia Muslim community which built the building are an ethno-religious community of East-African Asian migrants who trace their ancestry as a Khoja community to Gujurat.

Describing the architectural style of the building, a member explained: 'as a Shi'a community, we looked towards some of the Saudi Arabian, and more commonly, Iraqi and Persian styles of architecture and took a little bit from each of them. So our two minarets mirror the appearance of minarets in a mosque in Najaf, Iraq, the Imam Ali Shrine. The dome mirrors the Prophet's mosque in Medina – the Green Dome. The dome is actually solid copper, but we painted it green'. The mosque hall also includes fourteen windows, inspired by Persian stained glass, with Arabic inscriptions representing the Prophet, his daughters and progeny. The building is similar in intent then to the traditional-style Buddhist temples in evoking foundational religious buildings elsewhere but is more self-conscious in how this hybrid style is narrated. The community was also innovative in building a mezzanine floor to the main prayer hall which enabled women to hear sermons without compromising norms of gender segregation for worship practice. 'Essentially they're in the same room', suggested one respondent. There is, however, no particular attempt to engage with the specific local context and the resulting building is comparable to some of the other large 'boxy' religious institutions on the road.[10] For this Muslim community the building mirrors their transnational, diasporic identity evident in transnational worship practices including sermons conveyed via video link and the use of technology to link them to communities elsewhere.

Perhaps the only building on Number 5 Road whose architectural style is innovative and more explicitly engaged with its local context is the Dharma Drum Monastery. This modern, purpose-built Buddhist monastery was built in 2006, for a largely Taiwanese speaking community inspired by the traditions of Master Sheng Yen. The building was built by a Vancouver construction firm, Kindred, whose specialisms included exclusive private residences in Vancouver, innovative designs like the 'floating boathouse' for the University of British Columbia and renovation of the Dr Sun Yat-Sen Classical Chinese Garden in Vancouver's downtown Chinatown.[11] The monastery is a low building, described as 'California-style' by the Kindred building director. A spokesperson for Dharma Drum contrasts her own monastery with other Buddhist Monasteries on Number 5 Road: 'it's actually a bit different in terms of its construction. It's very simple. As opposed to the other temples they're more fancy, more decorative'. Describing the building, which is made of wood with intersecting verandas, she explains: 'In terms of the Dharma Drum Mountain, what we wanted to show was simplicity, more natural. The wood, it's all natural colours, having the windows so the natural light can come in. And that was important in terms of environmental issues'.

For the community at Dharma Drum, which has resident monks offering retreats for practitioners to learn ritual drumming and meditation, the architectural style of the centre is described as 'just a very simple, modern and clean. It's just what we practice here is basically, just making life simpler. So it's something very simple and hopefully with a design like this we hope to bring more calm and peace to us and maybe for people that come in and just experience that tranquility'. The Dharma Drum Monastery is an interesting hybrid space whose approach to religious and spiritual practice is articulated through an architectural style which is attenuated to a local vernacular and narrated through an environmental theology.

Analysing the architectural styles of religious buildings along Number 5 Road, suggests differences of aesthetics which may be described as 'functional', 'traditional' or more 'innovative' or 'hybrid', which were sometimes linked to particular discourses of belief or theology. The distinctive geographies of Number 5 Road in terms of its 'edge-city' location were engaged by the faith communities primarily in terms of the extent to which they facilitated larger facilities with plenty of space for car parking. The notion of a 'frontier space'[12] for new building echoes more secular discourses of speculative suburban-fringe development (Peck 2011) with large religious facilities consistent with existing suburban sprawl architecture. However, as I have suggested above, even the most functional of these religious spaces is rendered 'sacred', whether through ritual practice or the temporary presencing of the divine in collective worship. In the next section I explore further both how practices were understood by the faithful and how such 'performative presencing' creates the spiritual landscape of Number 5 Road.

Performing the suburban sacred: prayers, miracles and presence

Researching the histories and experiences of the diverse faith communities on Number 5 Road was often to experience a shift in register from the physical to the metaphysical as respondents shared their experiences and understandings of these religious spaces. Through these moments, I came to interpret Number 5 Road differently – as a landscape made spiritually meaningful through the performances of believers. This performativity is sometimes formal, part of regular authorised religious rituals, but also sometimes more spontaneous or unruly. Such moments, related through a series of vignettes from my fieldwork, offer an insight into informal and everyday experiences of faith communities on Number 5 Road.

As I suggested above the ordinary exterior of the Vedic Cultural Centre masks the dynamic spiritual activity of the Hindu temple within. Visiting the temple, an elderly devotee explains to me that the venue is animated both by the presence of the deities but also by the ritual pujas that took place the previous day. She explains:

> The energy stays, we have the energy from the Sunday and you can feel that energy, it stays here all week, it settles like the dust from construction work.

On several visits to the Subramaniya Swany Temple, a temple following South Indian traditions of Hinduism, devotees tell me about the miracles which have happened in the temple when the temple goddess has walked leaving footprints in the powder around the shrine or the guru has expelled a sacred stone. A young professional woman I talk to during an evening puja recounts her own experience:

> I know it's hard to comprehend when you're educated and everything, but once I was here, and the kitchen door just opened, and you felt a presence.

Such accounts by believers gave some insight into the 'affective atmospheres of the sacred or the divine' (Holloway 2013: 205) through which beliefs were strengthened and sacred sites were animated. When the deities (imported from India) were first installed in the Subramaniya temple, a forty-day cycle of rituals was required so that, as one participant explains to me, their 'faculties' might be enabled (see also Dempsey 2006). Once the gods are properly installed in the temple, a small wooden built temporary building dwarfed by its more imposing neighbours, their power and efficacy is evident for the faithful, as the miracles attest.

While buildings are animated, sometimes this spiritual presence can also be extended to the surrounding neighbourhood. Visiting the Lingyen Mountain temple, I witnessed an unusual performance which prompted reflection. Visitors to the temple typically enter the threshold of the temple space by making an offering and burning an incense stick. Instead a visitor took the proffered incense sticks and after he bowed first towards the Buddha shrine, then turned and bowed with his burning incense outward towards the Number 5 Highway. Witnessing his action I was intrigued, questioning whether this action was an inclusive gesture or suggested a more proselytising intent towards the suburban neighbourhood. Whichever was true, for my Chinese Christian companion this was an undeniably powerful gesture which risked 'opening up metaphysical or spiritual pathways which may not be benign', a recognition of the agency of these incense vapours as they travelled out into the highway.

In its different forms – spontaneous, read directly from a sacred text, chanted, sung, organised, collective or individual – prayer was central to religious life on Number 5 and was an important means through which the divine was made present. For worshippers at Richmond Evangelical Free Church, collective and spontaneous prayer produced the collective strength to send off a mission group gathered at a church service before setting off to Mexico in 2010. The efficacy of prayer can also extend beyond the church, and believers at the Richmond Bethel Church are encouraged to pray with and for neighbours and co-workers while a 'dinner ministery' at the church will provide a space of sharing Christian beliefs with visitors. All of the evangelical churches on the road were keen to develop networks of local prayer groups beyond the Sunday service as a means to support their ministeries.

An unguarded comment from some young people at one of the Chinese churches provided an intriguing example of prayer and performance – they admitted they had organised a secret 'prayer walk' along Number 5 Road as a means to symbolically reclaim the street for Christianity in the wake of the building of the new Thrangu Monastery. While their actions invoke a long tradition of prayer walking (Megoran 2010; Maddrell 2013; Middleton and Yarwood 2015), this was an unauthorised activity which they had concealed from the elders at their church since it would almost certainly have been forbidden. On another occasion visiting the Lingyen Mountain temple an older Chinese lady who acted as an usher at the shrine was keen to share the many CDs which were sold at the temple which had recordings of Buddhist chanting from the temple. She explained that she kept them in her car and would play the chant when driving to the temple as a means to both prepare herself for her visit and also to create a sacred space of a journey towards the temple. She raised the spectacle of many different drivers in cars which were mobile prayer spaces opening up an interconnected set of spiritual networks, trajectories and pathways converging on and emanating from Number 5 Road.

These diverse snippets from discussions with believers along Number 5 Road produce a more dynamic understanding of *Highway to Heaven* as a spiritual landscape animated through the religious rituals, beliefs and practices of its diverse faith communities. The spiritual processes and powers engaged by believers challenge the normative temporalities and spatialities of Number 5 Road both in extending the boundaries of the space of the spiritual and by animating the wider landscapes of the edge-city. Relating them engages the 'lived religion' of those who worship on Number 5 Road. Writing about belief in miracles, Robert Orsi (1997: 12) explains that 'religion comes into being in an ongoing and dynamic relationship with the realities of everyday life'. For some believers along Number 5 Road, the everyday suburban landscape is simultaneously imbued with the miraculous, the supernatural and the sacred. It was thus that emerging from the Plymouth Brethren Hall after staying late after a service and finding the carpark empty, my companions were to joke that such moments provoke the fear that 'the Rapture'[13]has taken place without their inclusion!

Conclusion: enchanting suburban geographies

My first encounter with Number 5 Road was an online blog written by a local tourist from Vancouver who expressed her joy and enchantment in a chance encounter with the unexpected religious diversity in her familiar urban environment. Her experiences were echoed in the narrative of a devotee at Lingyen Mountain Temple, a white Canadian businessman with no prior foundation in Buddhist practice, who explained: 'I was literally just driving by and thought: what a beautiful place'. Much of the public discourse about Number 5 Road which suggests that this is landscape of extraordinary and 'spectacular multiculturalism' (Dwyer *et al.* 2016: 17) also engages metaphors of enchantment. The philosopher Jane Bennett (2001: 5) defines: 'to be enchanted is to participate in a momentarily immobilising encounter, it is to be transfixed, spellbound . . . to be struck and shaken by the extraordinary that lives amid the familiar and the everyday'. Evoking Bennett's notion of 'enchantment', Julian Holloway argues that 'instead of focusing our accounts upon sacred space and times separate from the geographies and temporalities of our everyday . . . we should seek out the extraordinary as practised and sustained in the ordinary' (2003: 1961). While Number 5 Road is unusual in its religious diversity, this chapter has tried to illustrate the juxtaposition of the mundane and the miraculous in this functional edge-city landscape to illustrate a spiritual landscape where the 'extraordinary can be made out of the ordinary' (Gilbert 2012: 140).

The unusual proximity of different faith communities along Number 5 Road inevitably provoked reflection for its diverse inhabitants. The overarching narrative is of both peaceful co-existence (exemplified in the sharing of parking facilities) and of somewhat passive co-location. Exacerbated by an edge-city location heavily dependent upon automobility, our respondents told us: 'we're in our own little worlds'; 'we just drive in and drive out'. For some the proximity of other faiths is an opportunity to acknowledge shared or syncretic belief systems. The Buddhists, Sikhs and the Hindus were most likely to visit each other's place of worship while most of the non-Christian worship spaces participate in school visits or tours. There is a certain pride in ownership of the colloquial name for *Highway to Heaven*, as one of the Sikh elders suggests: 'God lives here in No. 5 Road'.

However, this presencing of the divine by so many different groups in close proximity can also provoke uncertainty. The impetuous prayer walk by young evangelical Christians would have been a source of embarrassment had it been made public, for it exposes the contradictions of Number 5 for some communities. Some respondents agreed that to find themselves beside so many other different faith traditions prompted reflection and even challenge. The pastor the Richmond Bethel Church explained:

> I enjoy being on the street. I love the fact [that] the gospel, from my perspective, can be spoken, practiced, lived in an avenue where there's variety. There's a smorgasbord of faith. I don't think that dilutes the gospel.

Another pastor found the juxtaposition more challenging and confusing in a telling comment which provided an insight into how he understood the dynamism of this unusual spiritual landscape:

> I sometimes think about the spiritual implications . . . there's so much airwaves in a sense if you think of prayers. . . . One can't help but wonder does that really affect our effectiveness? Just being surrounded by all the others?

This depiction of the overcrowded spiritual airways of Number 5 Road, filled with prayers and intercessions, drum beats and chanting, incense vapours and smoke, provides an intriguing metaphor for a distinctive spiritual landscape shaped by 'the agency of gods' (Chakrabarty 1997: 35). Number 5 Road is certainly an unusual and distinctive religious landscape shaped by planning, real estate and the aspirations of diverse transnational faith communities. In this chapter I have sought to trace the ways in which this is also a distinctive spiritual landscape of the edge-city animated by everyday lived experiences of the Spirit and encounters with the divine. Echoing the possibilities of finding the 'extraordinary in the ordinary' (Gilbert 2012; Holloway 2003), the mundane landscape of the 'edge-city' emerges as an extraordinary landscape of spirituality.

Notes

1 The starting point for this chapter was my keynote lecture to Geographies of Religion and Belief Systems Specialty Group, Annual Conference of the Association of

American Geographers, Seattle (April 2011). 'Encountering the Divine in W7 and off Highway 99: stories of the suburban sacred.' A later version of the paper was given at the session 'Investigating the Anthropo-Unseen: Mapping the Paranormal, the Extraordinary and the Unknown' at the Annual Conference of the RGS-IBG, Exeter, August 2015. I'm grateful to the audiences on both occasions as well as Justin Tse, Julian Holloway, David Gilbert, Betsy Olson and Steve Pile for helpful discussions of some of the arguments explored in this chapter.

2 Research was conducted between 2010 and 2012 and funded by Metropolis Canada (Grant Reference 12R47822). Work was undertaken in collaboration with Professor David Ley and Dr Justin Tse. The paper draws on archival and documentary sources, interviews and participant observation at all the religious buildings along Number 5 Road. See Dwyer *et al.* 2013a for more details of the methodology and research design.

3 Estimated population from 2011 Household Survey, which is a substantial increase on the Census total (2006) of 175,000. The 2016 census data will be released in 2017. (Richmond City Hall: www.richmond.ca/discover/about/profile.htm Accessed 20/10/16).

4 www.richmond.ca/__shared/assets/2006_Ethnicity20987.pdf Accessed 20/10/16. 'Visible minority' is the category used by Statistics Canada to refer to those 'who are non-Caucasian in race or non-white in colour and who do not report being Aboriginal'.

5 For a detailed history of the settlement of faith communities on Number 5 Road see Dwyer *et al.* 2013a.

6 'Say No to Buddha Disneyland' Letter from Carol Day to the Editor, *Richmond News*, 29 September 2010.

7 A further elaboration of this theological position in relation to architecture is provided by the Plymouth Brethren Community, who have a small meeting hall on Number 5 Road which has no external signage beyond a bible text and is named only as 'The meeting hall'.

8 Notes from fieldwork at opening of the Thrangu Monastery, 25 July 2010. *Richmond News*, 28 July 2010.

9 While its religious authentication may be confirmed by this process, the Thrangu Monastery may also be understood as a space which is more contested in geopolitical terms, serving as a space of religious and communal identity for both a Chinese-Canadian community of practitioners and an émigré Tibetan community who share very different backgrounds and political outlooks (see Dwyer 2017a).

10 There is an interesting contrast with another mosque built by a sister community in Harrow in north-London, where the same narratives of foundation in Persia are traced more self-consciously through India and Africa to London in the hybrid architectural style of the building (see Dwyer 2015).

11 Kindred also built the Thrangu Tibetan Monastery, partly as an outcome of their success at Dharma Drum monastery.

12 Entirely absent in any of our interviews was any discussion of the possibility of the land developed for new religious buildings having any prior sacred meaning. This was interesting given the contested status of 'Garden City Lands', a neighbouring plot of land owned by the City Council whose re-development had been challenged by First Nations Groups.

13 The Rapture refers to the belief (drawn from a reading of St Paul in the New Testament, 1 Thessalonians 4: 15–17) in an event when all born-again Christians will be gathered together to meet Christ on his return.

References

Agrawal, S. 2015. Religious clusters and interfaith dialogue, in M. Burayidi (ed.) *Cities and the Politics of Difference* (pp. 318–339). Toronto: University of Toronto Press.

Bartolini, N., Chris, R., MacKian, S. and S. Pile. 2017. The place of spirit: Modernity and the geographies of spirituality. *Progress in Human Geography*, 1–17.

Bennet, J. 2001. *The Enchantment of Everyday life: Attachments, Crossings, Ethics*. Princeton: Princeton University Press.

Chakrabarty, D. 1997. The time of history and the time of Gods, in L. Lowe and D. Lloyd (eds) *The Politics of Culture in the Shadow of Capital* (pp. 35–60). Durham NC: Duke University Press.

Della Dora, V. 2015. 'Sacred Space Unbound' introduction to virtual issue of *Society and Space* 13.

Della Dora, V. 2016. Infrasecular geographies: Making, unmaking and remaking sacred space. *Progress in Human Geography*, Published first online 0309132516666190.

Dempsey, C. 2006. *The Goddess Lives in Upstate New York*. Oxford: Oxford University Press.

Dempsey, C. 2008. Introduction: Divine proof or tenacious embarrassment: The wonders of the modern miraculous, in S.J. Raj and C. Dempsey (eds) *Miracle as Modern Conundrum in South Asian Religious Traditions* (pp. 1–23). Albany, NY: SUNY Press.

Dewsbury, J.D. and P. Cloke. 2009. Spiritual landscapes: Existence, performance and immanence. *Social and Cultural Geography*, 10(6), 695–711.

Dwyer, C. 2015. Reinventing Muslim space in suburbia: The Salaam Centre in Harrow, North London, in S. Brunn (ed.) *The Changing World Religion Map: Sacred Places, Identities, Practices and Politics* (pp. 2399–2414). New York: Springer.

Dwyer, C. 2017a. New religious architecture in the suburbs: Examples from London and Vancouver, in P.J. Margry and V. Hegner (eds) *Spiritualising the City* (pp. 115–129). London: Routledge.

Dwyer, C. 2017b. Transnational religion, multiculturalism and global suburbs: A case study from Vancouver, in D. Garbin and A. Strahn (eds) *Religion and the Global City* (pp. 173–188). London: Bloomsbury.

Dwyer, C., Gilbert, D. and B. Shah. 2013b. Faith and suburbia: Secularisation, modernity and the changing geographies of religion in London's suburbs. *Transactions of the Institute of British Geographers*, 38(3), 403–419.

Dwyer, C., Tse, J. and D. Ley. 2013a. *Immigrant integration and religious transnationalism: The case of the 'Highway to Heaven' in Richmond, BC* Working Paper 13-06, Metropolis, British Columbia, http://mbc.metropolis.net/assets/uploads/files/wp/2013/WP13-06.pdf

Dwyer, C., Tse, J. and D. Ley. 2016. Highway to heaven: The creation of a multicultural religious landscape in suburban Richmond, British Columbia. *Social and Cultural Geography*, 17(5), 667–693.

Gilbert, D. 2012. Sex, power and miracles: A suburban triptych, in A. Gibson and J. Kerr (eds.) *London From Punk to Blair* (pp. 131–140). London: Reaktion Books.

Gilbert, D., Cuch, L., Dwyer, C. and N. Ahmed. 2016. The sacred and the suburban: atmospherics, numinosity and 1930s interiors in Ealing, London. *Interiors: Design/Architecture/Culture*, 6(3), 211–234.

Holloway, J. 2003. Make-believe: Spiritual practice, embodiment and sacred space. *Environment and Planning A*, 35(11), 1961–1974.

Holloway, J. 2011. Spiritual life, in V. Del Cascino, M. Thomas, P. Cloke and R. Panelli (eds) *A Companion to Social Geography* (pp. 385–400). Oxford: Blackwell.

Holloway, J. 2013. The space that faith makes: Towards a (hopeful) ethos of engagement, in P. Hopkins, L. Kong and E. Olson (eds) *Religion and Place: Landscape, Politics and Piety* (pp. 203–218). Dordrecht, The Netherlands: Springer.

Kong, L. 2001. Mapping new geographies of religion: Politics and poetics in modernity. *Progress in Human Geography*, 25(2), 211–233.

Ley, D. 2010. *Millionaire Migrants: Trans-Pacific Life Lines*. Oxford: Wiley-Blackwell.

Li, W. 2009. *Ethnoburb: The New Ethnic Community in Urban America*. Honolulu, HI: University of Hawaii Press.

MacKian, S. 2012. *Everyday Spiritualities: Social and Spatial Worlds of Enchantment*. Basingstoke: Palgrave Macmillan.

Maddrell, A. 2013. Moving and being moved: More-than-walking and talking on pilgrimage walks in the Manx landscape. *Culture and Religion*, 14(1), 63–77.

Megoran, N. 2010. Towards a geography of peace: Pacific geopolitics and evangelical Christian Crusade apologies. *Transactions of the Institute of British Geographers*, 35, 382–398.

Middleton, J. and R. Yarwood. 2015. 'Christians, out here?' Encountering street-pastors in the post-secular spaces of the UK's night-time economy. *Urban Studies*, 52(3), 501–516.

Nayer, K.E. 2010. The making of Sikh space: The role of the Gurdwara, in L. DeVries, D. Baker and D. Overmyer (eds) *Asian Religions in British Columbia* (pp. 43–63). Vancouver: UBC Press.

Olson, E., Hopkins, P., Pain, R. and G. Vincett. 2013. Re-working the postsecular present: Embodiment, spatial transcendence and challenges to authenticity among young Christians in Glasgow, Scotland. *Annals of the Association of American Geographers* 103(6), 1421–1436.

Orsi, R. 1997. Everyday miracles: The study of lived religion, in D. Hall (ed.) *Lived Religion in America: Towards a History of Practice* (pp. 3–21). Princeton, NJ: Princeton University Press.

Peach, C. and R. Gale. 2003. Muslims, Hindus and Sikhs in the new religious landscape of England. *The Geographical Review*, 93(4), 469–490.

Peck, J. 2011. Neo-liberal suburbanism: Frontier space. *Urban Geography*, 32(6), 884–919.

Pottie-Sherman, Y. and D. Hiebert. 2015. Authenticity with a bang: Exploring suburban culture and migration through the new phenomenon of the Richmond Night market. *Urban Studies*, 52(3), 538–554.

Shah, B., Dwyer, C. and D. Gilbert. 2012. Landscapes of diasporic religious belonging in the edge-city: The Jain temple at Potters Bar, Outer London. *South Asian Diaspora*, 4(1), 77–94.

Wilford, J. 2012. *Sacred Subdivisions: The Postsuburban Transformation of American Evangelicalism*. New York: New York University Press.

Williams, A. 2016. Spiritual landscapes of Pentecostal worship, belief and embodiment in a therapeutic community: New critical perspectives. *Emotion, Space and Society*, published first online.

Woods, O. 2013. Converting houses into churches: The mobility, fission and sacred networks of evangelical house churches in Sri Lanka. *Environment and Planning D: Society and Space*, 31(6), 1062–1075.

8 Kendal Revisited

The study of spirituality then and now

Karin Tusting and Linda Woodhead

In the year 2000, as the new millennium began and before the twin towers of New York had fallen, a team of five researchers began a study of religion and spirituality in the market town of Kendal, Cumbria, in the north of England – population at the time 27,000. Two of that team, Karin and Linda, are responsible for this chapter.

We chose Kendal not because it was unusual, but because it was a rather typical market town in England outside of the South East. Sitting on the edge of the Lake District it is usually bypassed by tourists, but is remote enough to be self-contained and large enough to have its own amenities – schools, a college, and a hospital. In demographic terms like age and class it did not have any particularly unusual features compared with other English towns. In ethnic and religious terms, it was not very diverse and was very 'white British' with a Christian heritage and 25 functioning churches and chapels.

The Kendal Project took two years, and three years after that it resulted in a book, *The Spiritual Revolution* (Heelas and Woodhead, 2005). The book's subtitle – *Why Religion is Giving Way to Spirituality* – summed up the main findings: that organised religion as represented by the Christian churches and chapels of Kendal was declining and that new forms of holistic spirituality were growing. It was the sheer quantity and vitality of the latter which was the project's most surprising and important finding: there were 126 separate 'mind, body, spirit'-oriented groups and one-to-one offerings, even when we excluded those which said they did not have any spiritual purpose. They ranged from Yoga to Reiki to spiritual dancing to various forms of meditation.

The book predicted that if current trends continued, churchgoing in Britain would decline from over 7% to around 3% of the population by 2030. It said that active, regular participation in the 'holistic milieu' would increase from the roughly 1.5% of the population found in Kendal to about 3% of the British population in the same period. And so it predicted that by 2050 the 'holistic milieu' would start to overtake the 'congregational domain'.

Looking back from over a decade later (2017), these predictions have largely held good, and the significance of spirituality has been much more widely acknowledged. Churchgoing in Britain has declined as predicted. By 2015 it was down to 5% of the population, with no sign of bottoming out. Spirituality has achieved a much higher profile in Britain than when the study took place (today it is hard to credit what strong criticism the *Spiritual Revolution*'s findings about spirituality attracted – as documented by Woodhead, 2010). But the numerical predictions which *The Spiritual Revolution* made about the growth of spirituality are hard to test, because the evidence is lacking. The reasons for this take us to the heart of this chapter.

By going back to Kendal, both literally and by revisiting our research data, much of which remains unpublished, we discover some interesting things about how spirituality has changed in the intervening years – with important implications for the ways we should now study it. What we study and how we study it are never completely separate. The approach we used to research spirituality in Kendal was shaped by the way that 'alternative' spirituality was positioned at the time. Fifteen years later the situation has changed so much that the approach of the first Kendal Project would no longer be as appropriate. Reflecting on Kendal today, we find a situation in which the churches have become more marginal, and 'no religion' and spirituality more mainstream. We suggest that these changes are so significant that our original research design would need more than tinkering – it would need to be turned inside out. Just what that would involve is the subject of this chapter.

The street survey

One of the parts of the research project which was never written up was what we called 'the Street Survey'. In order to dip our toes into Kendal beliefs and religion outside of its organised forms we decided to do door-to-door interviewing in the most socio-economically varied street we could find. Karin, who undertook the research, takes up the story.

To one side of the town centre of Kendal is a small residential area. At one end is a lane of two-up two-down houses, and along from them slightly larger but still traditional terraced homes, set around a small green. Further up the hill the houses start to get bigger, until you turn into a private road of grand, imposing mansions, many of which have been split into two or more dwellings, and some into flats. At the end of the private road is a very large building which has been converted into luxury flats. Finally, the road drops down again into a modern development of detached and semi-detached homes.

For four months, from December 2001 to March 2002, I knocked on the door of each one of these 116 widely varied houses, to invite people to participate in short interviews about their attitudes to religion, their beliefs and their religious practices. Responses to this request varied widely, from those who politely refused the interview – some citing time pressures, others on the basis that 'we don't talk about that sort of thing' – to those who responded monosyllabically to the interview questions, through to those who invited me into the house and talked for an hour or more. The interviewees' beliefs, practices and affiliations with religious institutions varied enormously. Some of these interviews have stayed with me in the fifteen years since the street survey, a late part of the Kendal project, was carried out.

There was the practising Catholic who began the interview saying he attended Mass as a 'sacrifice of time', and prayed infrequently – then, almost out of the blue, shared a story of the powerful, personal mystical experience in an Italian monastery which had brought him back into the Church. There was the committed New Ager living in a small flat crammed with angel symbols, instructions for 'Emotional cleansing programmes' and books on topics like 'Natural Highs' who believed in a highly-organised life after death, a God made up of 'the combined power of every consciousness in the universe', Indigo children (a group of children from the 1970s onwards believed to be born with enhanced spiritual abilities), and, especially, channelling – but who, despite the absolute certainty of these beliefs, had never had what he would call a spiritual experience. There was the woman who nominally belonged to the Church of England but believed in reincarnation, karmic justice, and Spiritualist beliefs about life after death – and who had the vicar bless her home with incense to quieten the live-in ghost. I met an environmentalist who refused to have any belief in God at all, but rather believed in 'the power of Mother Nature', finding his community in environmental groups like Greenpeace and Friends of the Earth; a Pagan homeopath with a Wiccan altar in her upstairs windowsill, to be used for prayer and meditation when she needed to, without a belief in any kind of God; and a woman who, while expressing a definite lack of belief associated with what she called 'an essentially Darwinian understanding of life', had nevertheless had what she called 'inexplicable' out-of-body experiences and dreams of past lives which left her 'still havering' on questions of spirituality and faith.

In total, 56 individuals were interviewed, from two-fifths of the households (with a majority of the households not answering after three visits, and the rest refusing to participate in the survey). Respondents were drawn from all the different types of housing in the street.

The majority (55%) of the 56 people interviewed did not identify as Christian. Of these, just under a quarter expressed a definite spiritual or religious position which was important to them (like the Pagan Wiccan described above). Another half expressed a definite set of beliefs about what lies beyond the material world, but these beliefs did not play a large part in their lives and did not bring them into association with others to share those beliefs. The remaining quarter of the non-Christians said they did not believe in anything supernatural, but talked about there being 'something more to life', either locating this in a kind of humanism – 'I believe in people' – or in aesthetic experience. Only two people said they were definite atheists avowing a lack of any belief in anything beyond the material.

The remainder of the respondents (45%) identified as Christian and/or as belonging to one of the Christian churches. Of these, a majority (three-fifths) said they were churchgoers, but some may have meant they occasionally attended, because our meticulous church headcount – an important part of the original study – found that less than 8% of the town was in church on an average Sunday (Heelas and Woodhead, 2005). The next biggest group of Christians, just over a quarter, expressed an identification with Christianity and/or a church and belief in God but said they were not churchgoers. And the rest – about a tenth – were what we might call 'Christian atheists' who said that they belonged to a church or were Christian but did not believe in God.

With the benefit of hindsight we can see how important these findings were. They touched on what by the time of writing this chapter had become the most important new discovery about the religio-cultural landscape of Britain and some other countries: the rise of 'no religion'. 'No religion' is a category created almost by mistake – the category added to survey questions on religion (like the Census's 'what is your religion') for those who don't want to tick any of the existing boxes ('Christian', 'Muslim' and so on).

As Woodhead (2016) shows in detail, the 'nones' have only a few things in common besides their disaffiliation from a religious identity, strong commitment to liberal values and free choice in relation to personal moral issues, and relative youthfulness. In terms of their beliefs and practices they are varied, a large minority being atheist and the majority being believers in God or agnostics who are not sure. To most people's surprise, it is a category which has grown relentlessly to the point where, by 2017 it had just overtaken 'Christian' to become the single largest category of 'religious' affiliation in Great Britain, especially for people under 40 (the Census still finds 'Christian' to be larger, but there are reasons for thinking that on this question of religious affiliation the Census is less reliable than other surveys, including the British Social Attitudes Survey – see again Woodhead, 2016 for more discussion). The significance of 'no religion' becoming the majority is not just a matter of numbers: it means that 'Christian' has ceased or is ceasing to be the norm not just in terms of self-identification but for many institutions, rituals and activities, including weddings and funerals (by 2015 a third of people said they would like a non-religious funeral, a third a traditional religious funeral, and a third a mixture of the two). The normalisation of 'no religion', a category which is neither straightforwardly religious or secular, disturbs and displaces the old 'religious-secular' binary which was used to do so much work in the social sciences, and with it the theory of secularisation which still dominated the field when we undertook the Kendal Project.

As well as being early indication of the importance of 'no religion', the Street Survey also signalled the importance of 'non-institutional' or 'lightly-institutional' religion. The main study had focused primarily on institutions, organisations, communities and individuals clearly located within the fields of religion and spirituality and using their own demarcated spaces. We took a rather traditional, church-inflected, account of religion as our rule of thumb, looking for something organised, with a leader(s), boundaries, practices, teachings and a clear identity as 'religious' or 'spiritual'. However, the Street Survey shows that even at that time those institutions and the people actively engaging in them represented only a relatively small proportion of the population of Kendal, about 10%. Today that might be even smaller if we stayed with the same approach. What the Street Survey did was to show that the many people who are not actively involved in associational religion of a church-like type are nevertheless not necessarily atheists, nor indifferent to questions of meaning, cosmology, and transcendence, nor unreflective about them. We had fascinating glimpses of the thoughtful, reflexive, independent, and diverse ways in which many people generated meaning in their lives, even when not actively involved in any kind of organised religion or faith community.

Thus, were we to do a Kendal Project today, we would probably put such people and 'no religion' at the centre of the study. There were two main reasons why we left them out of our publications at the time. The first was that the street and its small sample were not representative, even of the town, and so it would be wrong to put it on the same footing as the other findings. The second was that, along with a lot of other data, it was not central to the main purpose of the book, which was to compare the fortunes of the 'congregational domain' with the 'holistic milieu' and explain why the latter was doing so much better. The only way we could make that comparison scientific was by comparing like with like. That is why we focused our attention in the spiritual milieu on 'church-like' organised, bounded, associational forms of spirituality, rather than on the wider manifestations of spirituality in the town.

This decision was the right one for the context in which we were working, mainly within the Sociology of Religion and in a climate which was still 'default Christian' (Luckmann, 1967; Woodhead, 2010). We succeeded in demonstrating that, even when looked at through a Christian lens, spirituality was a force to be reckoned with, and we were able to make sociologists who retained that church-inflected understanding of religion take our findings seriously, even when they argued against them. However, the downside of this decision was that we omitted something which would become increasingly important. The Street Survey revealed the green shoots which would grow so strongly that within just fifteen years they would push aside the existing paving slabs of institutional religion and leave a majority of people feeling they must dissociate themselves from the very category of 'religion', even though many of them retain belief in God or Spirit and some take part in spiritual practices (Woodhead, 2016). That small Street Survey turned out to be far more representative of what was happening in the wider population, not just of Kendal but of Britain, than we could have known at the time.

Going back to Kendal

As well as revisiting our data we can revisit the town in real life. We have been able to do this fairly regularly because we both continue to live and work nearby. This is the kind of informal research in which conversations and observations can lead to new insights and inform a new research design.

The holistic milieu

The original Kendal Project found 126 distinct holistic groups and one-to-one activities, and it had taken a full-time researcher well over a year to find them all. Although we couldn't possibly replicate this effort in a few short visits, the existence of the internet (only in embryo in 2000) and knowing who to talk to and what to look for allowed us to register some prominent developments.

We could see that one of the original hubs of such spirituality – places where a range of activities and several practitioners are housed together – was still active (the Fellside Centre), though we were told that it was increasingly hard to maintain the large and costly premises. Another, Rainbow Cottage, had closed many years ago, its owner having moved away, but there were at least two new hubs in its place, Staveley Natural Health Centre just outside Kendal and the Holistic Healing Centre. The latter offered massage, myofascial release, remedial massage, aromatherapy, reflexology, no hands massage, The Dorn Method, Gentle Touch therapy, Reiki, flower therapies, Indian Head Massage, Mindfulness, Emotional Freedom Technique, and shamanic healing.[1] Similarly, a shop offering spiritual healing of various kinds and selling a wide range of spiritual merchandise – from cheap crystals to expensive singing bowls – had closed, but another one offering some similar goods but with more of an emphasis on psychic readings had opened.

We gained further information from a long-term holistic practitioner, now in her mid-60s, who used to run the healing shop which had closed. Her personal career trajectory was instructive. When Linda last interviewed her fifteen years ago, she had been offering Reiki healing in the upstairs room of the shop she had opened. She closed the shop just a couple of years later because she realised she did not have the commercial skills or inclination to make it a success, even though customer demand was high. After that she had continued to practise, but in a very flexible way. She had practised out of a number of different rental premises in the intervening year, had had periods out of work to pursue other goals, and had continued to train and practise a variety of techniques.

Her original training had been as a nurse in the NHS (one of the refugees from church or Health Service or both we found to be so central in the holistic domain in our first study). While she was still a nurse she had tried out a strict Christian chapel but found it far too restrictive for women, and had moved onto a Spiritualist Church where she started to train as a healer. But it wasn't until she encountered Reiki that she really felt she had found her vocation. That didn't stop her experimenting with other techniques. Re-interviewing her we found that since the first Kendal Project she had been trained in Neuro-Linguistic Programming, which she now practised along with Reiki and other healing techniques, offering combinations tailored to individual client need. She explained to us how she had learned from all of these trainings and experiences to become a better healer. It was she – and 'the universe', including 'inspirational' people she met – who had shaped her career, rather than a career shaping her.

Piecing this all together we gain the clear impression that the holistic milieu has matured and become more mainstream in three senses. First, it has become less counter-cultural and 'alternative' and more culturally mainstream, widely accepted, and 'normal'; we argue below that it seems to have changed places with the churches in this regard.

Second, there appears to be more variety but also more integration.[2] Spirituality has shown itself to be extremely adaptable. When we were in Kendal at the start of the millennium we found that it was fairly common for clients to experiment with various therapies, often moving from one provider to another in doing so, or using several simultaneously. Some became expert consumers, able to rate and mix and match their treatments in reflective ways. Today it is clearer that this is also happening at the supply-side, not only because some practitioners work out of a single hub from which different activities can be accessed, but because some – like our key informant – build up a range of different skills and knowledges themselves, and are able to tailor their different offerings to the unique needs of each client. Individual journeys through the landscape of holistic activities are shaped by the needs, aptitudes, and biographies of both client and practitioner.

Third, spirituality has broken out of a bounded 'holistic milieu' understood as a sub-section of society shared with the churches and dedicated solely to religion. Already in 2000–2002 we noticed it was creeping into other sectors and institutions including the local College, Cancer Care Centre, and shops on the High Street. In subsequent visits we have noted how seamlessly it has inserted itself into the professionalised health and wellbeing scene of consumer capitalism: the holistic hubs, for example, present themselves as services offering clients the sort of satisfaction they could expect from other regulated health and wellbeing providers and therapists.[3] In 2016 we found a further powerful example of de-differentiation thanks to an invitation from a local primary school headteacher.

The schools

The school we visited is a Church of England state-funded primary school with 200 pupils. Such schools are common in England where over a third of primary schools are 'CofE' – a legacy of the time when the church was the major educational provider (Clarke and Woodhead, 2015, p. 17). The headteacher had made contact because he was interested in the Kendal Project and was beginning a study of spirituality in local schools. He was kind enough to talk to us, show us around his school, and share his findings.

It was immediately clear that spirituality was deeply embedded in the school, its ethos, and activities. The headteacher defined it in terms of 'Three Cs': consciousness, connection, and change. It was flexible enough to absorb many elements, including the CofE identity and a set of virtues and values drawn up by Archbishop Rowan Williams and prominently displayed on noticeboards in the school. To these Christian values the school had, more recently, added the 'fundamental British values' which schools now have a duty to uphold. We were told that the addition had been helpful and harmonious: the Christian values are chiefly about personal attitudes and virtues, whilst the 'FBVs' add a stronger social and political dimension. The values aren't 'imposed' on children, but made integral to the formal and informal curricula and allowed to shape the life of the school. In relation to 'Democracy' for example, we were shown a visual display made in lessons when children met and talked with their MP Tim Farron and local councillors and learned about the Prime Minister and opposition leaders of the day, David Cameron and Jeremy Corbyn.

A short survey on parental attitudes to spirituality in the school had discovered a very positive response. When asked 'How important do you think it is to promote children's spiritual development in school?' almost half said 'very important', nearly 40% 'quite important', and only a small number said that it was not important or they had no opinion. When asked which religion if any had an influence in their household, a sizeable majority said Church of England, with much smaller numbers (declining to single individuals) mentioning – in descending order – No Religion, Catholicism, Buddhism, Spirituality, Mormonism, and Paganism.

The research also involved asking eleven headteachers in Kendal about spirituality in their schools.[4] A couple were hesitant, feeling that their distance from religion also meant that they were unqualified to be experts on spirituality; they were more comfortable speaking about 'values', 'virtues', and school 'ethos'. Most of the heads, however, spoke fluently about 'spirituality', generally meaning something different from 'religion' and more inclusive, having an overlap with 'values', 'virtues', and school 'ethos' but not identical with them.

One head emphasised the importance of times and spaces for children to get away from noise and busyness during the school day. Another saw spirituality in a rather different way, emphasising that it helped children to know and understand their own culture and to be made conscious of history and tradition. Two schools placed a strong emphasis on the importance of the environment and connection with the natural world as well as human others. One thought that spirituality helped children to recognise the non-material aspects of the human experience, not least by way of the arts, as well as cultivating their critical awareness. In another, spirituality embraced fostering a critical approach by way of Socratic questioning and Philosophy for Children. And a school where a large proportion of pupils had difficult personal issues to deal with interpreted spirituality as the 'life force that pushes you through' and devoted time to helping pupils to make sense of what has happened to them and who they are. Only in a faith school which draws from a Christian faith community were religion and spirituality regarded as inseparable and virtually identical, with a strong emphasis on transmitting the faith, teaching scripture, and deepening children's relation to God and the wider global faith community.

In part these positive attitudes to spirituality can be explained by the legal requirement in England not only to hold daily acts of collective worship (assemblies) but also to attend to pupils' 'spiritual, moral, social and cultural development' – with 'SMSC' being inspected by OFSTED. But that duty has applied since 1944, and in the past had more to do with Christian instruction. What is striking in the interviews with heads isn't so much the fact that they are taking spirituality seriously, but that they use it in senses which draw more on the language and concepts of the holistic milieu than of the congregational domain, even when they are in Christian schools. Despite sharing much in common in their understanding of spirituality, each school has given spirituality a distinctive stamp of its own, tailoring it to particular needs, commitments, and ethos. Again we see clear signs of the maturation, adaptability, and normalisation of spirituality and the ease with which it can adapt to different settings.

The congregational domain

As well as finding that spirituality had become more mainstream, we found that Christianity had become more counter-cultural – a reversal of the situation in which Christianity had been central and spirituality marginal that had pertained for at least a century before (Woodhead, 2011). That situation was still evident when we undertook the first study: in 2000 the churches were the dominant players. They attracted more regular participants, had a much higher social and civic status, figured more prominently in the culture of the town and the media, and were generally more visible. Spirituality was a controversial 'alternative'. By 2016 that was no longer so true. It was not so much that the number of churches had declined – in fact only two had closed and one had opened – nor that attendance had continued its steady decline at a rate of around 1% per annum (we assume it has, in line with the rest of the country because when we counted typical Sunday attenders in 2002 Kendal was exactly 'on trend'): it was more that the profile, activities, and self-presentation of the churches had altered, as had the way they were culturally represented.[5]

Overall we found that the congregational domain as a whole had taken on a more evangelical-Protestant hue, with that particular churchmanship having extended its influence. In 2000, for example, the three Anglican churches each represented a different churchmanship: evangelical, Anglo-Catholic, and broad church. By 2016 the latter two, especially the parish church, had become more evangelical. The evangelical vicar of the parish church had made some dramatic changes, including sacking the choir in order to 'reclaim the sanctuary' as one evangelical church member told us, and initiating a mid-morning evangelical-style informal service with the singing of choruses displayed on TV screens placed in a side aisle of the magnificent medieval building. A more traditional main Sunday morning service remained in place, and the vicar told us that he continued to carry out some of the traditional civic duties of the parish priest – but the evangelical shift was evident. Its significance lies in the fact that evangelicals have a 'sectarian' rather than 'societal' ecclesiology which draws a much sharper line between church and society than traditional Anglicanism, and which values distinctiveness rather than integration (Brown and Woodhead, 2016). While the holistic milieu has de-differentiated and blended into various sectors of public life, the congregational domain appears to have differentiated and become more distinct from society.

This conclusion might seem to be challenged by something else we observed: the continuing role of the churches in social outreach and charitable action. However, despite the fact that some churches were actively involved in voluntary activities like food banks and helping people recover from flooding in Kendal in December 2016, we gained the clear impression in Kendal that the congregational domain had become less central to mainstream civic and local life than when we first studied the town. A prominent figure in the Kendal voluntary scene whom we interviewed reported that the churches had become 'less visible, prominent and confident in the last fifteen years', though they continued to be an important part of the voluntary sector. He also reported that 'I find fewer organisations are registering as primarily faith organisations' and 'the faith organisation that may lie behind the scenes can be somewhat hidden to the public that sees their works. . . . Much of the works being undertaken by them aren't thought of by the recipients as being carried out by faith organisations'.[6] The reason, he speculated, was that people could be ambivalent about using faith-based services, and that church involvement would not always be welcomed, something which a full-time employee of an ecumenical homeless shelter in the town confirmed to us. (Her explanation was that 'in the past the churches have not always been kind'.)

So declaring yourself Christian – even CofE – is no longer the normal, unproblematic marker of cultural identity, ethnic-majority-belonging, normalcy, and even good morals that it once was. In Kendal, as in the rest of the country, the churches have become more marginal to everyday life, ritual practice, and culture, whilst spirituality and 'no religion' have become more central.

This inversion, and its limits in media representation of religion, was given an interesting illustration by a newspaper story which occupied the front page of the local paper, the *Westmorland Gazette*, in February 2017, under the banner headline: 'Same Sex Duo Wed in Church First'.[7] The *Gazette* reported that 'Kass Conroy and Keysia Mattocks tied the knot in Kendal Unitarian Chapel, surrounded by family and friends, in a ceremony which is also thought to be the first of its kind in South Lakeland'. In 2000–2002 we had studied the Unitarian chapel now carrying out the wedding and had found it and the Quakers to be the only churches which had a significant overlap with holistic spirituality. Revisiting in 2016 we found it had continued to travel this path: it had appointed an Interfaith Minister rather than a Christian leader, and had commissioned a prominent new mural outside its entrance – the 'Spirit of Life' – displaying in full technicolour a multifaith, ecological, holistic, planetary spiritual orientation.

At one level this was not a 'story' at all, same-sex marriage having become commonplace in the UK since 2014. What made it news in Kendal was that it took place within the congregational domain, which has set its face against accepting wider social change in attitudes towards homosexuality. Had the wedding been undertaken by a holistic practitioner outside of a chapel context it would not have made the front page. Still, a traditional Christian framing lives on in the way that the story uses the wedding as a peg on which to hang a national story about a new report by Church of England bishops reaffirming their refusal to marry or even bless LGBTI partnerships. Solemn statements by local bishops are artfully juxtaposed with vox pop from Kendal: '"we sometimes still do experience discrimination, but we always will," Kass said. "I don't talk to my grandparents any more because they are Roman Catholic. But you can't change who you are."' The reader is left with the impression that it is the Unitarians who are normal and in touch with mainstream values and common decency – but the Christian framing remains. The old Christian 'establishment' may have become more questionable and open to challenge, but it continues to shape the newspaper's gaze.

Methods and theory then and now

Revisiting Kendal allows us to see what has changed and to consider how one would approach such a project today. The changes we have described suggest the answer: we would start from the outside and work inwards; rather than beginning in the official space of religion and working outwards to streets and other spaces, we would start with more and varied spaces of Kendal life and society and work inwards towards the explicitly religious.

The closest we came to this in the original project was with the Street Survey, which, as we have shown, proved surprisingly productive. It revealed that the main categories we were working with – religion/congregational domain, spirituality/holistic milieu – were even at that time only applicable to some respondents, and that even within these categorisations there were a very wide range of positions.

The churchgoers ranged from those who expressed a clear set of orthodox beliefs to those who expressed a fairly non-specific sense of God as 'something more' and saw church attendance as one way to explore this, and included those who prioritised belonging to the church community, or behaving in a particular moral or socially-acceptable way, over believing. This identification with Christianity or with one of the churches and chapels was not necessarily related to religious belief or spiritual practice.

The non-churchgoers were even more diverse in their accounts. Belief systems and practices were often put together from a wide range of cultural resources, and there is clearly no expectation that people hold similar beliefs to their neighbours. Combinations included a personal identification with the Church of England without a belief in God, a self-identification as a non-believer combined with an expressed belief in fate or destiny, and a fairly un-defined and definitely non-Christian belief in 'some kind of guiding Spirit' coupled with occasional church attendance. The only belief that was widely, though not universally, expressed was a belief in the positive value of tolerance of difference.

So where as recently as 30 or 40 years ago religion was still chiefly associated with one cultural domain and demarcated social space – the churches – and was accepted or rejected on that basis, we see here a much more flexible situation. Historians have reminded us that even when the church held maximum sway individuals often had diverse and heterodox beliefs and practices; what has changed is the sheer range of cultural resources on offer (especially because of the internet) and the diminishing cost of expressing them in public as Christianity has lost its social and cultural sway and centrality.

In a Kendal study of today an equivalent of the Street Survey would clearly have a very important role to play. By triangulating it with nationally-representative survey data on beliefs, values, and practices, it could be one useful starting point for a study which would seek to understand not just the beliefs and practices of churchgoers and 'alternatives' in a particular place, but those of the town as a whole. It would be better, however, to sample from across the town to represent different demographics (using a sample frame based on Census data) rather than focusing in on one small geographical segment. We would also need to be more careful to avoid the danger of undersampling those with no interest in the spiritual. In a climate in which 'no religion' is the norm, a restudy would have to focus in a much more open way on culture, values, beliefs, and practices which have to do with explicitly religious or spiritual meaning, identity, and significance.

As well as developing and using a more sophisticated Street Survey, a second Kendal Project would need to enter into many different spaces to see whether and how meaning and purpose is enacted, expressed, articulated, produced, or reproduced. We could once again rely on informal and short-term ethnography as the initial method, bringing in other methods like chats, interviews, surveys, or focus groups as and when needed, in relation to a flexible research design. We have illustrated how and why the domains of education, health and wellbeing, the local media, and voluntary services would have much to tell us. A wider study could step into many other spaces, such as the crematorium, graveyards, and a woodland burial site; places of work and leisure; or shops. It could talk to accountants, lawyers, social workers, and others who come into contact with large numbers of people dealing with difficult decisions – potential key informants.

As in the original Kendal Project we would use a carefully-combined mix of open and closed methods. The latter, mainly surveys, would be used to test how extensive certain beliefs and attitudes are and how they correlate with other factors (often useful at both the start and end of a piece of research – the first to get an initial map of the territory, the second to test out hypotheses arising). The more qualitative methods would be used to explore people's own approaches rather than forcing them to choose between pre-selected alternatives – enabling us to discover somewhat unexpected categories, such as those who identified as Christians without a belief in God. We know from the first Project that this allowed us to pay attention to the sorts of language people used to express their beliefs, the wide variety of cultural repertoires and meaning systems drawn on even within one small geographical area, and the varied frames of meaning within which people make sense of questions such as 'would you say you believe in anything?'. Following Geertz (1973, p. 5), this hermeneutical approach is situated within the tradition of cultural research which is not 'an experimental science in search of law but an interpretive one in search of meaning'. It does not assume coherence, and rather than looking for assumed correlations it can detect new ones – it does not assume, for instance, that 'lack of belief in an afterlife' is necessarily indexed with 'never having experienced contact with someone who has died'.

New technologies would make some aspects of this research much easier. While the initial Street Survey did not audio-record people's responses, a new study could easily incorporate this using a digital voice recorder on a phone – now so everyday as to be unobtrusive (whereas in 2000–2002 our recording equipment drew more attention to itself). Corpus linguistic analysis techniques could be used to identify keywords and clusters, with qualitative analysis of concordance lines developing a fuller picture of the different discourses people drew on in the interviews (Baker, 2006). Similar techniques and web-scraping could be used to identify the different discourses of spirituality being drawn on in the town, both by analysing religious texts such as sermons and church newsletters, and by analysing the more secular everyday texts such as local media or websites.

With the development and codification of research ethics in the last decade, however, some aspects of a new Kendal Project would have to be rethought. In relation to a new Street Survey, for example, a lone female researcher knocking on unknown people's doors and requesting consent on the spot would be likely to raise queries from university ethics committees – which, like smart phones, didn't really exist back in 2000!

Along with these changes, a restudy of Kendal would also be working with a very different repertoire of concepts and theories to the one which still held sway at the start of the century (Woodhead, 2009). The first project was firmly rooted in secularisation theory, even as it undermined it. It took the churches as its blueprint of 'real religion', even as it found that holistic spirituality was taking various forms, and placing more emphasis on body, emotions, practices, and rituals than on beliefs and belonging. Today, with a growing emphasis on lived and everyday religion; material objects and practices; spaces, places, and locations; and of course on 'no religion', the field has moved on. A new study would spot different things not just because things have changed on the ground, but because its tools, assumptions, and theoretical framing would focus attention in new ways.

Conclusion

The original Kendal Project put spirituality on the map, making it harder for scholars of religion who had ignored or dismissed it to continue to do so. It made a splash because it had a clear headline of obvious significance – 'spirituality taking over from the churches' – backed up with strong quantitative evidence. It even got a feature on BBC *Newsnight*, with a full camera crew arriving in the town – the first of several. The project became part of 'A' Level Sociology textbooks and launched many related student projects in towns across the country as well as studies by academics in other countries.

It worked because it took a widespread assumption – that church-based Christianity was mainstream religion in Britain and its decline meant secularisation – and challenged it on its own terms. By treating the 'spaces of spirituality' as analogous to the 'spaces of church-religion' and using very similar methods for interviewing and counting 'believers', it was able to present an alternative picture – religion wasn't just declining, it was changing, and secularisation wasn't the whole story. It offered a clear explanation for why this was happening, drawing on Charles Taylor's idea of a massive subjective turn in the culture of modern liberal democracies.[8]

This chapter and restudy shows just how much the picture has changed since the year 2000. Church Christianity no longer dominates social reality and imagination in the same way and no longer shapes the contours of 'alternative' spirituality. In many spheres it is now spirituality which is mainstream and Christianity which is 'alternative' and even counter-cultural. Religion has burst its boundaries both conceptually and empirically. The 'religious' and the 'secular' have been complicated and questioned and can no longer be used as straightforward alternatives with which to carve up the whole of culture. The growth of spirituality has certainly been an important element in this change, but the relentless rise of 'no religion' (not identical with 'secular' or 'spiritual') to displace 'Christian' as the majority self-identification is the broader change, and has been the subject of Woodhead's (2016) recent research.

The result is a much larger territory for the scholar of religion and spirituality to explore. Our forays into Kendal in the last few years illustrate how fruitful it is to step outside the traditional associational spaces of religion and spirituality to ask, with an open mind and open questions, how people are gathering, ritualising, making sense of life, death and suffering, sacralising, and drawing on the myriad resources now available to them in contemporary Britain.

What we report on in this chapter is a fascinating transitional situation in which the old religious centre has become increasingly marginal and the peripheries central. In many spheres spirituality is now more normal, mainstream, and 'unmarked' than Christianity. Where the latter used to shade into 'no religion' – 'I'm not religious, I'm CofE' – now spirituality does the same, and it is the churches, including the CofE, which have sharpened their boundaries and sense of distinctiveness.

This de-centering of church-like religion means we have to turn our methods inside out as well. Rather than starting with associational groups with shared beliefs and moving out to street and school, it makes more sense to start with street and school and move inwards towards dedicated forms of religion and spirituality. 'No religion' is a useful provocation, but is ultimately a placeholder which points beyond itself and cries out to be replaced once we have found more appropriate categories. For researchers who have been trained in the study of religion, values, and culture but who are open to reworking their methods and rethinking their approach, there is golden opportunity to make a major new contribution to cultural understanding.

Notes

1 www.holistic-healthclinic.co.uk/ Accessed 10–12–16
Staveley Natural Health Centre www.cumbriasupportdirectory.org.uk/kb5/cumbria/asch/service.page?record=FW8pcFbGcw8 Accessed 10–12–16

2 One informant thought the best indication of this in the UK as a whole was the 'IPTI List of Approved Treatments and Therapies' which has grown steadily; by 2016 it listed 164 different activities which it will insure, providing the practitioner can provide a separate diploma or certificate for each one. They range from Art Therapy to Face Reading, Baby Massage to Prana Healing, and four different kinds of healing for animals. http://www.iptiuk.com/treatments-covered-by-the-ipti-insurance-policy/ Accessed 12–12–2017

3 Quality marks and professional accreditations on spirituality are more evident than they used to be – the Kendal Holistic Health Centre, for example, is registered with the Complementary and Natural Healthcare Council (CNHC), the Federation of Holistic Therapists (FHT), and the Bach Foundation International Register. www.holistic-healthclinic.co.uk/About-us.html Accessed 18–2–17

4 The schools were all part of Kendal Collaborative Partnership, a company formed five years ago to enable the schools to work more closely together.

5 We revisited a handful of the churches and attended some morning and evening services at the Roman Catholic church, the main Anglican parish church, and Parr Street independent church (evangelical). We also spoke with some clergy, churchworkers, and others in the voluntary sector.

6 As examples he cited food banks, the work of Manna House with the homeless, and support for the formation of a local credit union.

7 *The Westmorland Gazette*, Thurs 27 February, pp. 1–2.

8 As Taylor puts it in *The Ethics of Authenticity*, 'I am called upon to live my life . . . not in imitation of anyone else's. But this gives a new importance to being true to myself. If I am not, I miss the point of my life, I miss what being human is for me' (1991, p. 29).

References

Baker, P., 2006. *Using corpora in discourse analysis*. London: Bloomsbury Academic.

Brown, A. and Woodhead, L., 2016. *That was the Church that was: how the Church of England lost the English people*. London: Bloomsbury.

Clarke, C. and Woodhead, L., 2015. *A new settlement: religion and belief in schools*. [online] Available at: <http://faithdebates.org.uk/wp-content/uploads/2015/06/A-New-Settlement-for-Religion-and-Belief-in-schools.pdf>

Geertz, C., 1973. Thick description: towards an interpretive theory of culture. In *The interpretation of cultures: selected essays*. New York: Basic Books. pp. 3–30.

Heelas, P. and Woodhead, L. with Seel, B., Szerszynski, B. and Tusting, K., 2005, *The spiritual revolution: why religion is giving way to spirituality*. Oxford, UK, Malden, MA: Blackwell.

Luckmann, T., 1967. *The invisible religion*. London: Collier-Macmillan.

Taylor, C., 1991. *The ethics of authenticity*. Cambridge, MA: Harvard University Press.

Woodhead, L., 2009. Old, new and emerging paradigms in the sociological study of religion. *Nordic Journal of Religion and Society*, 22(2), pp. 103–121.

Woodhead, L., 2010. Real religion, fuzzy spirituality. In D. Houtman and S. Aupers, eds. *Religions of modernity: relocating the sacred to the self and the digital*. Leiden: Brill. pp. 30–48.

Woodhead, L., 2011. Christianity and spirituality: untangling a complex relationship. In G. Giordan and W. Swatos, eds. *Religion, spirituality and everyday life*. Chicago: Springer. pp. 3–21.

Woodhead, L., 2016. 'No religion' in Britain: the rise of a new cultural majority. *Journal of the British Academy*, 4, pp. 245–261.

9 The small stuff of barely spiritual practices

Jennifer Lea, Chris Philo and Louisa Cadman

Introduction

The spiritual sector is growing in economic, social and cultural significance in the UK. Particularly significant are those practices grouped under the term 'New Age' or 'spiritualities of life', such as yoga, massage, reiki and meditation (Sointu 2006). At the same time as the sector is growing, the practices that constitute it are changing (Carette and King 2004) and new geographies of spiritualities are emerging. This chapter draws on a wider research project that attempted to trace the formation of some elements of these new geographies on the ground, taking Brighton and Hove (a south coast UK city, home to many spiritual practitioners) as a case study for the emergence of an 'everyday urban spiritual' landscape. The broader project asked how far, and in what ways, the spiritual comes to matter both in explicitly spiritual spaces (e.g. Buddhist centres, Natural Heath centres), and also across the kinds of mundane spaces of everyday life that are often seen as resolutely non-spiritual, notably workplaces and homes. The chapter draws on extracts from diaries completed by research participants which offer an understanding of spiritual practices (here chiefly yoga) as constituted by the broader contexts within which they are pursued. In enabling us to develop an understanding of how such spiritual practices relate to other aspects of people's lives, the chapter contributes to wider debates emerging in response to the growth and proliferation of the spiritual sector, as well as to the small body of geographical work on spiritualities (e.g. Bartolini *et al.* 2013, 2017; Conneely 2003; Holloway 1998, 2000, 2003, 2011; MacKian 2011, 2012).

Our project moves beyond the concern shown by geographers of religion (e.g. Park 1994; Sopher 1967; Stump 2008) for the most obvious, self-proclaimed sites expressing – enabling, bearing witness to – faith in godheads of one stripe or another, what Lily Kong (2001: 228) calls the 'officially sacred' spaces of (organised) religious observance. Instead, we reach out to all manner of practices and attendant spaces, places, environments and landscapes, many of which strike an ambiguous pose with reference to both conventional notions of 'belief' and geographical interest in 'sacred space'. If the field was originally animated by the 'big relations' of person, world and divinity, as mediated through small numbers of core spaces anchoring what Peter Berger (1967; also Wilford 2010) terms the 'sacred canopy' of religious guiding principles, then we are excited more by 'the "small stuff" of spirituality' (Bartolini *et al.* 2017: no pagination). Hence, in line with newer emphases in the geographies of religion and spirituality (Gökariksel 2009; Henderson 1993; Holloway and Valins 2002; Philo *et al.* 2011), our attention instead turns to multiple micro-instances of 'other ways' for being-in-the-world, however localised or momentary, where a spiritual charge arises in isolated patches and along dangling threads (scantly referencing any overarching 'canopy'). More specifically, we explore the intimate 'small relations' between snatched spaces of yoga practice and how they 'rub up' against everyday workaday and personal lives, thereby opening up the quite mundane, often unexciting, undramatic and, echoing Sara MacKian (2012: Chap.2), 'spirituality lite' geographies of spirituality lived by our case study participants.[1]

Thinking the spiritual

Perhaps due to the proliferation of the spiritual sector, many of the existing academic accounts have tended toward defining or quantifying the sector to lend it some coherence (e.g. Chandler 2008; Glendinning and Bruce 2006). One lens frequently applied to the spiritual is consumption, with authors suggesting that New Age spirituality is a form of bricolage in which a variety of (previously separate) traditions, practices and objects are combined together and repackaged as commodities to be bought/consumed. This has led other writers, such as Jeremy Carette and Richard King, to critique the sector in strong terms because it has become a kind of 'pick and mix', allegedly undermining the integrity of the practices and producing narcissistic and self-serving identities (2004: 21–22).

Much of the research reaching such conclusions has deployed a rather abstracted form of discourse analysis, and therefore has not engaged in detailed empirical investigation of the actual happening of spiritual practices in particular contexts. Work on 'spiritualities of life' (Heelas 2008; Heelas and Woodhead 2005), emerging from a disciplinary framing in Religious Studies, has been more attuned to such grounded practices as arising in the spiritual milieu of Kendal, UK. This work has broadly concerned the processes of identity formation and the connection with a 'life force': an experiential quality of being in the 'here and now' that the practices were seen to afford (Heelas 2006: 224), one seen to make a substantial difference to the participants' subjective lives (Heelas 2008: 33). This idea of 'life-itself' has led Paul Heelas and others to situate the practices within a broader 'subjective turn of modern culture' which authors such as Charles Taylor (1992) argue has been present in Western societies since the 1960s, a 'subjective turn' involving a turn away from 'external' roles and obligations, such as those traversing the 'sacred canopy', towards a 'life lived in reference to one's own subjective experiences' (Heelas and Woodhead 2005: 2; also Heelas 2006, 2008).

This is life lived in deep connection with the self, through the cultivation of attentiveness to 'states of consciousness, states of mind, memories, emotions, passions, sensations, bodily experiences, dreams, feelings, inner conscience, and sentiments' (Heelas and Woodhead 2005: 3). An exploration of such inner subjective experience as afforded by spiritual practices has been taken up by some geographers, as they have begun to examine practices that can be understood as 'spiritualities of life', including yoga (Lea 2009) and meditation (Conradson 2008, 2010). Conradson's work, most notably, follows up some of the questions asked by the spiritualities of life literature, and delves into the qualities of the subjective experiences that might be opened up through encounters with specific spaces, places and environments. These inner states are conceptualised through the idea of 'stillness', which Conradson (2008, 2010) particularly addresses in his empirical research on so-called 'retreat' settings. Rather than stillness being defined by corporeal inactivity, stillness here refers to a kind of inner experience which is characterised: firstly, by a present focus (awareness of what is happening here and now, rather than thinking about what might be happening in other times or spaces); secondly, by an internal state of calm wherein the mind becomes less active and less engaged in ruminative thought; and thirdly, by the kind of clarity of thought that might emerge through stilling the mental habits of distraction and dissection.

Conradson considers the emergence of stillness in retreat settings by two routes: firstly, focussing on the 'experiential economies' of stillness using a 'therapeutic landscapes' perspective to understand the kinds of spaces in which stillness might potentially emerge (2008); and secondly, looking at the 'orchestration of feelings' of stillness via 'mind-body shift technologies' such as meditation (2010). These moves allow Conradson variously to analyse how stillness as a mind-body state is valued and how it occurs across different spaces, with a particular focus on why retreats offer settings in which experiences of stillness might emerge. Notably, he pinpoints factors such as the distance between the retreat and everyday working and caring responsibilities, the fact that the meditation practices are supported by the retreat's regime and routine as expressly designed to allow stillness, and also the situated social context of being surrounded by others also seeking experiences of stillness.

The distance between the retreat and everyday life is understood to allow the possibility of thinking and feeling 'differently', and for many the retreat setting was experienced as 'counterposed to the stress of demanding and busy lives' (Conradson 2010: 72). Looking at these practices in separation from 'everyday life' draws similarities with scholarship on religion and pilgrimage (e.g. Maddrell and Della Dora 2013), but contrasts with the majority of other geographical enquiries into spiritual practices, which have precisely sought to approach them when set in the bustling scenes of everyday life. Julian Holloway's work on the spiritual milieu of Glastonbury (2003), MacKian's work on spirits and enchantment (2011, 2012) and our own work on the spiritual everyday (Lea *et al.* 2015a, 2015b; Philo *et al.* 2015) all assert that spiritualities *must* be addressed as contextual and embedded in such scenes, while still maintaining a relationship, more-or-less muted, with the 'otherworldly' aspects of such practices. This kind of research thus starts to show how 'sacred space' has to be seen as not solely inhering in obviously religious buildings and sites, but as leaking into all sorts of spaces (some of which are indeed very ordinary and everyday).

This chapter, drawing on accounts offered of how individuals manage to integrate 'spiritual practices' into their everyday lives, attempts to address spiritualities as contextual and embedded through a focus on the inner life of participants (as outlined in work on spiritualities of life, as well as in Conradson's work on stillness). The chapter considers the inner states of individuals who attend yoga classes as part of their weekly routine, taking seriously the implications of striving to 'fit in' these classes alongside work and/or caring responsibilities. These inner states are those that might be 'orchestrated' through organised practices in which the individuals are taking part (in this case, yoga classes), but they also entrain those states that come about in other parts of the participants' lives (e.g. anxiety, stress and worry). The space-time data from the diaries that our participants completed allow us to trace the emergence, change and endurance of various bodily states (including stillness) across transitions between the everyday spaces of home, work and leisure, the workaday sites of busy urban lives, and the yoga classes that they elect to attend. What begins to emerge is an understanding: firstly, of how experiences of stillness, and perhaps then a more all-embracing spirituality, might be enjoyed – and, to an extent, actively made – in much closer proximity to the everyday than proposed by Conradson; and secondly, of the relationships that emerge between spiritual and what we might (perhaps too simplistically) understand as 'non-spiritual' arenas of a life.

Space-time diary-keeping

The data here derives from a project which explored spiritual practices (yoga and meditation) in Brighton and Hove, which has a reputation for a high density of spiritual practices and associated 'alternative' lifestyles. Central to the project was an interest in the everyday nature of spiritual practices: how participants used them in their working and home lives, as well as in the more 'formal' spiritual spaces that they might frequent, such as natural health centres and classes. While the broader project used other methods – in-depth interviews with participants, teachers and centre owners; participant observation in yoga classes and on meditation courses – this chapter is based on the space-time diaries that we asked participants to complete. Based loosely on the 'diary: diary-interview' method (Latham 2003), we asked participants who practised yoga and/or meditation to create a written record of their practices in the context of their wider lives; and in so doing to become observers of their own practices, thoughts, feelings and sensations while engaged in (and also while *not* engaged in) spiritual practices.

A detailed consideration of our diary methodology is given in Louisa Cadman *et al.* (2017), but in summary we asked diarists to record their activities on five days when they practised yoga or meditation. We asked them to give details of their practice, where and when they practised, but also of what *else* they did, where and when, to gain a picture of how people fit their engagement with spiritual practices into (often hectic) daily schedules. We were also interested in whether these practices had longer-term effects throughout their days (or maybe longer) and across a range of other sites and activities, thereby asking about what

MacKian (2012: 3) describes as 'a tendency [of spiritualities] to spill out into the broader fabric of everyday life'. While the practices in question often happen in separate space-times – participants take time out of their day to visit a class in a yoga studio, gym, village hall or a natural health centre; they often change their clothes and may use equipment such as yoga mats to transform the space – this does not mean that these space-times are experienced in isolation from other aspects of the participants' lives. By situating these practices in the space-times of the broader day, we were able to develop an understanding that departed from typical assumptions about the separation of the 'sacred' from the 'profane' aspects of people's lives.

More specifically, we gave participants a time-space diary template with each day divided into timeslots at two-hour intervals, and with columns for the recording of key activities and locations, including but not restricted to spiritual practices, and a column for participants to reflect on these practices and their 'fit' into the day. (Figure 9.1 includes the precise instructions provided at the head of each column.) We explicitly invited participants to elaborate here on the feelings, sensations and experiences of the spiritual practices involved, as well as on any broader elements of the participants' lives that they felt were being touched by these situated practices. We also requested some background information, including basic demographic details, and also asked them to complete a section where they could reflect on their relationship to their practices, working lives, health, where they lived and so on. We offered some examples of how the diaries might be completed, but we also indicated that the diary template was only loosely formatted and that participants were entirely welcome to use it as seemed best for them.

Diarists were recruited through various means: we put flyers and posters up in the 'associational domains' (Heelas *et al.* 2005) where yoga and meditation took place or were advertised (e.g. natural health centres, cafés, health food shops, community noticeboards); we asked yoga and meditation teachers (who were also interviewed for the project) to pass on flyers to their students; and we used our own participation in yoga and meditation classes to recruit, either by the teacher giving us time to make a call-out at the end of the class or via informal socialising after classes. This third route was the most successful. Of the 23 diaries completed, ten used an electronic template, eleven a paper template and two their own formats; most completed five entries within a timescale of six weeks.

This chapter focuses on the diaries of two participants – diarist 9 and diarist 11 (these numbers being our identifiers for them) – permitting in-depth focus homing in at the grain of individual experiences. These particular diarists were chosen because they attended weekly yoga classes alongside their working and home lives, and because their diaries show a clear attention to their thoughts, feelings and the interwoven nature of their practice with their work commitments. Both are female and live with their partners (without children). Diarist 9 is in the age range 35–45, and works as a psychologist within a mental health team; she attends a weekly yoga class and also has a regular evening home practice of yoga and meditation (four or five times a week). She also uses, in her own words, 'meditation/

breathing and mindfulness techniques either in a structured or interwoven into daily routine and activity (particularly if I feel stressed or anxious)' (*diary introduction*). Diarist 11 is between 56–65 and works as a tutor in higher and further education; she attends classes once a week and, in contrast to diarist 9, does not do any practice outside of the sessions. In what follows, we roughly arrange our treatment of the two diarists' contributions according to the temporalities and spatialities of attending a yoga class: firstly, thinking about going to the class; secondly, thinking about the class itself; and thirdly, thinking about what happens after the class. We should underline the importance of reading carefully through the 'raw text', unedited by us, in the diary entries shown (in Figures 9.1–9.7) below: this text, arguably more than our own interpretation, is what really gets at the 'small stuff' of everyday spiritual practices which is the beating heart of this chapter.

Beginning the class

The diaries showed the spatio-temporal proximity of the classes to the working and home lives of participants, demonstrating how they assembled the classes alongside their other commitments. This proximity meant that, most often, the participants brought themselves to the class in whatever state they were; bringing bodily sensations, feelings and thoughts from work or home. Participants as a whole were more or less able to 'let go' of these thoughts and feelings, certainly as a class progressed, but there were different ways in which they linked back to the world 'outside' of their class. For example, each of diarist 11's entries note some kind of difficulty experienced during her day, including: feeling 'rather miserable' that her son would not be visiting her for another six weeks (*entry one*);[2] feeling 'rather redundant and demotivated' that not many people had attended a charity drop-in session that she helps to organise (*entry four*); and experiencing difficulties relating to work (*entry four*). Despite these contextual starting-points, in each of the entries she describes starting the class and being able to access feelings of stillness almost immediately. This access was even the case when there were potentially disruptive circumstances, such as having a different teacher from normal (*entry two*) and being away from class for weeks because of holidays (*entry one*).

In contrast, diarist 9 tends to experience a more difficult transition to the class, being less able to detach from her thoughts and feelings from 'outside', and therefore bringing such extraneous influences into the class. The following extract from her diary (Figure 9.1) reveals her very busy day before attending a class, one in which almost every moment is accounted for, from the moment she leaves for work to the time she gets changed for class in the evening. She is constantly at work, doing household chores or eating – or even doing more than one of these at once (lunch is eaten during a work meeting). While she might anticipate the positive effects of attending the yoga class, which provides the motivation to get up and get changed and go to the class, it is not surprising that she feels reluctant and rushed. This is indeed not an isolated occurrence, and *entry five* describes a similar

Time	*Main activity/activities during this time block (please note the times you practised yoga or meditation [e.g. meditation 12.30-13.30])*	*Where were you? (where possible, please give the postcode or area [e.g. Kemptown], or street name. Also include where you practised yoga and/or meditation)*	*Reflections on yoga and/or meditation (for example: did you enjoy the practice? how did it affect you physically and emotionally? how did the practice fit into your day? what impact did the practice have on your day)*
5-7 am	Sleep.	Home.	
7-9 am	Get up at 7.30am; shower and dress; eat cereal breakfast at 8.30am, whilst flicking through house magazine; leave house at 8.40am; travel in car along Shoreham Road to work and arrive at 8.55am.		
9-11 am	Attend morning meeting with [colleague] at 9am. She is late (annoying) and end up meeting at 9.20am instead. Sit outdoors in cafe with coffees, because lovely weather. Makes up for lateness. Go to Library in Education Centre, on site to look at Psychology and Therapy resources. 10.30am Client paperwork and G.P. letters.	Work.	
11 am – 1 pm	Individual supervision at 11am. 12 noon client work/assessment with new client (as CBT specialist psychologist). Then rushing to get to Supervision meeting at 1pm.		

Figure 9.1 Diarist 9, extract of entry from day one.

Time	*Main activity/activities during this time block (please note the times you practised yoga or meditation [e.g. meditation 12.30-13.30])*	*Where were you? (where possible, please give the postcode or area [e.g. Kemptown], or street name. Also include where you practised yoga and/or meditation)*	*Reflections on yoga and/or meditation (for example: did you enjoy the practice? how did it affect you physically and emotionally? how did the practice fit into your day? what impact did the practice have on your day)*
1-3 pm	Running 10 mins late. Arrive for 2-hour Supervision Group at 1.10pm and eat lunch during process which I hate doing (ready prepared salad which I have bought in - nice - some consolation)		
3-5 pm	3pm Client work. 4pm Trying to catch up on day's clinical notes and data entry.		
5-7 pm	5pm. Leave 15 mins late for home.[3] Not too bad. Home by 5.15pm. Prepare dinner. Eat dinner early at 5.30pm, a light supper, as going out to Yoga later. 6.15pm. Attend to personal emails, make some phonecalls and use the internet.	Home.	
7-9 pm	7.30pm. Change for Yoga. 7.50pm, travel to Yoga Class, walking. 8pm. Attend Yoga Class.	Yoga class	Reluctant to go to yoga, feel rushed (rushing). Frustrated during yoga by my attempts to constantly monitor thoughts and affect and how body responding to yoga postures and breath. Thought processes taking me out of the mindfulness of the moment.

Figure 9.1 Continued

pattern, wherein the residual traces of this diarist's day shape the beginning of her class in a negative fashion (Figure 9.2).

There are a number of things worth noting here. First is that the persistent and problematic bodily rhythms and sensations are seen to arise from the weather, physiology and the diarist's stress levels, offering a clear example of how the practice within the class is enmeshed in wider relations with the 'natural' environment as well as in the social roles, relationships and tasks that encompass lives outside of the classes. A (perhaps) lifetime of monitoring and working on her breath because of an asthmatic pathology, compounded by the weather and high stress levels, gives rise to the negative orientation of diarist 9, who is well used to, and indeed highly sensitised to, differences in her breathing. This negativity might be compounded because the (restricted) breath and (tight) chest are central to practising the kind of modern postural yoga being studied at her class. The breath

Time	*Main activity/activities during this time block*	*Where were you?*	*Reflections on yoga and/ or meditation*
7-9 pm	6pm Snack supper as attending Yoga and hate having full stomach as interferes with process. Spend next 1 and 45 mins cutting and pasting Yoga diary entries from computer notepad and completing and finishing diary. 7.50pm drive to Yoga, as in a rush. 8pm. Yoga for one and half hours.	Home and yoga studio.	Stress levels have been high today and generally really struggling with breath (not helped by weather). Unsure if emotional or asthmatic. I rarely use my inhaler because I don't believe that it is a good habit but prefer to work with my breath to deepen, but notice chest feels tight and restricted all day to day (breath feeling rushed, like me). My heart is pounding, with rapid shallow breathing and pounding head - difficult to rest as mind racing and then reluctance to begin postures - want to stay on back. Postures are easier in heat, but motivation lessened.

Figure 9.2 Diarist 9, extract of entry from day five.

is often the first point of reference in a yoga class: both for the student who lies or sits on the mat before the class, checking in with their breath and body, and for the teacher who often directs the students to attend to their breath in the opening moments of the class. The breath is a foundational aspect of yoga, and hence it is unsurprising that diarist 9 is dispirited about their restrictive shallow breathing in the light of this practice, finding it hard to shift, via a focus on the breath, to the kind of present focus and inner state of calm described by Conradson (2008, 2010).

The second thing to note is the relationship between the mind and body discussed by this diarist. Even though she has arrived at the class, she is not able simply and straightforwardly to engage with the practice, because of the feelings of stress and the physical sensations of a pounding head and heart. While the heat might make the body more supple and yielding, the diarist reports having to meet with, and overcome, resistance in order to participate in the class. While she wants to do the practice, it is hard to exert the will over the body in order to start the postures and to elevate the body from the floor. Stillness does not straightforwardly emerge here, then, precisely because of the agitated mind-body relationship that has been established during the rest of the day.

From this look at these couple of diary extracts, we can see how traces of the surrounding world push into the classes that the participants attend in a number of ways. Transitions between the different spaces of work/home, or the broader city, and the class – both of which have different norms around what forms of embodied existence are appropriate – can be hard to enact successfully due to the persistence of bodily and mental states from outside of the class. In this context, these practices rub up against the rest of the participants' lives in a close fashion. For the diarists, their spiritual practices are not enacted in some kind of vacuum, and the relationship to the self is constituted and mediated via the embodied traces of work or home life, ones manifested through niggly corporealities, habits of thought or a combination of both.

During the class

It is important to note, however, that once in a class participants generally managed to effect a kind of change in 'subjectively experienced state of consciousness characterised by calmer mental rhythms and a shift in attention from other places and times (the "there" and "then") towards the present moment (the "here" and "now")' (Conradson 2010: 72), at least to some degree. This section of the chapter looks at the kinds of experiences that the participants reported emerging from their practices as they were conducted: the space-time situated changes in mood, sensations and feeling. Some reported a straightforward shift in sensation. Considering *entry four* from diarist 11's diary, it can be seen that her afternoon consisted of worrying about work. Her mind was 'taken off' it by chatting with her hairdresser, and then the yoga class allowed her to gain some new perspective on the problems that she was experiencing, as can be seen in Figure 9.3.

Time	*Main activity/activities during this time block*	*Where were you?*	*Reflections on yoga and/or meditation*
1-3 pm	Drove home and checked emails. Rather alarming message from college re a change of class at the last minute. Felt stressed and put upon. Trawled through several emails with instructions, lists, forms, docs relating to new term. Wished I wasn't working for the college. Rang line manager to discuss issues but she couldn't talk till tomorrow. Printed out pages and pages of notes from college website. Miserable about amount of paperwork.		
3-5 pm	Drove in for haircut. Unusually didn't turn on radio to think through what to do about problems at the college. Forgot worries whilst chatting with hairdresser.	Hairdresser's flat.	
5-7 pm	Dropped off at home briefly to check more emails and got back into car to drive to yoga class for 5.30.	Yoga class	More emphasis on breathing exercises though from the start 'to bring our minds into our bodies' – this is just what I needed tonight. I was conscious of needing relief from the work issues and to get some sense of perspective on the importance of these new developments that had come to light today. Relaxation was, as always, lovely. This week the addition of small eye 'cushions' filled with lavender that the tutor placed over our eyes was particularly welcome. Shutting out the light and giving off a lovely perfume.

Figure 9.3 Diarist 11, extract of entry from day four.

The teacher explicitly attempted to orchestrate a change in feeling and attention by asking the participants in the class to shift *from* their minds *into* their bodies. The breath was foregrounded right at the beginning of the class to 'kick-start' a feeling of stillness in the participants. This feeling came together for this diarist with a desire to get some respite from thinking about work issues, and to gain perspective on the problems upon which they were dwelling; almost a textbook description of the kinds of 'lucid' or calmer thoughts that are seen to be part of becoming still. As Conradson reflects, 'the mind becomes less busy and attention is drawn towards the scale of the body. Cycles of mental rumination may soften or begin to dissipate, enabling calmer and more lucid states of consciousness to emerge' (2010: 72). In this case, diarist 11 successfully manages to achieve this change in feeling, shifting her attention away from work and to the present moment. She understands the class as being 'just what [she] needed', partly because the shift to the body allows some lucidity to emerge around work issues. Physical objects and pleasant smells (scented eye cushions), '[s]hutting out the light and giving off a lovely perfume', were also gathered into the situation, 'circumstantial' accumulations (McCormack 2016) that assisted in releasing a positive affect for this diarist.

Diarist 9 describes a somewhat more problematic route towards stillness in the following diary entry (Figure 9.4), which continues from an extract shown previously (Figure 9.2). Here, the diarist fleshes out the transition made in the felt sense of the self during the class. The participant brings the traces of a busy day into the class, meaning that, at first, she finds it hard to experience the 'shift in consciousness' that Conradson (2010: 72) suggests constitutes stillness. The 'then' and 'there', as Conradson puts it, stubbornly enter into the class, rendering it difficult to connect with the 'here' and 'now'. The habits of thought endure from her everyday life elsewhere and elsewhen, draining into the class and exerting a lasting hold over the diarist even when she has entered the class and started her yoga practice. Yet, within the space of the class, she still experiences quite profound transformations in her 'subjective experience' (Conradson 2010: 72) which can readily be equated with the kinds of transformations that Conradson describes as stillness. It is maybe telling that geographical metaphors abound here: the diarist's eventual transformation moves her into 'an even deeper place of mindfulness', the yoga postures and breathing 'creat[ing] an internal space . . . for the chi energy to habit' which has been 'shower[ed]' and 'cleansed' – the everyday of elsewhere and elsewhen has now been substituted for an interior bubble of space-time 'stilled and slowed', as Ben Anderson (2004) has claimed in a different context (to do with the 'boredom' of home music-listening).

While some of the mind-body states that comprise barriers in the way of accomplishing stillness endured for longer and were more persistent than others, all of the entries suggest that there was some kind of shift in feeling towards stillness during the class. The diary methodology puts these shifts of feeling into a processual frame, allowing us to see what changes occur in the yoga classes, almost despite the things that the participants might bring in with them. If the spiritual

Time	*Main activity/ activities during this time block*	*Where were you?*	*Reflections on yoga and/or meditation*
7-9 pm	8pm. Yoga for one and half hours.	yoga studio	I again notice (like previous yoga class) that noticing and monitoring too carefully my breath and physical sensations is distracting me from . . . engaging . . . with my breath - not allowing me to engage more holistically, in that I am constantly questioning how am I feeling, what am I experiencing etc.-Reminds of when I first began yoga and the process of conscious incompetence to conscious competence. Eventually I notice and let go and this seems to move me into an even deeper place of mindfulness with the yoga. It feels as though the postures and breathing create an internal space, like a shower cleansing internally and allowing expansion and a space for the chi energy to inhabit. Feels really good.

Figure 9.4 Diarist 9, extract of entry from day five.

practice here is changing feelings, shifting bodies and offering new, more lucid, perspectives on the kinds of everyday problems that the diarists report, then this finding goes some way towards answering the question of why people carve out time, space and energy in their already overfull days to go to their yoga classes; and why they exert the will over the self to lift the body off the mat and to begin to move the body as instructed by the yoga teacher.

Leaving the class

The chapter will now look at how feelings and sensations *from* the class may endure, or otherwise, after the class has ended. Again, participants reported a range of experiences, from positive effects which endured over a range of time-scales, to less straightforwardly positive effects. Diarist 11 uses her journey home from yoga to describe various states of energy, ranging from feeling 'much more energetic than earlier in the day and walked fast, noticing the strength in my legs as I strode up the hill' (*entry one*, 7–9pm); driving home 'feeling loose and nicely weary' (*entry three*, 5–7pm); to walking home and 'feeling rather heavy and tired. Would have liked a lift today' (*entry two*, 7–9pm). Diarist 9 focussed her description on her senses rather than her 'energy' (and see Philo *et al.* 2015 for more discussion of energy in relation to these practices), as disclosed below (Figure 9.5).

Time	*Main activity/activities during this time block*	*Where were you?*	*Reflections on yoga and/or meditation*
9-11 pm	Yoga Class finishes at 9.30pm, leave by 9.40pm and walk home.		Feeling calm, happy, grounded. Glad came and thinking about it has helped to deepen my practice in being more thoughtful and attentive to breath and positions and effect on body, instead of doing automatically when familiar with poses. Feel happy, grounded and body feels more supple/freer.
11pm -1am	Home at 9.50pm. Feel grumpy. Hypersensitive to mess in kitchen that Husband has created in my short absence. irritated.	Home.	Hyperarousal and sensitivity of senses in concert with sense of calm, really normal response for me after yoga so easily irritated/annoyed ironically. During the walk home always more sensitive to light and noise. I feel more incongruent with my environment. I have a desire to surround myself with peace and calm.
1-5 am	Bed at 11am and turn lights off at 11.30pm. Sleep really well,		Feel slightly stimulated so takes a while to fall asleep, but fall into deep and replenishing sleep (wake up next day and feel refreshed and calm).

Figure 9.5 Diarist 9, extract of entry from day five.

While she had achieved a state of calmness and happiness during her practice, when she stepped outside of the yoga class she reports this sensibility changing into a feeling of being ‘out of step’ with the outside world – of being ‘incongruent with [her] environment’ – set apart in a state of hyperarousal and hypersensitivity. She expresses a desire to surround herself with peace and calm, but on arriving home was confronted by a messy kitchen. The state of hypersensitivity experienced after yoga then became manifest as irritation, followed by stimulation, which hindered the onset of sleep for her. This re-telling shows how the state of stillness can be changed as bodies travel through different settings, suggesting too that stillness can occasionally morph into agitation or turbulence, stillness’s opposite. Once this diarist had slept, though, she still woke up feeling refreshed and calm, perhaps regaining some of her feelings of stillness after their slightly bumpy transition back into ‘normal’ life. This outcome points towards a lengthier and more enduring set of feelings that may arise from the yoga practice.

Time	*Main activity/activities during this time block*	*Where were you?*	*Reflections on yoga and/or meditation*
5-7 pm	Dropped off at home briefly to check more emails and got back into car to drive to yoga class for 5.30.	Natural Health Centre	. . . It did help – I came away feeling more distanced from all the chaos and pressures of the last couple of weeks spent at the college. Writing this 1.5 hrs later I still feel quite detached and able to keep it in proportion.
7-9 pm	Received call from friend who needs help tomorrow morning when I was looking forward to some space in which to tackle the college problem and speak to my manager on the phone. Found myself explaining that I wouldn't be available at that time – untypically putting my needs before hers. I wonder whether the yoga session had any bearing on my decision to not offer to drop everything and help her out.		
9-11 pm	On reflection I realised there would be plenty of time in the morning to help out my friend. Rang her back. Went to bed 11.30 fell asleep quite quickly.		
11 pm - 5 am			
1-5 am	Woke up while still dark thinking about work. Fell asleep again and woke again about 7 still thinking about work. . . .		

Figure 9.6 Diarist 11, extract of entry from day four.

We can usefully trace out feelings and how they change over time and space through diarist 11's entries. In the diary extract shown in Figure 9.6 this diarist suggests that after a class she was able to maintain changed feelings about the things that had been problematic beforehand, now feeling more distanced from work. In addition to these changes in her relationship to her problems, she also notes a changed relationship to herself and her friend, noting that it was 'untypical' for her to put her own needs in front of others. She wonders whether it is the

Time	*Main activity/activities during this time block*	*Where were you?*	*Reflections on yoga and/or meditation*
7-9 am	Get up early at 7.00am; shower and dress; eat cereal breakfast at 8.00am, whilst looking at hotmail account; leave house at 8.20am; travel in car along Shoreham Road to work and arrive at 8.30am.	Home and Work (as above).	Wake up early refreshed after previous evening Yoga practice. Also good mood because Friday. Notice how much more expansive my breath feels even now and this escalates my sense of relaxation. Notice more than usual the profound lasting effect that the yoga has today, although feeling really stressed and anxious about new job on one level, breathing ok, less difficult.
9-11 am	Individual Clientwork and accompanying paperwork.	Work (as above).	I'm more attentive to my breath and body today (because of the diary) and notice when I am feeling stressed/ tense and consciously breath abdominally to calm me which works really effectively.
11am-1pm	Individual Clientwork and accompanying paperwork.		Same as above.
1-3 pm	2.30-3pm. Take a late lunch break.	Meadow at back of work in Sunshine.	Despite breathing, beginning to feel stressed because have been unable to have a break with lots of new clients and paperwork to complete. I feel that I shouldn't take a break and I know if I eat in the office that I'll rush, so I go outside instead to eat and after I lie in the sunshine in 'lieing pose' breathing for 20 mins I feel completely rejuvenated and destressed. Breathing expansively and abdominally again. It is as [if] I have breathed in the sunshine and it is relaxing me and melting away my tensions from inside. I notice that it is easier to activate this sensation after yesterday's yoga class (I must remember the benefits of longer practice)

Figure 9.7 Diarist 9, extract of entry from day two.

yoga session that has brought about this change in orientation to her friend? The next timeslot shows her changing her mind, but in a measured, reflective and spacious manner, and not as a result of feeling guilty or as if she had put their friendship under pressure. All the same, she still wakes up early the next morning thinking about work, showing the complex shifts between states of stillness and everyday feelings that happen.

Longer-lasting effects are detailed by diarist 9, who identifies feelings and sensations that lasted overnight. *Entry two* (Figure 9.7) comes the day after the entries detailed prior (Figure 9.5) where the diarist had felt grumpy with the messy kitchen. She wakes up feeling refreshed, experiencing her breath as more expansive and less constrained. While she feels the lasting stress about her job, she can still maintain good feelings from the yoga, and she reiterates the importance of her breathing being good. Later in the day, however, she offers a less clear-cut positivity and shows that other events had interfered in the maintaining of these positive feelings. Here, therefore, we can see the stress taking over after a rushed morning at work where there has been no opportunity to stop, but significantly she recognises her ability to intervene in the process by going outside to eat and to do a yoga pose which returns her to being able to breathe less restrictively and to relax. She identifies that this ability is linked to the class attended the previous day, perhaps because the sensations and feelings were closer temporally and experientially, and she was therefore able to access the feelings again more easily and, moreover, to remember the benefits of doing so. There is something to note here also about the wider context – that the diarist has access to a sunny space to lie down in without fear of being disturbed – central to a low-key but helpful micro-geography of yogic practice.

Conclusion

The chapter has paid close attention to the shifts in feelings and sensations that arise as bodies move between everyday worlds and snatched space-times of yogic/meditative practice. Our diary methodology has enabled us to begin to develop an understanding of the relationship between these different spaces: to address how the one influences, inflects, enhances or sometimes compromises the other. In this case, we have approached these contexts through an intimate acquaintance with the kinds of feelings and sensations that were registered by the bodies of our participants, and which variously lingered and/or dissipated once the yoga class began. The species of close attention paid here to the mind-body and its feelings and sensations, as constituted in and by the diary entries, lends us clear indications about the modes of corporealities that 'we' bring with us to yoga classes, what we work with when we work upon the mind-body in this way, and the imbrications or foldings of minds and bodies that might be created through spiritual practices such as yoga.

This close attention to experiences, feelings and sensations also gives some pointers towards what it is about these snatched time-spaces that matters – the qualities of experience that emerge (stilling, slowing, calming, thinking) even when participants have come into a class full of worry or anxiety. At the same time, though, the data indicates that a more diverse range of outcomes might emerge – stillness does not always straightforwardly emerge, and it does not always give rise to some kind of congruence between the participant's mind-body and its environing world. Nonetheless, the spiritual practices in play are often used to 'press back' against the demands of everyday life, as figured through work and care. Here we present small accounts of everyday worries, anxieties and bothers stalled, if momentarily (and possibly even just countered in a more positive 'spirit'). These are sustained for shorter or longer periods, across smaller and larger spaces. On re-entering the more everyday contexts of home or work, however, these worries and anxieties themselves 'press back' against the experience of stillness that might have come about thanks to the yoga practice.

If the kinds of things described here are, in some way, 'new geographies of spiritualities' then we accept that they are not really geographies of enchantment, revelation or other 'big stories', but rather geographies of the everyday, the ordinary and the mundane: the 'small stories' (Lorimer 2003) of 'small stuff' spiritualities that are, in many respects, barely even spiritualities at all. These are geographies full of church halls, community centres, leisure centres, health centres, alternative cafés and bookshops, as well as fine-detailed micro-geographies of where bodies can find room to do yoga at home, at work, in the park, on the beach, on public transport – which are themselves set in the everyday fabric of settlements big and small. Nonetheless, these geographies really matter: the diary extracts underline the fact that, when we act upon the self in order to 'orchestrate' some kind of feeling (through a practice such as yoga), our action (and indeed agency) is always variously shaped (constrained/enabled) by the contexts in which we are situated. These everyday geographies offer vital information in our understanding of the 'place' of yoga, and indeed of other associated spiritual or barely spiritual practices, carved from the maelstrom of people's lives and the chaos of broader settings.

Notes

1 Actually, whereas MacKian (2012: 2) 'focussed on those spiritual experiences and practices which have a distant air of enchantment about them', many of the spaces intriguing us have few trappings of such enchantment: a few did, but many are utterly prosaic and ostensibly just flotsam of the secular world.
2 *Entry one* refers to the first day for which the participant made an 'entry' in their diary.
3 These timings don't add up. The text is taken directly from the diary entries. This reflects the difficulty in accounting for practice within the linear temporal framework that the diary provides.

Acknowledgements

We are very grateful to the AHRC-ESRC Religion and Society Research Programme, which provided funding for this project (award number AH/H009108/1).

References

Anderson, B. (2004) Time-stilled, space-slowed: how boredom matters. *Geoforum*, 35: 739–754.

Bartolini, N., Chris, R., MacKian, S. and Pile, S. (2013) Psychics, crystals, candles and cauldrons: alternative spiritualities and the question of their esoteric economies. *Social and Cultural Geography*, 14: 367–388.

Bartolini, N., Chris, R., MacKian, S. and Pile, S. (2016) The place of spirit: modernity and the geographies of spirituality. *Progress in Human Geography*, Online First.

Bartolini, N., MacKian, S. and Pile, S. (2017) Personal communication to authors of the chapter (e-mail).

Berger, P.L. (1967) *The Sacred Canopy: Elements of a Sociological Theory of Religion*, Stanford, CA: Stanford University Press.

Cadman, L., Philo, C. and Lea, J. (2017) Using time-space diaries and interviews to research spiritualities in an 'everyday context'. In Woodhead, L. (ed.), *Innovative Methods in the Study of Religion*, Oxford: Oxford University Press.

Carette, J. and King, R. (2004) *Selling Spirituality: The Silent Takeover of Religion*, London: Routledge.

Chandler, S. (2008) The social ethic of religiously unaffiliated spirituality. *Religion Compass*, 2: 240–256.

Conneely, J. (2003) *New landscapes of self-creation*. Unpublished PhD thesis, School of Geographical Sciences, University of Bristol, UK.

Conradson, D. (2008) Experiential economies of stillness: the place of retreat in contemporary Britain. In Williams, A. (ed.), *Therapeutic Landscapes*: 33–48, Aldershot: Ashgate.

Conradson, D. (2010) The orchestration of feeling: stillness, spirituality and places of retreat. In Bissell, D. and Fuller, G. (eds.), *Stillness in a Mobile World*: 71–86. London: Routledge.

Glendinning, T. and Bruce, S. (2006) New ways of believing or belonging: is religion giving way to spirituality? *The British Journal of Sociology*, 57: 399–414.

Gökariksel, B. (2009) Beyond the officially sacred: religion, secularism and the body in the production of subjectivity. *Social and Cultural Geography*, 10: 657–674.

Heelas, P. (2006) The infirmity debate: on the viability of New Age spiritualities of life. *Journal of Contemporary Religion*, 21: 223–140.

Heelas, P. (2008) *Spiritualities of Life: New Age Romanticism and Consumptive Capitalism*. Oxford: Blackwell.

Heelas, P. and Woodhead, L. with Seel, B. Szerszynski, B. and Tusting, K. (2005) *The Spiritual Revolution: Why Religion Is Giving Way to Spirituality*, Oxford: Blackwell.

Henderson, M.L. (1993) What is spiritual geography? *Geographic Review*, 83: 469–474.

Holloway, J. (1998) *Sacred space: a study of the New Age Movement*, Unpublished PhD thesis, School of Geographical Sciences, University of Bristol, UK.

Holloway, J (2000) Institutional geographies of the New Age movement. *Geoforum*, 31: 553–565.

Holloway, J. (2003) Make-believe: spiritual practice, embodiment and sacred space. *Environment and Planning A*, 35(11): 1961–1974.

Holloway, J. (2011) Spiritual life. In Del Casino, V., Thomas, M.E., Cloke, P. and Panelli, R. (eds.), *A Companion to Social Geography*: 385–401, Chichester: Wiley-Blackwell.

Holloway, J. and Valins, O. (2002) Placing religion and spirituality in geography. *Social and Cultural Geography*, 3: 5–80.Kong, L. (2001) Mapping 'new' geographies of religion: politics and poetics in modernity. *Progress in Human Geography*, 25: 211–233.

Latham, A. (2003) Research, performance, and doing human geography: some reflections on the diary-photograph, diary-interview method. *Environment and Planning A*, 35(11): 1993–2017.

Lea, J. (2009) Liberation or limitation? Understanding Iyengar Yoga as a practice of the self. *Body and Society*, 15: 71–92.

Lea, J., Cadman, L. and Philo, C. (2015a) Interventions in habit: mindfulness meditation. *Cultural Geographies*, 22: 49–65.

Lea, J., Philo, C. and Cadman, L. (2015b) 'It's a fine line between . . . self-discipline, devotion and dedication': negotiating authority in the teaching and learning of Ashtanga yoga. *Cultural Geographies*, 23: 69–85.

Lorimer, H. (2003) Telling small stories: spaces of knowledge and the practice of geography. *Transactions of the Institute of British Geographers*, 28: 197–217.

MacKian, S. (2011) Crossing spiritual boundaries: encountering, articulating and representing otherworlds. *Methodological Innovations Online*, 6(3): 61–74.

MacKian, S. (2012) *Everyday Spirituality: Social and Spatial Worlds of Enchantment*. Basingstoke: Palgrave Macmillan.

Maddrell, A. and della Dora, V. (2013) Crossing surfaces in search of the holy: landscape and liminality in contemporary Christian pilgrimage. *Environment and Planning A*, 45: 1105–1126.

McCormack, D.P. (2016) The circumstances of post-phenomenological life. *Transactions of the Institute of British Geographers*, Online First.

Park, C. (1994) *Sacred Worlds: An Introduction to Geography and Religion*, London: Routledge.

Philo, C., Cadman, L. and Lea, J. (2011) *The new urban spiritual? Tentative framings for a debate and project*. The New Urban Spiritual AHRC Project, Working Paper No.1, available through Glasgow University Library ENLIGHTEN system at: http://eprints.gla.ac.uk/96750/

Philo, C., Cadman, L. and Lea, J. (2015) New energy geographies: a case study of yoga, meditation and healthfulness. *Medical Humanities*, 36(1): 35–46.

Sointu, E. (2006) The search for wellbeing in alternative and complementary health practices. *Sociology of Health and Illness*, 28: 330–349.

Sopher, D. (1967) *Geography of Religions*, Upper Saddle River, NJ: Prentice Hall.

Stump, R.W. (2008) *The Geography of Religion: Faith, Place, and Space*. Lanham, MD: Rowman and Littlefield.

Taylor, C. (1992) *Sources of the Self*, Cambridge, MA: Harvard University Press.

Wilford, J. (2010) Sacred archipelagos: geographies of secularisation. *Progress in Human Geography*, 34: 328–348.

10 Rethinking youth spirituality through sacrilege and encounter

Elizabeth Olson, Peter Hopkins and Giselle Vincett

Introduction

Contemporary interest in Western young people's spirituality experienced a notable uptake in interest in the early 21st century (Cusack 2011), and within geography, the role of religion in young people's and children's lives has formed an important axis for the rejuvenation of the study of religion (Kong 2010; Olson and Reddy 2016). Yet within this work, examinations of youth spirituality have been relatively modest when compared to studies of religious identity and agency within schools, youth groups, and communities (see Hemming and Madge 2012). This contrasts with a growing geographic interest in a range of spiritual practices and approaches toward studying spirituality amongst adults, such as through therapeutic communities and emotion (Finlayson 2012; Williams 2016), spiritualism (Bartolini *et al.* 2013; Holloway 2006), pilgrimage and landscape (Maddrell 2009; della Dora 2016), ethics (Cloke 2002), and activism (Pulido 1998). In the fields of sociology and psychology, previous framings of youth spirituality as shrouded in mystery and secrets, to be carefully extracted by meeting children in their own worlds (e.g. Hart 2003), have given way to surveys detailing individual youth perspectives on concepts of God, afterlife, the occult, and transcendence.

The purpose of this chapter is to experiment with a different approach for researching youth spirituality in geography, one which might respond to Bartolini *et al.*'s (2017) observation that our current theories and questions remain insufficient for understanding emerging spirituality beyond our modernist conceptions. Our experiment thus entails attuning ourselves to everyday encounters that reveal youth engagements with spirituality that have previously been excluded or sidelined in youth spirituality research. Specifically, we focus on performances of sacrilege through blasphemy, its discursive practice, in order to think differently about relationships between practices and beliefs in contemporary spirituality. We draw our data from a research project designed to explore youth religiosity in areas of urban economic deprivation, focusing on the work we conducted in Glasgow, Scotland. Sacrilege was a common practice amongst our young participants, but as we explain below, has not been taken up broadly in studies of youth spirituality.

Accounting for the spirituality of youth

Youth religiosity has been researched through diverse disciplinary perspectives and methodologies, but as Hemming and Madge (2012) suggest, existing frameworks can often be inappropriate for understanding how young people engage with belief and religion. The same might be said for research on youth spirituality, which has acknowledged the need for better methodologies to categorize and describe changes in spiritual practices and perspectives amongst young people (Singleton *et al.* 2004). In this section, we briefly outline research trends in two fields – developmental psychology and sociology – which have made the most robust contributions to research on youth spirituality. These trends and their resulting frameworks have influenced our own research questions and methodologies, but they also illustrate the barriers to researching spirituality in a way that avoids reproducing existing modernist categories. We suggest that alternatives might be found through methodological experiments that train our focus on encounters and discursive practices.

Until the start of the 21st century, youth spirituality tended to preoccupy the attention of religious and moral educators working in the context of a secularizing Europe. However, with a growing recognition of new forms of spiritualism in the West coupled with claims of social 're-enchantment', and broader population trends embracing claims of 'spiritual, but not religious' (Fuller 2001) or 'believing without belonging' (Davie 1994), other researchers began to pay attention to the function and practice of spirituality in the lives of young people. Within developmental psychology, youth spirituality emerged as a potentially important variable in explaining adolescent 'moral' behavior such as voluntary sexual activity (Holder *et al.* 2000) or drug consumption (Belgrave *et al.* 1997). Lerner *et al.* (2008), for instance, found that professed spirituality had positive impacts on self-esteem, community membership, and pro-health behaviors. However, many of these early attempts at capturing spirituality were critiqued for their tendency to define spirituality from the perspective of a world religion, rather than as something that might be distinct from religion or monotheistic religious doctrines (e.g. King *et al.* 2014; Zinnbauer *et al.* 1997). Subsequent research has attempted to address these biases by capturing 'the internal, personal, and emotional expression of the sacred' (Cotton *et al.* 2006, p. 273). Surveys such as the Measurement of Diverse Adolescent Spirituality (MDAS), evaluated and tested by King *et al.* (2016) in the context of youth in Tijuana, Mexico, and the Youth Spirituality Scale (Sifers *et al.* 2012), thus emphasize higher powers and ultimate realities, relationships, and wellbeing.

These studies are relevant to our interest in this chapter not for their conclusions or content, but because they illustrate how researchers have grappled with the methodological challenges of classifying and analyzing youth spirituality. Overall, critiques of methodologies point to two main challenges: the challenge of adequately defining youth spirituality in ways that avoid cultural and age-oriented biases, and the lexical challenges associated with the study of spirituality (Savage *et al.* 2006). The first challenge is most frequently concerned with the conflation of spirituality with religion, and with the prominence of definitions of spirituality which reflect a bias toward religious dogma. As Ezzy and Halafof (2015) suggest, a focus on spirituality is more common in work that engages with traditional world religions than with spiritualism or those of the occult, which can tend toward describing practices rather than beliefs about the sacred. Studies from across Europe and the US describe young people embracing a traditional religious identification while also engaging spiritualist or new age practices (Vincett *et al.* 2015). Nonetheless, the common presence of questions about belief in a 'Higher Being' or an afterlife in social science studies of youth spirituality may fail to capture the spirituality of young people who are not traditionally religious or who might be secular. Studies of raves and witchcraft (Ezzy and Halafoff 2015), and of satanism and vampirism (Cusack 2011) illustrate this point, and suggest that more open questions might be necessary to capture occult spirituality. In the case of indigenous spirituality, Christianity sits in very different relation to other culturally-embedded ontologies for indigenous youth than it might for non-colonized youth (Collard and Palmer 2015). Questions about sacredness can be normatively biased, assuming that the absence of the sacred is associated with an absence of spirituality.

The second challenge is more sharply concerned with the reliability of elicited information about spirituality. Singleton *et al.* (2004, p. 250) propose using a stipulative definition of spirituality as 'a conscious way of life based on a transcendent reference' in order to avoid lexical associations with religious institutions or other philosophical touch-points. They identify ten dimensions of spirituality including themes such as salience and authority, as well as more agential possibilities through categories such as eclecticism and expression. Critical youth scholars have followed suit by opening up the entanglements of religious symbols as central to spirituality (Ezzy and Halafoff 2015). Others have reflected new and emerging arrangements through the language of religiosity (e.g. Olson and Reddy 2016) or lived religion (e.g. McGuire 2008), both of which point toward the assemblage of practices, beliefs, representations, and institutions that produce spirituality as a discursive practice as well as a feeling or a belief. Nonetheless, the overwhelming approach toward studying youth spirituality treats it as a matter of personal choice or agency (Flory and Miller 2000), or as Lynch (2010, p. 37) cautions, as 'an unquestioned view of the importance of metaphysical belief for individuals'.

Our own research has pointed to the importance of opening up the scope of analysis beyond traditional religions or religious spaces and seeking to describe the boundaries and the fuzzy edges of faith and belief. We find common ground in Bartolini *et al.*'s (2017, p. 14) call for work that engages 'forms of spirituality, spiritual practices and spiritual experiences that do not look like a religion or a religious practice or a religious experience'. We have also clarified that research claiming declining religion amongst young people is often simply dismissing young people's religious categories because they don't fit neatly into existing frameworks (Olson *et al.* 2013; Vincett *et al.* 2012). Nonetheless, having attempted to do this kind of work for over ten years with diverse groups of young people, we admittedly have found it a difficult task; looking beyond existing framings of spirituality and the sacred could make the concepts too broad and thus meaningless descriptions of anything or nothing at all.[1] We are also aware that broadening the category could have a disingenuous and distorting effect if it forces spirituality upon diverse practices that might resist the label. For example, de la Cadena's (2015) work with Quechua 'speaking men' provides alternatives to the language of spirituality for practices that might be interpreted from a Western theoretical/theological position as religious practice. Deeming certain practices spiritual or religious potentially reproduces the categorical and analytic schemes that are historically rooted in Eurocentric hegemony (Asad 2009). Finally, though our research on youth religiosity has always been open to spiritual practices including ghosts, spiritualism, and afterlife, we have dismissed other evidence about spirituality when it is not equivalent to personal/individual belief. Lynch warns that this approach,

> . . . can obscure the possibility that issues of existential meaning may only be important for young people in specific moments, that young people may only learn to become 'believing subjects' through particular social contexts, and that assent to metaphysical or existential beliefs may play a relatively unimportant role in the day-to-day conduct of many young people's lives.
>
> (2010, p. 38)

In the remainder of this chapter, we look to the day-to-day discursive and performative practices that fall well outside of the categories and engagements normally associated with youth spirituality. Specifically, we consider performances which, though clearly about spirituality, may or may not be about belief. To do this, we focus on practices that might be described as sacrilege – actions and practices which take sacred things for secular use (St John 2006, p. 180) – in modern studies of religion and the sacred. Our data is drawn from *Marginalized Spiritualities*,[2] a project examining the spiritual and religious experiences of young people in areas of urban economic deprivation. Our overall research design was inspired by Kim Knott's work on religion and space, for we hoped her locational approach could 'reconnect "religion" with those other categories – "society", "politics" and "economics" – from which it has been separated for the purpose of classification and study' (Knott 2009, p. 159). However, in order to move away from the classifications discussed above, we also incorporated in-depth ethnographic and participatory work, including filmmaking with teams of young people and spending time at a youth club in a Glasgow neighborhood. While the locational approach allowed us to avoid bias toward religious or spiritually-articulate participants, we found that everyday encounters were important for exposing 'unmarked non-religious cultures' (Lee 2015, p. 20) that might better describe the emergence of youth spirituality.

Sacrilege

We are riding on a very slow train which jolts us as it takes in all the uneven surfaces of the tracks, killing time on our journey to the town center. There are four of us – one researcher, and three girls between the ages of 14 and 16. The girls know each other well. As young carers, they frequently make use of the services provided by the care support services in their ward, and this includes trips for the exceptional (boating trips) and the necessary (registering for ID cards for things such as public transportation access). We decide to go into town to collect footage for a movie we are making about spirituality, and to maybe interview some people in shops on camera if they agree. The trip is a relatively rare social outing for the girls, a chance to hang out where other kids hang out and get moved on periodically by police or shop owners, just like other kids. We had been working hard out of the cramped space of the neighborhood youth club for several months, and we all thought of this as a treat. Our research budget meant that we could safely deliver everyone home by taxi. Being in the center of town and having dinner in a restaurant also meant that the girls could relax and wander, something that was difficult to do in their own neighborhoods which were known for gang activity, distrust between different ethnic groups, aggressive policing, and a drug trade that would ebb and flow into public space in unpredictable ways.

Because we are filming, taking pictures, and recording audio, we are dressed in multi-colored hoodies with the names of our film crew and an image of a film clapper on the back. Our research team hoped to avoid suspicion by being conspicuous in our intent; an Italian art student had been assaulted and arrested for filming in public space recently, and though the right to film in public had been reasserted, there were too many questions being raised after 7/7.[3] The final film that we produced, 'Being', drew together a series of short, edited interviews with adults speaking about the difference between religion and spirituality. The film team liked the topic because they claimed to have never spoken about spirituality before involvement in our project, though they had been in compulsory Religious Education classes since primary school. They tripped over the word itself when we first began to talk about it as if it were being presented to a foreign tongue; 'spirituality' is not an easy word to manipulate with the Glaswegian dialect if unpracticed.

The researcher holds the camera during the train ride, in case the girls want to include some of the footage in the final film (they don't). When they begin to speak about 'organized' and 'unorganized' religion, one of the members of the team asks the researcher to turn the camera on and record what they are saying. The conversation battles against the cacophony of the train, of screeches and bangs, and in doing so, draws out a dynamic that might have otherwise been passed over in only a few seconds:

B (researcher): (Camera pointed at L) So what do you think about organized versus unorganized religion?
L: (Looking very bored, staring out the window) I don't have a clue what they are, really.
C: (To L, sitting beside her) A church, a Mass, a people.
L: (Turns to C) Ah, right. So that wouldn't be Dude-ist.
C: Taoist? No.

L: *Dude*-ism (with emphasis, hand cupped to her mouth)
C: Dude-ism?
A: (from off camera) Taoism?
L: Taoism? Taoist?
C: Dude-ism!
L: Dude-ism. (L and A reiterate the word several times. B laughs.)
B: Define Dude-ism for us.
L: (with an ironic smile, head tipped back, playful) It's Dude-ism. Where you're a dude. You're a Dude-ist, and then Dude-ism religion. Pretty cool religion. I'm a part of it.
C: Oh. (deadpan and looking straight into the camera). Tell me about it.
L: You just be a dude.
C: So you just exist? (very skeptical look, again directly at the camera.)
L: (looking at L and the camera interchangeably) yeah, and you relax and be cool to others and be kind and friendly and be cool and that makes you popular and no harm comes to you.

Interpreted from the literature of youth spirituality, L's description of 'Dude-ism' echoes many of the findings encountered with middle-class youth in the U.K. and the U.S. Sometimes referred to as 'moralistic therapeutic deism', this youth perspective is interpreted as replacing traditional Christian teachings of a powerful God who punishes evil with a benevolent God that watches over people and aims for 'personal happiness and interpersonal niceness' (Smith and Denton 2005, p. 171). Dean (2010, p. 28) refers to this trend amongst young Christians in the U.S. as the 'cult of nice', and defines it as perhaps the most important category of contemporary youth spiritual understanding. This positioning of youth spirituality is interpreted against an adult and institutionalized religious theology that is set in contrast with Moralistic Therapeutic Deism, which Smith (2011) considers a degraded organized theology that collapses into confused transitions into adulthood.

Indeed, L's commentary could be fixed as a nearly textbook example of the cult of nice – relaxing and being cool. However, to interpret it as evidence of the cult of nice would require ignoring the intermittent mocking and sincere manner that signals L as engaging in something much more akin to sacrilege. Dude-ism has a doctrine, following, and website based on the movie character from which it comes.[4] It is satirical, but also content-heavy. It belongs in the category of other modern satirical traditions evoked by post-boomer generations including the Church of the Flying Spaghetti Monster, but rather than being overtly secularist in its intent, it also condones a form and structure of spiritualism that is playful but also ambiguous. Isn't Dude-ism relevant in that it could produce a kinder world than we have now? Couldn't it be taken seriously as a spiritual practice, just like the film it is drawn from? But even L delivers her sacrilege with a recognition that she is committing it; the purpose of the performance is to make a point about herself as a subject that can be in formation, and also about religion and her ability to satirize it. Hers is not a claim to the sacrilege of the Protestant Reformation, but more of a reform without replacement. L is not a Dude-ist – or maybe she is. What is important for us as researchers is her eagerness not only to confront traditional forms of religious subjectivity, but also the normality of committing sacrilege, one which her friends find amusing but also interesting and confusing.

This discursive practice of sacrilege could be dismissed as banter on a train, but it illustrates a much wider trend that religious scholars have to work increasingly hard to ignore or dismiss. The emergence of distinctively modern secular humanist movements which have also taken to acts of sometimes wide-scale sacrilege in order to create space for 'being godless' (Blanes and Oustinova-Stejepanovic 2015). In Australia and the United Kingdom, the movement to write in 'Jedi Knight' as a religious category in the national census confounded state governments and researchers. Initially praised by atheist activists, they later responded with a more serious call to record 'no religion' as it became clear that Jediism would go on to claim religious status.

Academic accounts of religious change often relegate the episode to footnotes that simultaneously acknowledge and dismiss its presence and significance – it requires explanation, but only in as much as it artificially manipulates our descriptions of religion. Interpretations include speculations that respondents 'may be being humorous', or trying to make the point of 'just how inadequate the categories provided on the census are' for categorizing religiosity and spirituality (Ezzy and Halafoff 2015, p. 850). Since 2001, the date of the first significant census write-in movement, Temple of the Jedi Order has become a tax-exempt non-profit in the United States[5] with an articulated doctrine and method for developing as a Jedi spiritual leader. An application in 2016 by The Temple to be recognized as a religion by the UK Charity Commission failed because of the Commission's assertion that charity law defines religion as 'belief in one or more gods or spiritual or non-secular principles or things . . .' (The Temple of the Jedi Order – Full Decision 2016, p. 3) Referring to the *Hodkin* case, the Commission based its decision on the assertion that Jediism, though 'open to spiritual awareness', could nonetheless 'be advanced and followed as a secular belief system' (*ibid* p. 4).

If the census protest and L's playful performance as discursive and material practices are considered as part of the spiritual landscape of Western societies, rather than abnormalities, we might begin to describe youth spirituality differently. Both L's Dude-ism and Jediism suggest a blending of spiritual types, but a transcendence that emerges from this act of assembly and play across the religious and the secular, or if ascribing to the argument advanced by the British state, an act of sacrilege. They are expressed through discursive practices carried out in informal and formal everyday spaces, sometimes on a train, other times through governmental institutions that challenge the boundaries of definitions and categories of religion and spirituality. Though the encounter described above can't be conflated with an elaborated belief or practice, it should not be entirely discounted from our descriptions of youth spiritual practice and belief. As a discursive practice, it suggests that sacrilege should be thought of as a part of, rather than apart from, the production of youth spirituality; it requires explanation rather than dismissal.

Encounter

The performance of sacrilege on the train serves as a check upon our assumptions of how young people navigate religious categories and their significance. In describing Australian secularism, Coleman and White (2006, p. 3) explain that although the space for the sacred is ubiquitously assumed to demand societal respect, 'blasphemy and sacrilege are both affronts to this value: they are acts of disrespect, irreverence, or destruction'. The implication is that sacrilege is antithetical to a secular society because of its divisiveness and role in perpetuating social prejudice. It can be carried out deliberately and on a wider social order, such as the desecration of Muslim sacred objects and Muslim bodies in the clandestine prisons maintained by the United States since September 11th, 2001. However, this kind of sacrilege is different from that expressed on the train encounter, which we would describe as more creative than destructive, carried out haltingly and singularly, without clear conviction except a dedication to committing the sacrilege. Here we describe two more expressions that might be described as sacrilege, one which illustrates a more concerted effort to articulate spirituality through practice, and a second which turns to an individual account of encounter.

On a day when we were scheduled to brainstorm a fictional short film about spirituality, C arrived at the youth club visibly energized, and excitedly presented a flier she had been handed with the word 'Spirituality' printed boldly on the front. The flier advertised a workshop and event described as 'new age spirituality', with opportunities for attendees to experience meditation, shamanism, yoga, and reiki. None of the young women were familiar with any of these practices or knew what they were, and we looked some of them up on the computer and talked through them, toying with the idea of paying to go to the event. As we sat around a large piece of newsprint in the youth club meeting room to try and think of a storyline that would explore their ideas about spirituality, the girls moved quickly away from the new-age practices advertised in the flier, and instead gravitated toward what a spiritual person would do. Spiritual people would be good people, 'better', generous, and caring. One of the screenplays that resulted from these exercises – their favorite, and the one we would have filmed had it not been for our inability to procure the central prop – focused on a brief interaction between an older man making his way down the pavement on a motorized mobility scooter, and a teenage girl walking in the opposite direction. Rather than passing the girl, who they decided would project visible characteristics of distress (head down, walking slowly, looking dejected), the old man would stop and speak with the girl and ask her how she was feeling, and if she was OK. In just a few moments their exchange would be over, but the girl would look different – thoughtful, reflective, reconnected, and looking at the world around her. Hopeful.

The film team's version of spirituality is better described as care and connection across difference, less 'interpersonal niceness' than encounter, or what Wilson (2016) identifies as 'the unpredictable ways in which similarity and difference are negotiated *in the moment*' (p. 5, emphasis in original). Wilson points to the uncanny sense of interruption that is the outcome of encounter across difference, or 'meetings where difference is somehow noteworthy' (p. 14). Encounters of spirituality thus imply a very different process than the individualistic and possessive framings discussed in the youth spirituality literature; whereas the latter assumes that spirituality is embodied and thus measurable as an attribute of more or less spiritual, encounter implies that difference, time, and place are central to spiritual experiences. The time and place of our filmmaker's lives was 'broken Britain', a phrase offered by Sir Iain Duncan Smith and others in the Conservative Party as a broad-brushed description of the decline of working class neighborhoods. Youth, who were viewed as products of irresponsible and morally-bereft parents, were the outcome of the social pathology of poverty. Then Prime Minister David Cameron painted neighborhoods like the one where our film crew lived as 'the seed bed of crime', emerging from the tragic and unruly consequence of failed families.[6] To find the sacred in a pensioner who will notice and speak with a teenager tells us as much about their context as it does their spiritual beliefs.

Explaining and understanding youth spirituality thus might require more explicit groundings in encounter than belief, even when seeking to understand individual perspectives and experiences. To illustrate this final point, we turn to a 17-year-old young woman, 'Z', who was interviewed for our project. As a frequent fixture in the youth club she was often listening in, though not participating, in our other activities. She was born in Pakistan and moved with her family to Glasgow when she was just a year old. She is fond of her neighborhood because she knows most of the people who live around her, though she worries about how many young people do drugs, drink, and get into trouble. She describes herself as always the most interested in speaking about religion amongst her peers, always wanting to help others understand God as she did but recognizing that most people her age weren't interested, including her boyfriend. She states confidently, 'I believe in God with all my heart, I've been brought up as Muslim but. . . (pause) my heart doesn't really belong there, if you get me'. Z is therefore difficult to categorize in our existing interpretations of spirituality beyond the individualism that is often affixed to post-boomer generations. Still, she defines herself as 'more on the Christian side than I am on the Islam side. I think all religions are basically the same. It's just the small differences that set them apart'. Her certainty about God is buttressed by her ability to successfully cope with her long-term depression. 'It's mostly because I trust God that I see life in a different way now'.

Meeting across difference, and holding difference in tension, is central to Z's sense of the spiritual and the sacred, and it is also what drives her to piece together practices that would be considered sacrilege, blasphemy, or 'fuzzy theology' from the perspective of either Christianity or Islam. Spirituality for Z is located both in and through difference, and she insists that '[. . .] so many places that are spiritual, but in their own way. I would just love, you know, to go around the world and just see what everybody is like'. The discovery of the sacred through encounter of spiritual places is a grounding characteristic of her spirituality, and she explains that this is a long-standing practice for her. 'It sounds really strange, but sometimes I used to just like skip off school and like I would just sit outside the church just to find, you know, comfort . . . just to feel like I'm closer to God in a sort of way'. Though Z confesses lacking words to describe the power of church buildings, she thinks that it must have to do with 'so many people that have been there baring their souls out to God'. Her practices might be described according to existing theories of youth religiosity as taking on a 'candy-jar' or 'spiritual seeker' role. However, these explanations would mischaracterize her certainty about her belief, and perhaps also her reasons for her sacrilegious practices, which are not intended as desecration or destruction. Instead, her sacrilege allows for the folding together of difference, revealing not an individualistic form of spirituality but a complementary place of peace, saturated with souls.

Conclusion

To conclude, we would like to highlight three ways that attention to everyday encounter, captured outside the practices, discourse, and spaces normally associated with religion and spirituality, might lend new insights into youth spirituality. Firstly, this approach provided relief from existing categorical biases and from the assumption that spirituality is an analytic category best studied by asking about individual beliefs and then describing trends. Instead, our analysis suggests that there are other processes we should be paying attention to and discussing. Focusing on practices of sacrilege reveals some problems with our own assumptions as scholars when we categorize things that fall outside of our existing definitions as inaccuracies, or as not quite serious. Secondly, and relatedly, our focus on everyday encounters revealed a range of possibilities for studying youth spirituality: as engagement and play with popular culture, as attempts to articulate spirituality within context and that which is missing, and also as the more common scholarly reflections on novel combinations of practice-led (rather than theologically driven) belief. Furthermore, encounters don't have to be lasting to be meaningful to our study of youth spirituality; L doesn't have to become a Dude-ist, and Z doesn't have to articulate a personal theology to correspond with sitting outside of churches, in order for us to pay attention to their experiences. Thirdly, research that focuses on spirituality through the collection of beliefs will provide only a limited understanding of both the construction and the meaningfulness of youth spirituality. Here we have experimented with looking across encounters that would traditionally be described as sacrilege or its related practice of blasphemy in order to look at instances of encounters with spirituality that may or may not be felt or expressed as belief. Though tentative and limited, it illustrates the importance of both moving beyond existing discursive categories (Bartolini *et al.* 2017), and considering approaches and methodologies that encourage researchers to engage youth spirituality as not just existential or metaphysical (Lynch 2010), but also as encounters that produce and reflect important signals in spirituality more broadly.

We would like to end by briefly reflecting on sacrilege and blasphemy as potentially important qualities of contemporary spirituality. We have resolved very few questions about its function in late modernity, and it requires more research. Our analysis suggests that exploring who is able to commit sacrilege, what is condoned or condemned, in what context and for what ends are all worthwhile questions that may help us understand contemporary spirituality with new insights and new categorical emphases. Understanding that which used to be deemed sacrilege or blasphemy, and tracking both continuity and new emergences of its discursive performance, could be an important undertaking for describing contemporary spirituality. We find room, and perhaps even an urgency, for moving further from our traditional scripts and into areas that are still partly formed and tentative, revealed through close attention to everyday encounters.

Notes

1 See, for instance, discussions surrounding Taves' (2009) recommendation to shift religious studies toward 'experiences deemed religious.'
2 The project was funded by the AHRC-ESRC Religion and Society Research Programme (www.religionandsociety.org.uk/)
3 www.youtube.com/watch?v=nNEgLLGLL18 and www.theguardian.com/law/2011/aug/31/do-we-have-right-to-film-police
4 http://dudeism.com
5 www.templeofthejediorder.org/
6 http://news.bbc.co.uk/2/hi/5166498.stm

References

Asad, T. (2009). *Genealogies of Religion: Discipline and Reasons of Power in Christianity and Islam*. Baltimore, MD: Johns Hopkins University Press.

Bartolini, N., Chris, R., MacKian, S., and Pile, S. (2013). Psychics, crystals, candles and cauldrons: Alternative spiritualities and the question of their esoteric economies. *Social and Cultural Geography*, 14(4), 367–388.

Bartolini, N., Chris, R., MacKian, S., and Pile, S. (2017). The place of spirit: modernity and geographies of spirituality. *Progress in Human Geography*, 41(3), 309132516637763. https://doi.org/10.1177/0309132516637763

Belgrave, F. Z., Townsend, T. G., Cherry, V. R., and Cunningham, D. M. (1997). The influence of an Africentric worldview and demographic variables on drug knowledge, attitudes, and use among African American youth. *Journal of Community Psychology*, 25(5), 421–433.

Blanes, R., and Oustinova-Stjepanovic, G. (2015). Godless people, doubt, and atheism. *Social Analysis*, 59(2), 1–19.

Cloke, P. (2002). Deliver us from evil? Prospects for living ethically and acting politically in human geography. *Progress in Human Geography*, 26(5), 587–604.

Coleman, E., and White, K. (2006). Negotiating the sacred in multicultural societies. In E. Coleman and K. White (Eds.), *Negotiating the Sacred: Blasphemy and Sacrilege in a Multicultural Society*. ANU Press: Canberra. 1–14. Retrieved from www.jstor.org/stable/j.ctt2jbjjq.5

Collard, L., and Palmer, D. (2015). Koorlankga wer wiern: indigenous young people and spirituality. In J. Wyn and H. Cahill (Eds.), *Handbook of Children and Youth Studies*. Singapore: Springer Singapore. 875–888.

Cotton, S., Zebracki, K., Rosenthal, S. L., Tsevat, J., and Drotar, D. (2006). Religion/spirituality and adolescent health outcomes: a review. *Journal of Adolescent Health*, 38(4), 472–480.

Cusack, C. M. (2011). Some recent trends in the study of religion and youth. *Journal of Religious History*, 35(3), 409–418.

Davie, G. (1994). *Religion in Britain Since 1945: Believing Without Belonging*. Oxford: Blackwell.

Dean, K. C. (2010). *Almost Christian: What the Faith of Our Teenagers Is Telling the American Church*. Oxford: Oxford University Press.

de la Cadena, M. (2015). *Earth Beings: Ecologies of Practice Across Andean Worlds*. Durham and London: Duke University Press.

Della Dora, V. (2016). *Landscape, Nature, and the Sacred in Byzantium*. Cambridge: Cambridge University Press.

Ezzy, D., and Halafoff, A. (2015). Spirituality, religion, and youth: an introduction. In J. Wyn and H. Cahill (Eds.), *Handbook of Children and Youth Studies*. Singapore: Springer Singapore. 845–860.

Finlayson, C. C. (2012). Spaces of faith: incorporating emotion and spirituality in geographic studies. *Environment and Planning A*, 44(7), 1763–1778.

Flory, R.W., and Miller, D.E. (2000). *GenX Religion*. London and New York: Routledge.

Fuller, C. (2001). *Spiritual, But Not Religious: Understanding Unchurched America*. Oxford and New York: Oxford University Press.

Hart, T. (2003). *The Secret Spiritual World of Children: The Breakthrough Discovery That Profoundly Alters Our Conventional View of Children's Mystical Experiences*. Makawao, HI: Inner Ocean.

Hemming, P. J., and Madge, N. (2012). Researching children, youth and religion: Identity, complexity and agency. *Childhood*, 19(1), 38–51.

Holder, D. W., DuRant, R. H., Harris, T. L., Daniel, J. H., Obeidallah, D., and Goodman, E. (2000). The association between adolescent spirituality and voluntary sexual activity. *Journal of Adolescent Health*, 26(4), 295–302.

Holloway, J. (2006). Enchanted spaces: the séance, affect, and geographies of religion. *Annals of the Association of American Geographers*, 96(1), 182–187.

King, P. E., Clardy, C. E., and Ramos, J. S. (2014). Adolescent spiritual exemplars. *Journal of Adolescent Research*, 29(2), 186–212.

King, P. E., Kim, S.-H., Furrow, J. L., and Clardy, C. E. (2016). Preliminary exploration of the Measurement of Diverse Adolescent Spirituality (MDAS) among Mexican youth. *Applied Developmental Science*, 0(0), 1–16. https://doi.org/10.1080/10888691.2016.1203789

Knott, K. (2009). From locality to location and back again: a spatial journey in the study of religion. *Religion*, 39, 154–160.

Kong, L. (2010). Global shifts, theoretical shifts: changing geographies of religion. *Progress in Human Geography*, 34(6), 755–776.

Lee, L. (2015). Ambivalent atheist identities: power and non-religious culture in contemporary Britain. *Social Analysis*, 59(2), 20–39.

Lerner, R.M., Roser, R.W., and Phelps, E., Eds. (2008). *Positive Youth Development and Spirituality: From Theory to Research*. Conshohocken, PA: Templeton Foundation Press.

Lynch, G. (2010). Generation X religion: a critical approach. In S. Collins-Mayo and P. Dandelion (Eds.), *Religion and Youth*. Farnham: Ashgate. 33–38.

Maddrell, A. (2009). A place for grief and belief: the Witness Cairn, Isle of Whithorn, Galloway, Scotland. *Social and Cultural Geography*, 10(6), 675–693.

McGuire, M. B. (2008). *Lived Religion: Faith and Practice in Everyday Life*. Oxford and New York: Oxford University Press.

Olson, E., Hopkins, P., Pain, R., and Vincett, G. (2013). Retheorizing the postsecular present: embodiment, spatial transcendence, and challenges to authenticity among young Christians in Glasgow, Scotland. *Annals of the Association of American Geographers*, 103(6), 1421–1436.

Olson, E., and Reddy, S. (2016). Geographies of youth religiosity and spirituality. In T. Skelton and S. Aitken (Eds.), *Establishing Geographies of Children and Young People*. Singapore: Springer Singapore. 1–23.

Pulido, L. (1998). The sacredness of "Mother Earth": spirituality, activism, and social justice. *Annals of the Association of American Geographers*, 88(4), 719–723.

Savage, S., Collins-Mayo, S., Mayo, B., and Cray, G. (2006). *Making Sense of Generation Y: The World View of 15 to 25 Year-Olds*. London: Church House Publishing.

Sifers, S. K., Warren, J. S., and Jackson, Y. (2012). Measuring spirituality in children. *Journal of Psychology and Christianity*, 31(3), 205–214.

Singleton, A., Mason, M., and Webber, R., 2004. Spirituality in adolescence and young adulthood: a method for a qualitative study. *International Journal of Children's Spirituality*, 9(3), 247–262.

Smith, C. (2011). *Lost in Transition: The Dark Side of Emerging Adulthood*. Oxford, UK: Oxford University Press.

Smith, C., and Denton, M. (2005). *Soul Searching: The Religious and Spiritual Lives of American Teenagers*. Oxford and New York: Oxford University Press.

St John, E. (2006). The sacred and sacrilege – ethics not metaphysics. In E.Coleman and K. White (Eds.), *Negotiating the Sacred: Blasphemy and Sacrilege in a Multicultural Society*. ANU Press: Canberra. 179–190.

Taves, A. (2009). *Religious experience reconsidered: A building-block approach to the study of religion and other special things*. Princeton, New Jersey: Princeton University Press.

The Temple of the Jedi Order – Full Decision 2016. The Charity Commission. 19 December 2016. Retrieved from www.gov.uk/government/publications//the-temple-of-the-jedi-order

Vincett, G., Dunlop, S., Sammet, K., and Yendell, A. (2015). Young people and religion and spirituality in Europe: a complex picture. In J. Wyn and H. Cahill (Eds.), *Handbook of Children and Youth Studies*. Singapore: Springer Singapore. 889–902.

Vincett, G., Olson, E., Hopkins, P., and Pain, R. (2012). Young people and performance Christianity in Scotland. *Journal of Contemporary Religion*, 27(2), 275–290.

Williams, A. (2016). Spiritual landscapes of Pentecostal worship, belief, and embodiment in a therapeutic community: new critical perspectives. *Emotion, Space and Society*, 19, 45–55.

Wilson, H. F., 2016. On geography and encounter: bodies, borders, and difference. *Progress in Human Geography*.

Zinnbauer, B. J., Pargament, K. I., Cole, B., Rye, M. S., Butter, E. M., Belavich, T. G., Hipp, K. M., Scott, A. B., and Kadar, J. L. (1997). Religion and spirituality: unfuzzying the fuzzy. *Journal for the Scientific Study of Religion*, 36(4), 549–564.

11 Transnational religion and everyday lives

Spaces of spirituality among Brazilian and Vietnamese migrants in London

Olivia Sheringham and Annabelle Wilkins

Introduction

In a study of irregular migration from Mexico and Central America to the United States, Hagan (2008: 7) argues that 'religion permeates the entirety of the migrant experience'. Migrants draw upon their faith for guidance before embarking upon migration, as well as turning to religious objects, practices and institutions for material, emotional and spiritual support during frequently dangerous journeys (Hagan 2008). Once they have reached their destination, migrants engage with local religious sites and practices that enable them to feel a sense of belonging in an unfamiliar, often hostile environment (Hagan 2008, see also Sheringham 2013). Religion is also trans-temporal, connecting migrants' memories and traditions with ideas of the future, including potential returns to the homeland (Vásquez 2016). However, while there has been increasingly widespread recognition of the significance of religion within migrant experience, few studies have examined the connections between home, migration and spirituality in the city (Wilkins 2016; Blunt and Sheringham 2015). This chapter explores everyday urban and transnational spiritualities, with a particular focus on religious and spiritual practices, objects and spaces among Brazilian and Vietnamese migrants in London.

Drawing upon qualitative research with people who have migrated to London over a range of time periods from 1979 to the present day, the chapter explores the everyday urban spaces in which religion and spirituality are practised, as well as the relationships between material and spiritual worlds.[1] This chapter also extends research on transnational religion in its attention to relationships between the domestic and the urban as sites of religious and spiritual experience, both of which are mediated by transnational connections. Throughout our discussion, we develop the idea of the migrant home in the city as a site of connection between domestic, urban, transnational and spiritual realms. We also consider the ways in which spiritual objects and practices are present in workplaces and those that travel with migrants on their journeys around the city. The chapter draws upon these objects and practices to theorise how everyday spiritualities unsettle the boundaries between the home, the city and worlds beyond. We emphasise the importance of locality and migrant home-making alongside transnational networks, showing how religious practices can contribute to the shrinking of space between home and (imagined or remembered) homeland. We argue that migrants' everyday spiritual practices not only contribute to understandings of transnational religion, but also articulate broader debates within geographies of home and migration, including what home is and where it might be located.

Through its examination of religious and spiritual practices in the lives of Brazilian and Vietnamese migrants in London, this chapter develops a holistic perspective on the significance of transnational religion in relation to home, work and the city. Previous research has revealed how individual migrants draw upon religious and spiritual practices in multiple ways, including as resources for navigating and coping with the challenges of everyday life (Vásquez and Knott 2014), as strategies for ensuring success in work (Hüwelmeier and Krause 2010; Wilkins 2016), or in maintaining relationships across transnational space (Sheringham 2013). Religious practices are also significant in terms of building a sense of home, identity and belonging (Tolia-Kelly 2004). In emphasising material, affective and intangible aspects of religion, our approach highlights the ways in which religious practices operate at multiple scales, from the individual body to the dwelling, the workplace, the city and across transnational space. It takes into account the ways in which religion is reconfigured by and through mobility, as migrants find new and innovative ways to adapt their practices to new contexts. This is not without significant challenges, as everyday spiritual practices are influenced and constrained by material, economic and personal dimensions of life in the 'super-diverse' city.

Our analysis is informed by a 'lived religion' approach, in which the emphasis is shifted from institutional or formalised religion towards a focus on how religion is actually practised in the everyday lives of individuals and communities (Ammerman 2007; McGuire 2008; Orsi 2003). Ideas of lived religion have been incorporated into migration studies, as research has explored the roles of spiritual practices, sacred objects and spaces in enabling senses of belonging, home and identity among migrant communities (Sheringham 2013; Vásquez and Marquardt 2003; Vásquez and Knott 2014). Alongside broader understandings of lived religion, this chapter draws upon ideas that are embedded within particular cultural contexts, such as concepts of the Vietnamese home as a site of connection between material and spiritual worlds (McAllister 2012; Jellema 2007a). In addition to exploring the multi-scalarity of religion and spirituality among Vietnamese and Brazilian migrants in London, we propose that spiritual practices are important elements of home and belonging in the context of migration and urban super-diversity. Furthermore, we argue that a focus on everyday spiritual practices not only enables new understandings of home and migration, but also contributes to knowledge on the complex relationships between these domains. The city is emphasised as a crucial site for transitions and transformations in religious and spiritual practices in the everyday lives of individual migrants and communities. Before examining the particular spiritual practices, sites and objects that are significant for participants, the following section situates the chapter within broader geographies of religion and migration.

Religion and transnationalism: from institutional religion to everyday spiritualities

Relationships between religion and migration have been explored by increasing numbers of scholars across a range of disciplines (Sheringham 2010; Vásquez and Dewind 2014; Wong 2014). The globalisation of migration focused attention on the ways in which religious and spiritual practices travel and change through mobility (Levitt 2007), the formation of transnational religious networks (Ebaugh and Chafetz 2002), and the heightened visibility of diasporic religious identities in urban space (Garnett and Harris 2013; Hüwelmeier and Krause 2010; Vásquez 2016). Religious institutions have been theorised as liminal spaces in which migrants can experience a sense of belonging in what can be an exclusionary environment (Vásquez and Marquardt 2003). However, others have argued that focusing solely on institutions and adopting a 'functionalist' or 'materialist' perspective on religion have led scholars to overlook personal and embodied aspects of religion and spirituality within the migrant experience (Dwyer 2016; see also Mazumdar and Mazumdar 2009). Similarly, Sheringham (2010: 1689) calls for geographers to examine 'how religious beliefs and practices travel across borders not just through institutions and formal networks, but also as an integral part of the identities and experiences of many migrants.' This perspective regards migration and religion as inseparable from broader practices and processes of mobility, dwelling and everyday life, not only for particular migrants or religious communities but also for those who stay put (*ibid*).

An emphasis on the intertwining of religion and everyday life draws upon ideas of 'lived' or 'everyday' religion (Ammerman 2007; McGuire 2008; Orsi 2003), whereby attention is focused on embodied practices, sacred objects and spaces in homes and places of work, as well as public rituals or performances of faith that take place during religious celebrations (Dwyer 2016, MacKian 2012; Vásquez and Knott 2014). Both the home and the city have been explored as important realms of religious and spiritual experience and practice. Fewer scholars, however, have focused on how migrants' religious practices can create connections *between* these domains as well as with spaces beyond (see however Wilkins 2016; Tolia-Kelly 2004). Moreover, as we argue below, migrants' religious and spiritual practices in urban and domestic spaces can foster temporal connections, including relationships with deceased ancestors and aspirations for the future.

A substantial body of research draws upon domestic material culture to explore emotional dimensions of home, migration, memory and identity (Hurdley 2006; Miller 2001; Walsh 2006). Studies have also examined the role of domestic objects and images in creating and remembering homes across diasporic space (Burrell 2014b; Parrott 2014; Walsh 2006). Recent scholarship argues that domestic possessions are not simply 'identity markers' or reminders of a homeland, but involve layers of emotion and sensation that are interwoven with cultural and personal values (Parrott 2014). Other studies have examined the ways in which the material culture of diasporic homes is imbued with religious and spiritual meaning (Tolia-Kelly 2004; Mazumdar and Mazumdar 2009). Drawing on research with South

Asian women in London, Tolia-Kelly (2004) discusses domestic religious spaces including *mandirs* (home altars or shrines), exploring their material and sensory significance in relation to feelings of home, belonging and identity in the context of migration. Tolia-Kelly's theorisation of shrines as objects of 're-memory' points to the ways in which these sacred spaces change over time, as objects and images symbolising personal and collective histories are continually added (*ibid*: 319). Discussing the significance of photographs of the Dalai Lama among members of the Tibetan diaspora, Harris (2001) notes that these images do not only evoke religious feelings, but are drawn upon to generate a sense of solidarity and resistance against Chinese politics. Parrott (2014: 51) notes that objects that were intended to bring a sense of comfort may evoke feelings of loss or isolation in a new location.

Within the growing research field of geographies of religion (Dwyer 2016), there has been an increasing interest in the materialisation of religion in urban landscapes, sometimes involving contestations over public space and identity (Naylor and Ryan 2002; Smith and Eade 2008). Recent work has also highlighted the intersections between religious identities, space and power, often, as Dwyer (2016: 2) suggests, producing 'richer accounts of the intersectionalities of social formations, power and resistance.' A number of studies have examined the roles of faith-based organisations (FBOs) in the 'post-secular city' in addressing social and welfare-related issues (Beaumont 2008; Jamoul and Wills 2008). Religious and spiritual practices can therefore be understood as resources for building a sense of belonging and community in what can be exclusionary contexts. However, a focus on the particular buildings and functions of religion risks overlooking the embodied aspects of spirituality and the diverse spaces in which they take place. In this chapter, we respond to scholarship that recognises how spiritualities are 'infused into ordinary spaces', including homes, workplaces and public spaces in the city (Bartolini *et al.* 2017), while also examining spiritual and otherworldly experiences as important topics of study in their own right (Holloway and Valins 2002; MacKian 2012).

A spatial perspective on religion, as Knott (2005) argues, encompasses not only physical and cultural spaces, but also the wider expressions and practices of religiosity and spirituality across and within multiple domains. This chapter responds to calls for scholars to adopt an embodied and spatial approach to transnational religion. However, this chapter also contributes to geographies of spirituality by broadening spatial perspectives on religion to encompass the spirit world alongside domestic, urban and transnational dimensions of space. We also demonstrate the ways in which spiritual practices connect multiple temporalities of home and relationships between living relatives, spirits and ancestors. We explore the significance of spiritual objects found in the home, as well as objects that migrants carry with them on their journeys around the city. These objects are examined in relation to the multiple ways of 'doing religion' in the city. Migrants' everyday spiritualities encompass face-to-face and virtual dimensions, including online apps that facilitate worship, alongside objects and rituals that are important resources in work and home-making practices. These practices and objects generate new connections between migrants' homes and mobilities in the city, as well as forming translocal connections between London and their countries of departure. We argue that attending to everyday religious practices and spaces offers new ways of theorising home beyond the material, urban and domestic, incorporating spiritual and temporal worlds.

It is important to clarify how we are using the terms religion and spirituality, which whilst overlapping, also refer to distinct concepts and realms of experience. Several of our interviewees did not describe themselves as religious, but emphasised important spiritual aspects of their everyday lives. Conversely, several participants referred to the social aspects of their affiliation to a formal religious denomination as an important site of belonging in the city, without highlighting the spiritual dimension. Here we understand spirituality as often encompassing, but also moving beyond, formal and institutionalised religion. Drawing on MacKian's (2012) work on 'everyday spiritualities', we point to the importance of taking seriously the 'agency and salience of the spiritual' (Dwyer 2016: 758–759). Our approach to home is informed by established frameworks conceptualising home as a 'spatial imaginary': not only a physical location, but also a site of emotions, memories and imagination that is intertwined with power relations (Blunt and Dowling 2006; Blunt 2005; Brickell 2012). Finally, we build on recent work that examines the intersections of home, city and migration, where the city is understood as an important site of home for migrants and diasporic groups (Blunt and Bonnerjee 2013), and the home is revealed as a site of connection between the urban and the domestic (Burrell 2014a; Blunt and Sheringham 2015). Drawing together the literature on home, city and migration with perspectives on lived religion enables a comprehensive understanding of the importance of everyday spiritualities within migrant home-making, as well as the broader role of the city as a site in which religious practices are maintained, adapted and transformed. Before discussing the spiritual dimensions of home, migration and the city in relation to our empirical material, the following section briefly establishes the background of Brazilian and Vietnamese migration to London.

Brazilian and Vietnamese migrants in London

Brazilian and Vietnamese migrants are two relatively recent migrant groups in London, whose migratory trajectories follow different patterns to those of migrants with particular colonial links to the UK. While the groups have very different backgrounds and circumstances of migration, they share a number of important features with regard to their internal diversity and visibility in the city. While earlier waves of Brazilian migrants to London included those who came in the 1960s to escape the military dictatorship in Brazil, it is widely acknowledged that the most significant inflow of Brazilian migrants to the UK, and to London in particular, occurred in the early 2000s. Unofficial estimates put the number in London alone at somewhere between 150,000–300,000 (Evans *et al.* 2015), while estimates from the Brazilian Ministry of External Affairs put the figure in the UK at around 120,000 (MRE 2014). Existing studies of Brazilians in London highlight the internal heterogeneity of Brazilians with regards to region of origin, race, sexuality, class background and migration status as well as the fairly wide dispersal of Brazilians across the city (McIlwaine 2011; Evans *et al.* 2015; Sheringham 2013). Despite such internal diversity, a high proportion of the recent flow of Brazilians to London are 'economic migrants', the majority employed in low skilled, low paid service sector jobs (Evans *et al.* 2011). London's sizeable Brazilian presence is reflected in the large numbers of shops, restaurants, beauty salons and publications that exist, established predominantly by Brazilians, to serve Brazilians. More significant, perhaps, are the growing number of churches – of varying denominations – that seem to represent crucial spaces of support for Brazilian migrants as well as seeking to attract followers from other denominations (Sheringham 2013). The empirical discussion in this chapter is based on a wider study – conducted in 2011–2012 in Brazil and London, with follow-up interviews in London conducted in 2016 – of the role of religion in the everyday lives and migration experiences of Brazilian migrants in London and those who return home (see also Sheringham 2013). The research involved a total of 78 interviews with men and women from a range of backgrounds who were either currently living in London or had returned to Brazil after a period of time living in the city.

During the 20 years following the takeover of South Vietnam in 1975, over two million refugees left Vietnam. This mass exodus was triggered by the reunification of North and South Vietnam in 1975, when Communist Vietnamese forces took control of Saigon and thousands of people fled the city, many of them in small fishing boats (Chan 2011). Other refugees left as a result of conflict between China and Vietnam after 1979. Many refugees who were rescued at sea endured long stays in refugee camps in Hong Kong (Hale 1993). The first Vietnamese refugees were accepted by the UK from Hong Kong in 1979, when Hong Kong was under British rule (Sims 2007). Around 22,000 refugees were resettled in Britain between 1979 and 1988. Alongside those who came as refugees, the Vietnamese population includes people who have migrated for work or education in more recent years. The 2011 census estimated around 30,000 people living in England and Wales who were born in Vietnam, of whom around half live in London. However, these official totals are markedly lower than the 55,000 Vietnamese in the UK estimated by community organisations (Sims 2007: i). While Vietnamese communities have formed across East and South East London, Hackney is commonly regarded as an important location for the city's Vietnamese population, evidence of which can be seen in the presence of Vietnamese restaurants, shops and nail salons. The Vietnamese diaspora is diverse in terms of ethnicity, political affiliation and religious practice, and includes followers of Buddhism, Catholicism and ancestor veneration. The research on which this chapter is based draws upon qualitative research involving 22 men and women from a range of backgrounds, all of whom were born in Vietnam and now live and/or work in London. Alongside repeated semi-structured interviews, the study used ethnography and visual methods to examine material, emotional and imaginative dimensions of home, work and the city.

As relatively 'new' migrant groups without colonial links to the UK and characterised by internal heterogeneity with regard to class, migration status and religion, both Brazilian and Vietnamese migrants could be seen as contributing to the increasing intensification of diversity in London that has been termed 'super-diversity' (Vertovec 2007). Both groups have made their mark on the urban landscape through the emergence of shops, restaurants, nail salons and other businesses, as well as public cultural events such as Brazilian Day or the Vietnam Discovery festival. Yet these visible traces of ethnicity to some extent mask the more complex and contested ways in which migrants inhabit and experience the city (Knowles 2013). In the following sections, we examine migrants' everyday religious and devotional practices – which span domestic, urban and spiritual worlds – to contribute to a more nuanced and layered understanding of home, transnational religion and super-diversity.

Religion in the home

The home was a significant space of religious and spiritual practice among Brazilian and Vietnamese migrants in our case studies. The presence of home shrines or altars, for example, facilitated contact with deities, ancestors and spiritual beings, as well as contributing to a diasporic identity. Domestic altars can be observed in the homes of many Vietnamese people in London, and include those that venerate the Buddha, others that are directed towards Taoist deities and altars to familial ancestors. Alongside the major religious faiths of Buddhism and Christianity, ancestor worship is widely practised in Vietnam, including among people who would not define themselves as religious (McAllister 2012). Ancestor worship is part of a belief system in which spirits of the deceased are considered to exist alongside the living, and is related to the on-going repayment of 'moral debt' to parents for the sacrifices they have made in raising their children (Jellema 2007a). Practices of worship include the offering of food, water and other material goods, and have been understood as a means of connecting the living in this world with the dead in the other world (Di Gregorio and Salemink 2007). Several participants described ancestor veneration as a private and intimate practice, and many rituals are undertaken in the home.

Son left Vietnam as a refugee in 1979 and had lived in Hackney for over 20 years at the time of the research. His altar was given a prominent position in the main living room and included statues of the Buddha and pictures of his deceased parents. Son described how the ancestors are considered to return home to receive offerings from their relatives. He noted that the altar must be kept brightly lit so that the spirits can find their way home:

> The spirit wanders, the spirit comes to the house, they have to know where to go . . . when you make the offering they come back and sit in the picture . . . then they start to enjoy what you offer them . . . you offer the incense stick, you tell them who you are and you ask them for whatever you want – for a better career, to find a partner, whatever you want them to bless you to have.

Several participants noted the importance of keeping the altar clean and regularly replenishing the offerings as a mark of respect to the ancestors. As he described his altar, Son admonished himself for not keeping it clean and for using dried flowers instead of fresh ones, frequently pointing out differences between the placement of his altar and how it would be positioned in a Vietnamese home. Despite these challenges, Son described the altar as the focal point of his home and a valuable connection to his ancestors and to the Buddha. However, several other participants were unable to house an altar because of a lack of space in their rented accommodation. Many younger interviewees shared flats and sometimes rooms in East London and changed their accommodation frequently, making it difficult to maintain an altar. Indeed, migration impacts upon participants' religious practices in multiple ways, including through the effects of constraints on space, differences in housing design, access to places of worship and living costs. However, Ngọc, a Vietnamese student who lived in a shared house near Mile End, had adapted the material culture and rituals surrounding ancestor worship to her East London home. Her shrine resembled a traditional altar, but was focused upon a depiction of the Buddha and did not venerate any of Ngọc's ancestors. It did not include any pictures or shelves that were fixed to the wall, making it more practical for living in rented housing. Ngọc explained that she consulted her mother for advice on the practicalities and demands of worship in her new location:

> She said, well, basically, when I was asking her, do we need that? As in, do we need to have a picture of my grandfather or something? And then she said that it's too complicated for you, so instead you can have a picture of Buddha, it's a good blessing for the house.
>
> (Ngọc)

Like Vietnam, the Brazilian religious landscape is marked by diversity. Despite being the world's largest Catholic country, a vast array of other religions and spiritual movements have emerged and proliferated in recent decades that have challenged the hegemony of the Catholic Church as well as fostering change – new 'creolisations' (Rocha 2006) – within it. These religious shifts are also bound up with processes of migration and mobility, making Brazil an important node in diasporic religious networks (Rocha and Vásquez 2013). Among Brazilians in London, religious practices both reflect the dynamic religious field of Brazil, as well as demonstrate distinct and innovative ways of practising religion in a new and unfamiliar environment. Within such a context, Brazilian migrant homes became important sites for these complex religious and spiritual practices, which often reveal affiliation to major religious denominations as well as creativity and adaptability in response to the challenges of migration. Several interviewees had domestic shrines, crucifixes and other devotional objects in their homes, which, whilst not necessarily regarded as creating links to ancestors, were emphasised as providing spiritual support as well as enabling ties with people and places back in Brazil.

Vera, for instance, a practising Catholic, lived in a small flat in East London with nine other people, including her son-in-law, granddaughter and other family members. As well as working as a cleaner with shifts in the early morning and late at night, Vera worked from home as a manicurist and beautician, and as a child-minder. The flat was thus a hub of activity: Vera's clients would come to have their nails painted or legs waxed, while others would drop off their children for her to look after. At the same time, the other inhabitants of the flat would come and go, often to eat or sleep before heading out to their next workplace. Yet within this unadorned space in which the often mundane, practical realities of everyday life were carried out, there was a small shrine to the *Nossa Senhora de Aparecida*, the Patron Saint of Brazil, and to Jesus, positioned on a shelf in the kitchen. As well as offering everyday protection and being a symbol of God's presence, this small shrine took on great significance when Vera's husband was run over by a car and suffered severe injuries.

Looking at the shrine, Vera explained how it was thanks to God that he had not been killed, and how as a family their faith had helped them get through this difficult time. Ivone, aged 25, who had lived in London since she was 14, had recently purchased a flat with her husband on the outskirts of the city. She explained how important it was to bless their home 'just asking God to protect, bless and keep this house' as well as 'when something is important to us.' Eventually they had it officially blessed by a Priest with a prayer called 'Minor Exorcisms' which, she explained, 'frees that house from any bad and makes the house God's property.' Yet unlike her home in Brazil where she and her family had had a room in the house dedicated for prayer, where they had 'photos, some small sculptures, rosaries, candles and the Bible', Ivone and her husband didn't have such a room or even an altar as they did not have space in their London flat. She talked about their plans to rearrange furniture and redecorate so as to incorporate space for more 'religious signs like photos or sculptures'. Enrique, aged 30, didn't see himself as belonging to a particular religious affiliation but talked about his 'personal faith' which, he said, didn't require him to go to a specific church or temple, but could be practised in the space of his home. Enrique explained how he would often sit in silence, light a candle and feel a sense of comfort, as well as a connection to absent people and places.

These examples demonstrate not only the links between religious objects, memories and lives pre-migration (Tolia-Kelly 2004), but also highlight the adaptability and mobility of religion and spirituality and how they can be shaped to respond to shifting needs and demands in new environments. Domestic shrines are also examples of the transnational movement and circulation of spiritual objects. Items used in worship, such as Buddha figures and incense holders, could be purchased from Vietnamese shops in London which import the objects from Vietnam. Other people obtained religious objects on visits to Vietnam or had them brought to London by visiting relatives. Buddhist objects, incense holders and other items were also obtained from temples around the city. Son noted, however, that a Buddha obtained from the temple must be blessed by a monk before it enters the home.

Relationships between the Vietnamese home and the wider world become particularly apparent during *Tết*, the Lunar New Year festival, which is seen as a time to renew social bonds and to prepare for the coming year. Many of the rituals involved take place within the domestic space, and include the cleansing of the home as families prepare to welcome living and dead relatives (McAllister 2012; Jellema 2007a). These activities also link the home to other sites in the city, including markets, shops and pagodas. McAllister (2012) draws upon the Foucauldian concept of heterotopia to theorise the Vietnamese home as a place that connects to other sites through the activities of its inhabitants:

> The home connects to the heavens, to an ancestral graveyard, to the ancestors in the underworld or in heaven, to pagodas or churches, and to other sites associated with the supernatural, in various ways – spatially, temporally, materially and spiritually.
>
> (McAllister 2012: 120)

Practices involved in ancestor veneration are closely tied to particular notions of home. Ancestor worship summons ancestors and the living to return to their 'native homeland' (*quê hương*), the ancestral home that is associated with birth and childhood. This highlights the complex relationships between home and mobility, presence and absence. While Vietnamese people overseas are urged to return to the homeland to venerate the ancestors, Jellema (2007b) argues that these returns are accepted as temporary and occasional, describing a flexible 'coming and going' relationship with the nation. Migrants' practices of worship also link the Vietnamese migrant home with the wider world. Minh described how she includes her current location and Vietnam in her prayers, connecting her London home with broader ideas of home and identity on a transnational scale.

In some cases, the home becomes a significant site of migrant religious practice for practical reasons. Places of worship may be located far from people's homes or workplaces, and their worship may be affected by the demands of work and family life or the absence of religious services. Brazilian and Vietnamese participants demonstrated how they adapt their practices to the new environment, leading to a reconfiguration in practices of worship over space and time. Several individuals described how the act of worship provided a sense of comfort and guidance, helping them to establish a sense of home in their new environment. However, this relationship also placed practical and emotional demands upon worshippers, particularly among those in temporary or transient housing conditions. In addition to the religious significance of objects located within the home, several participants referred to spiritual objects that they would carry with them in their everyday lives. In the next section, we examine the portability of these objects and how they enable new connections between domestic, urban and spiritual worlds.

Connecting home, city and worlds beyond: portable and virtual spiritualities

Participants' narratives emphasised the ways in which spiritual practices extend beyond the home and how the meanings of objects can be altered in contexts of mobility. Ty, a Vietnamese student in London, described the emotional and spiritual significance of a Buddhist amulet given to him by his mother for protection on his journey from Vietnam. Alongside its protective capacities, Ty cherished the amulet because it was a tangible reminder of his home and family. Ty kept the amulet in his wallet and carried it with him around the city, enabling a sense of spiritual presence as he navigated urban space. Fernando, an undocumented migrant from Brazil, also showed an amulet that he carried with him in order to feel 'a bit protected' in what could be an unwelcoming environment. Lucia, who attended a Brazilian Catholic church in London, carried an image of the Portuguese 'Our Lady of Fatima' in her purse, which her sister had bought her from Portugal, along with one of Jesus. She explained how she would look at them and touch them when she was feeling 'alone, or homesick', describing how these images reminded her of her faith as well as her sister who now lives back in Brazil.

Similarly, Ivone carried in her purse an image of Jesus and a prayer, which she drew comfort from during her journeys around the city. Yet what seemed more significant for Ivone were the various religious apps that she had downloaded onto her phone, such as the Bible, the liturgy and the Eucharist, and various prayers. She explained how they allowed her to study religion, as well as to pray, when she was travelling on the underground or at home. Minh, a Vietnamese woman studying in London, used a mobile app based on the lunar calendar that would help her determine the optimal day or time for particular life events, such as travelling or planning a wedding. These examples demonstrate the portability and adaptability of migrant everyday spiritual practices and the ways in which religious objects take on particular significance while navigating the challenges of urban life. Furthermore, the portability of many religious objects and practices enables worshippers who migrate to feel a sense of belonging as part of transnational religious networks (Hüwelmeier and Krause 2010).

Religion's adaptability is increasingly incorporating new technologies, forging virtual as well as physical connections. Several scholars have explored migrants' use of the Internet to access religion through the use of online chat rooms, prayer groups or the live broadcasting of religious services (Levitt 2007; Oosterbaan 2011; Sheringham 2013). Indeed, for many participants who were unable to attend services because of their working hours, the Internet was a crucial resource for allowing participation in a religious community – either in London or transnationally – and the practice or experience of religion in the home. Another example of the intersections of the virtual and the religious came in the form of a Brazilian soap opera – or *telenovela* – called 'The Ten Commandments' that many Brazilian families watched.[2] As one participant explained: 'The TV channel had done many other religious series in the past, different stories from the different characters in the Bible. But for some reason this one was special. Maybe because it was about Moses and his journey to become the leader and take people to the Promised Land.' Yet it wasn't just the religious content that was important for Brazilians living in London. Rather, for Ivone, it was the one time in the week when she and her family came together without fail and all sat down together at home, huddled around the TV to watch this important programme. In this sense the spiritual, the virtual, the domestic and the familial are all intertwined, the home becoming a space of connection with people who are present and absent, as well as a sense of connection with a wider religious community.

For many Brazilian and Vietnamese migrants, spiritual practices are not only associated with the home, family and domesticity, but are also connected to the domains of work, business and urban public life. Thus, for example, altars to the God of Fortune (who is associated with prosperity and success) are widely found in Vietnamese restaurants, shops and nail salons in London, illustrating further connections between spirituality, home and the city. Several Brazilians also talked about the connections between their spirituality and their economic successes in London and, subsequently back home. As one return migrant exclaimed as she showed her newly built house in Brazil: 'this was possible thanks to London and to God' (Anete).

The city figures throughout our research as both enabling and constraining migrants' spiritual practices. While some religious objects and activities are mobile and adaptable, those that require a large public space or visible presence in the city can be more difficult to sustain transnationally. Son, for example, explained that many Vietnamese Buddhists would prefer to have a temple located in Hackney, but the community faced barriers including the cost of land and obtaining planning permission. While some participants visited temples elsewhere in London and beyond, the need for transport to and from the temple posed a barrier to regular public worship. Son also described how the Linh Son Buddhist temple in South London had been the target of complaints by some local residents, who objected to parking problems and the sounds of early morning ceremonies. In contrast to the Vietnamese Catholic community, who have a large congregation in a church in East London, Buddhist worshippers must travel further afield. Several Brazilian and Vietnamese participants said that the timings of religious services were incompatible with their working hours. In this context, worship in the home and on the move can be understood as practical as well as spiritually-based practices that are influenced by the challenges of everyday life in the city.

Conclusion

This chapter has demonstrated the significance of everyday spiritual practices and spaces within and beyond the migrant home, offering new understandings of religion and spirituality in contexts of urban migration and mobility. We have explored the multiple ways that migrants draw upon religious and spiritual objects as resources for navigating the challenges of everyday life in the city, in maintaining relationships across transnational space and time and in building a sense of home and belonging. These objects operate at multiple scales, from the individual body to the dwelling, the workplace and the city, as well as circulating within transnational networks. Yet the chapter also uncovers the challenges of maintaining spiritual practices in contexts of migration and urban super-diversity. Religious practices are reconfigured in movement and are frequently adapted to the new location. People, of course, also face barriers to mobility and settlement that will impact upon their spiritual practices. Participants in this research develop pragmatic and creative solutions to the challenges of practicing their religion or spiritual beliefs in the city, yet these practices cannot be separated from the structural inequalities and power relations involved in housing, work, immigration status and (im)mobility.

This chapter contributes to geographies of spirituality in broadening spatial perspectives on everyday religion to include the spiritual world alongside homes, workplaces and public spaces in the city as sites of spiritual practice and significance. Exploring the place of spirit within contexts of modernity, Bartolini *et al.* (2017) argue that 'we need to rethink the lines drawn between the secular and the religious' towards 'those sites and spaces where the fuzzy and fluid boundaries between superstition, religion and modernity are evident . . . where new forms of modern spirituality are being created.' In addition to examining how migrants adapt their spiritual practices to the urban environment, this chapter demonstrates how everyday spiritualities unsettle the boundaries between the home, the city and worlds beyond. We emphasise the importance of locality and migrant home-making alongside transnational networks, showing how migrants' religious practices can contribute to the shrinking of space between home and (imagined or remembered) homeland.

Migrants' everyday spiritual practices not only contribute to understandings of transnational religion, but also speak to broader debates in geographies of home and migration. The home itself becomes a site of spatial and temporal connection with people and places in Brazil and Vietnam, for example, as well as deities and the spirit world. Spiritual practices are important ways in which migrants generate a sense of home in the city, yet these practices are constrained by the spatial and economic conditions of urban life. Understanding religion in contexts of mobility requires us to explore how practices are lived across borders, as well as the power relations that come into contact with ideas, practices and people as they move.

Notes

1 This chapter draws upon empirical data taken from two research projects, both of which incorporated in-depth interviews and ethnographic research. Olivia Sheringham's study explored transnational religious practices among Brazilians in London and on their return to Brazil, and was based on multi-sited ethnographic fieldwork in London and Brazil (Sheringham 2013). Annabelle Wilkins' research examined home and work among Vietnamese migrants in East London, drawing upon interviews, photography and ethnographic research with Vietnamese people who migrated to London in diverse circumstances, including individuals who arrived as refugees following the Vietnam War, as well as participants who migrated for work or education in recent years (Wilkins 2016).

2 'Os Dez Mandamentos' is a Brazilian telenovela that was produced and broadcast by the TV channel *Rede Record* in 2015.

References

Ammerman, N. (ed., 2007) *Everyday Religion: Observing Modern Religious Lives*. Oxford: Oxford University Press.

Bartolini, N., Chris, R., MacKian, S. and Pile, S. (2017) 'The place of spirit: modernity and the geographies of spirituality'. *Progress in Human Geography*, first published April 26, 2016.

Beaumont, J. R. (2008) 'Faith action on urban social issues'. *Urban Studies* 45(10), pp. 2019–2034.

Blunt, A. (2005) 'Cultural geographies of home'. *Progress in Human Geography* 29(4), pp. 505–515.

Blunt, A. and Bonnerjee, J. (2013) 'Home, city and diaspora: Anglo-Indian and Chinese attachments to Calcutta'. *Global Networks* 13(2), pp. 220–240.

Blunt, A. and Dowling, R. (2006) *Home*. London: Routledge.

Blunt, A. and Sheringham, O. (2015) 'Domestic urbanism and urban domesticities in East London'. Paper presented at Migrants in the City conference, Sheffield, October 2015.

Brickell, K. (2012) '"Mapping" and "doing": critical geographies of home'. *Progress in Human Geography* 36(2), pp. 225–244.

Burrell, K. (2014a) 'Spilling over from the street: contextualising domestic space in an inner-city neighborhood'. *Home Cultures* 11(2), pp. 145–166.

Burrell, K. (2014b) 'The objects of Christmas: the politics of festive materiality in the lives of Polish immigrants', in M. Svašek (ed.) *Moving Subjects, Moving Objects: Transnationalism, Cultural Production and Emotions*. Oxford: Berghahn, pp. 55–74.

Chan, Y.W. (2011) 'Revisiting the Vietnamese refugee era: an Asian perspective from Hong Kong', in Y.W. Chan (ed.) *The Chinese/Vietnamese Diaspora: Revisiting the Boat People*. Abingdon: Routledge, pp. 3–19.

Di Gregorio, M. and Salemink, O. (2007) 'Living with the dead: the politics of ritual and remembrance in contemporary Vietnam'. *Journal of Southeast Asian Studies* 38(3), pp. 433–440.

Dwyer, C. (2016) 'Why does religion matter for cultural geographers?'. *Social and Cultural Geography* 17(6), pp. 758–762.

Ebaugh, H.R. and Chafetz, J.S. (2002) *Religion Across Borders: Transnational Immigrant Networks*. Walnut Creek, CA: AltaMira Press.

Evans, Y., Dias, G., Martins Jr., A., Souza, A. and Tonhati, T. (2015) *Diversidade de Oportunitidades: Brasileiros em Londres*. Londres: GEB/Goldsmiths/Queen Mary/Oxford Brookes.

Evans, Y., Tonhati, T., Dias, G., Brightwell, M.G., Sheringham, O., Souza, A. and Souza, C. (2011) *For a Better Life: Brazilians in London*. London: GEB/QMUL/RH/Goldsmith's. www.geog.qmul.ac.uk/docs/staff/87292.pdf.

Garnett, J. and Harris, A. (2013) *Rescripting Religion in the City*. Farnham: Ashgate.

Hagan, J. (2008) *Migration Miracle: Faith, Hope and the Undocumented Journey*. Cambridge, MA: Harvard University Press.

Hale, S. (1993) 'The reception and resettlement of refugees in Britain', in V.C. Robinson (ed.) *The International Refugee Crisis: British and Canadian Responses*. London: Macmillan, pp. 273–294.

Harris, C. (2001) 'The politics and personhood of Tibetan Buddhist icons', in C. Pinney and N. Thomas (eds.) *Beyond Aesthetics: Art and the Technologies of Enchantment*. Oxford: Berg, pp. 180–200.

Holloway, J. and Valins, O. (2002) 'Editorial: placing religion and spirituality in geography'. *Social and Cultural Geography* 3(1), pp. 5–9.

Hurdley, R. (2006) 'Dismantling mantelpieces: narrating identities and materializing culture in the home'. *Sociology* 40(4), pp. 717–733.

Hüwelmeier, G. and Krause, K. (2010) 'Introduction', in G. Hüwelmeier and K. Krause (eds.) *Traveling Spirits: Migrants, Markets and Mobilities*. New York: Taylor and Francis, pp. 1–16.

Jamoul, L. and Wills, J. (2008) 'Faith in politics'. *Urban Studies* 45(10), pp. 2035–2056.

Jellema, K. (2007a) 'Everywhere incense burning: remembering ancestors in Đổi Mới Vietnam'. *Journal of Southeast Asian Studies* 38(3), pp. 467–492.

Jellema, K. (2007b) 'Returning home: ancestor veneration and the nationalism of post-revolutionary Vietnam', in P. Taylor (ed.) *Modernity and Re-Enchantment: Religion in Post-Revolutionary Vietnam*. Singapore: Institute of Southeast Asian Studies, pp. 57–89.

Knott, K. (2005) *The Location of Religion: A Spatial Analysis*. London: Equinox.

Knowles, C. (2013) 'Nigerian London: re-mapping space and ethnicity in superdiverse cities'. *Ethnic and Racial Studies* 36(4), pp. 651–669.

Levitt, P. (2007) *God Needs No Passport: Immigrants and the Changing American Religious Landscape*. New York: New Press.

MacKian, S. (2012) *Everyday Spiritualities: Social and Spatial Worlds of Enchantment*. Basingstoke: Palgrave Macmillan.

Mazumdar, S. and Mazumdar, S. (2009) 'Religion, immigration and home-making in diaspora: Hindu space in Southern California'. *Journal of Environmental Psychology* 29(2), pp. 256–266.

McAllister, P. (2012) 'Connecting places: constructing *Tết*: Home, city and the making of the lunar New Year in Urban Vietnam'. *Journal of Southeast Asian Studies* 43(1), pp. 111–132.

McGuire, M. (2008) *Lived Religion: Faith and Practice in Everyday Life*. Oxford: Oxford University Press.

McIlwaine, C. (ed., 2011) *Cross-Border Migration Among Latin Americans: European Perspectives and Beyond*. New York: Palgrave Macmillan.

Miller, D. (2001) *Home Possessions: Material Culture Behind Closed Doors*. Oxford: Berg.

MRE (Ministério das Relações Exteriores) (2014) *Estimativas Populacionais das Comunidades*. www.brasileirosnomundo.itamaraty.gov.br/a-comunidade/estimativas-populacionais-das-comunidades/estimativas-populacionais-brasileiras-mundo-2014/Estimativas-RCN2014.pdf (accessed 14/09/16).

Naylor, S. and Ryan, J.R. (2002) 'The mosque in the suburbs: negotiating religion and ethnicity in South London'. *Social and Cultural Geography* 3(1), pp. 39–59.

Oosterbaan, M. (2011) 'Virtually global: online evangelical cartography', *Social Anthropology* 19(1), pp. 56–73.

Orsi, R.A. (2003) 'Is the study of lived religion irrelevant to the world we live in?' Special presidential plenary address, Society for the Scientific Study of Religion, Salt Lake City, November 2, 2002. *Journal for the Scientific Study of Religion* 42(2), pp. 119–174.

Parrott, F. (2014) 'Materiality, memories and emotions: a view on migration from a street in South London', in M. Svašek (ed.) *Moving Subjects, Moving Objects: Transnationalism, Cultural Production and Emotions*. Oxford: Berghahn, pp. 41–54.

Rocha, C. (2006) *Zen in Brazil: The Quest for Cosmopolitan Modernity*. Honolulu, HI: University of Hawaii Press.

Rocha, C. and Vásquez, M. (2013) *The Diaspora of Brazilian Religions*. Leiden: Brill.

Sheringham, O. (2010) 'Creating alternative geographies: religion, transnationalism and everyday life'. *Geography Compass* 4(11), pp. 1678–1694.

Sheringham, O. (2013) *Transnational Religious Spaces: Faith and the Brazilian Migration Experience*. Basingstoke: Palgrave Macmillan.

Sims, J.M. (2007) *The Vietnamese Community in Great Britain: Thirty Years On*. London: The Runnymede Trust.

Smith, M.P. and Eade, J.P. (eds., 2008) *Transnational Ties: Cities, Migrations, and Identities*. New Brunswick, NJ: Transaction Publishers.

Tolia-Kelly, D. (2004) 'Locating processes of identification: studying the precipitates of re-memory through artefacts in the British Asian home'. *Transactions of the Institute of British Geographers* 29, pp. 314–329.

Vásquez, M. (2016) 'Religion, globalization and migration', in L. Woodhead, C. Partridge and H. Kawanami (eds.) *Religions in the Modern World.* Routledge: Abingdon, pp. 431–452.

Vásquez, M. and DeWind, J. (2014) 'Introduction to the religious lives of migrant minorities: a transnational and multi-sited perspective'. *Global Networks* 14(3), pp. 251–272.

Vásquez, M. and Knott, K. (2014) 'Three dimensions of religious place-making in diaspora'. *Global Networks* 14(3), pp. 326–347.

Vásquez, M.A. and Marquardt, M.F. (2003) *Globalizing the Sacred: Religion Across the Americas*. New Brunswick, NJ: Rutgers University Press.

Vertovec, S. (2007) 'Super-diversity and its implications'. *Ethnic and Racial Studies* 30(6), pp. 1024–1054.

Walsh, K. (2006) 'British expatriate belongings, mobile homes and transnational homing'. *Home Cultures* 3(2), pp. 119–140.

Wilkins, A. (2016) 'Home, work and migration for Vietnamese people in East London'. Unpublished PhD thesis, Queen Mary, University of London.

Wong, D. (2014) 'Time, generation and context in narratives of migrant and religious journeys'. *Global Networks* 14(3), pp. 306–325.

12 Life cycles of spirituality, religious conversion and violence in São Paulo

Kim Beecheno

Introduction

In this chapter, I explore the role of violence in the religious conversion of women to Pentecostalism[1] in Brazil. Drawing on ethnographic data and interviews with female converts from a low-income, high-crime area of São Paulo, as well as literature analysing religious conversion in the Americas (Brenneman, 2012; Brusco, 1995; Freston, 2008; Lehman, 1996; Mariz and Machado, 1997; Martin, 1993; Rostas and Droogers, 1993; Smilde, 2007; Stoll, 1990 among others), this chapter finds that some women use religious conversion and continued spiritual practice as a strategy for dealing with everyday violence and especially domestic violence.

This study employs Gooren's (2007) concept of *conversion careers*, a life-cycle approach to the examination of religious conversion, which highlights how women use various levels of religious adherence over time to deal with the violence of everyday life (Scheper-Hughes, 1993) and domestic violence. Although Pentecostalism is generally considered a patriarchal and conservative form of evangelical Protestantism, this study demonstrates that some women feel empowered by their conversion and religious adherence, which allows them to create spaces of safety in which they negotiate and ultimately escape the violence they are experiencing. This also underscores a spatialized understanding of conversion, the effects of which are played out in different 'spaces', notably in the street and in the home.

Data for this chapter was collected in the low-income, periphery city of Mauá, São Paulo metropolitan region, much of which is *favela* (slums). It suffers from high rates of urban violence, including 10.4 homicides per 100,000 inhabitants, considered epidemic levels by the United Nations Office on Drugs and Crime (UNODC) (Waiselfisz, 2012a). In addition, statistics reveal high levels of robbery, car theft, unemployment and drug abuse, particularly in the form of crack cocaine (Waiselfisz, 2012a). While São Paulo is the 26th most dangerous state for women with 3.1 femicides per 100,000 inhabitants, statistics show that these numbers are unequally distributed, with metropolitan areas including Mauá reaching femicide rates above 10 per 100,000 (Waiselfisz, 2012b). Although statistics are hard to gain, rates of domestic violence are believed to be very high and Brazil is the 5th deadliest country in the world for women (Waiselfisz, 2015). There is a plethora of Pentecostal churches which have grown significantly in the last two decades, demonstrating the ease of access potential converts have to Pentecostalism and highlighting the tendency of these churches to grow in impoverished areas (Freston, 2008; Garmany, 2013).

I lived near Mauá for two years and returned to conduct the study over two months in April and May 2012. I had intimate knowledge of the area and the difficulties faced by its residents which allowed an entry point as well as access to contacts. I conducted in-depth semi-structured interviews[2] with female converts to Pentecostal churches from Mauá (n=15). In addition, I spoke to many people in the area such as residents, church members and leaders (n=46), and attended numerous church services while living with a family in the area. Several interviewees (n=8) worked as assistants in a local health centre earning Brazil's minimum salary of R$600 per month (around US$300) and the other women were from favela *Pedreirinha* in Mauá (n=7) with no fixed income.

The chapter is divided as follows: first, a brief look at current literature on conversion to Pentecostalism to set the scene and highlight the importance of this study. Next, I will turn to a section describing the reasons for women's conversion, in which domestic violence was found to be the overarching reason for conversion. This is followed by analysis of the subsequent effects of women's conversion from a life-cycle perspective and the way in which religious conversion and spiritual practice were used in relation to violence, before turning to the conclusion.

Religious conversion in Latin America

Despite important contributions examining the high rates of religious conversion to evangelical Protestantism in Latin America, few studies have linked it to violence. Researchers have suggested that conversion was linked to processes of US acculturation and cultural imperialism (Lehman, 1996; Martin, 1993) or even a natural effect of the 'religious economy' whereby adherents are likened to consumers in a religious market (Chesnut, 2007). Studies have often focused on conversion as a strategy for dealing with poverty (Stoll, 1990); a short-term, problem-solving strategy giving converts a sense of empowerment (Rostas and Droogers, 1993) or a form of cultural agency that allows converts to gain control over personal, economic and social aspects of their lives (Smilde, 2007). These important and varied theories demonstrate the complexities around religion and spirituality, highlighting that reasons for conversion are not isolated to specific 'sacred' spaces or moments (Garmany, 2013).

Conversion has sometimes been considered within the context of the high rates of urban violence that plague the region, leading Brenneman (2012) to describe religious conversion as an exit strategy for men from drug gangs. Goldstein (2003) took a different stance, calling it a gendered form of oppositional culture for women against gang membership and participation in urban violence. Several authors pointed out that the conservative dress converts wear sends out a visual message that they are not part of the violence around them (Abi-Eçab,

2011; Goldstein, 2003). Garmany (2013) suggested that in low-income areas evangelical pedestrians help to break down spatial barriers induced by fear of public spaces at night. These theories also demonstrate the spatial effects of conversion, which in these cases are played out in public spaces, in the city and in the street. This is pertinent in the context of Brazil, where conversion levels have increased concurrently with mounting levels of urban and interpersonal violence: evangelical Protestants increased from 6.6 per cent of the population in 1980, to 20.2 per cent of the population in 2010. Over the same period, there were more than a million homicides in Brazil, which is an average of around 36,000 deaths a year (Waiselfisz, 2012a).

However, statistics demonstrate that while Brazilian men are by far the greatest victims of homicide, women are the overwhelming victims of domestic violence.[3] A national phone line set up for victims of domestic violence receives around 175 calls a day, and data suggests that a woman is beaten every two minutes and one woman is killed every 1.5 hours in Brazil (Agencia Patricia Galvão, 2016). Domestic violence is committed mainly against women or children by an intimate partner or family member and occurs predominantly, although not exclusively, in the home. In Brazil, domestic violence includes physical and sexual violence as well as verbal and psychological violence such as swearing, threatening or humiliating someone (ibid). Therefore, for millions of Brazilian women, the spatial distinctions of home/safety, street/danger, are not applicable. Conversion theories analysing urban violence may therefore be less pertinent for women than studies exploring the role of domestic violence. This underscores the importance of this study, which found domestic violence to be the overarching reason for conversion, although other forms of violence were also found to exist.

Not often considered in conversion theories are the intertwining effects of *everyday violence* (Scheper-Hughes, 1993). This includes structural violence, e.g. Brazil's historical, political and economic oppression, creating significant socio-economic inequality, as well as institutional violence, e.g. created by agents acting on behalf of the State, such as the police and those who oppose its authority, such as armed gangs (Scheper-Hughes, 1993). Everyday violence is also the normalisation of interpersonal aggression in communities and individually lived experiences. This includes drug abuse, delinquency, domestic and sexual abuse. These forms of violence lead to the creation of an ethos of violence (Scheper-Hughes, 1993) and a culture of fear (Kruijt, 2001).

One of the few authors to link conversion and domestic violence, Burdick (1996) suggested that Pentecostalism and Umbanda or Candomblé, Afro-Brazilian Spiritist religions, would be more appealing than Catholicism to women seeking help for domestic conflict. This is because they are 'cults of affliction' (153) whose clienteles are drawn through the experience of suffering. These cults of affliction created spaces of social privacy, where blame for conflict could be safely articulated and projected onto spiritual 'others' (Burdick, 1996). It has also been suggested that male conversion elevates domesticity by limiting traditional aspects of 'macho' male behaviour, such as drinking, smoking, gambling and extramarital relations (Mariz and Machado, 1997; Drogus, 1997; Brusco, 1995).

This led Brusco (1995) to call evangelical conversion a 'strategic women's movement, like Western feminism, because it serves to reform gender roles in a way that enhances female status' (1995: 6).

These theories demonstrate clear benefits for conversion although they fail to highlight how women negotiate everyday violence through religious conversion and spiritual practice. In addition, the concept of 'conversion' suggests a one-time event. I believe that a temporal analysis of women's conversion and continued religious practice is important for understanding why and how it could be used as a strategy for dealing with violence, particularly within the private space of the home and within intimate relationships. Using a temporal lens helps to explain how women engage in the church in different ways and why they leave if they do not find an answer.

Therefore, I employ Gooren's (2007) concept of *conversion careers* to analyse women's conversion and continued use of religion and spirituality in relation to violence, using a life-cycle framework. Gooren (2007) identified five levels of higher or lower religious participation during a person's life, the first of which is *pre-affiliation*, when a potential member makes their first contact to see if they would like to affiliate themselves on a more formal basis. All the women interviewed in this study went through this step as they sought a solution to their problems. The second step is *affiliation*, where the person refers to being a formal member of a religious group, but the membership does not form a central aspect of one's life or identity. The next step is *conversion*, which all the women interviewed achieved, referring to a radical change of worldview, and attribution of religious identity not only from members of the church, but also non-members outside the group. The fourth level identified is that of *confession*, a theological term for core member identity, denoting a high level of participation within the church and a missionary attitude towards non-members, a position that many of the women achieved within their respective churches. Finally, *disaffiliation* refers to those who have rejected membership of the church *or* inactive members who still self-identify as believers (Gooren, 2007), examples of both of which were found in this study. This theory is helpful for analysing religious adherence particularly in relation to violence, as Brazil's high levels of religious syncretism mean that conversion from one religion to another and often subsequent disaffiliation is relatively common (ibid). The following section will address women's initial reasons for conversion before turning to the consequences of their conversion from a life-cycle perspective.

What were the women's reasons for converting to Pentecostalism?

A variety of push and pull factors explain the reasons the women in this study had for converting to Pentecostalism. However, this study found domestic violence to be the overarching reason for women's conversion. All the women turned to the church during a moment of crisis in their lives, directly or indirectly affected by alcoholism and drug abuse, domestic violence, depression, illness or bereavement.

Several elements attracted them to convert: a sense of community, friendship, advice from the pastor, a safe space in which to verbalise problems, miracle cures, divine intervention and predictive visions from other members.

Conversion to Pentecostalism was offered as the solution to all their problems, and they were told it would give them a personal relationship with God once they had accepted Jesus as their saviour, allowing the Holy Spirit to act in their lives. Pentecostal churches are therefore postulating their services as a one-stop-shop for the kind of poverty-related issues common to periphery urban areas and favelas, suggesting a lack of institutional or state-run services, which churches have been quick to fill. None of the women interviewed converted at a time when they were happy or when their lives were going well, hence conversion to Pentecostalism can be seen as a survival strategy (Rostas and Droogers, 1993; Smilde, 2007; Stoll, 1990).

The violence of everyday life is often exacerbated by poverty. In Mauá and Favela Pedreirinha in particular, the violence of everyday life is played out through the indignity of living in makeshift housing inadequate for human living, a lack of basic services from the State and vulnerability to urban violence. One of the interviewees, Jacinta, 48 years old, suffered from these forms of everyday violence and had converted to Pentecostalism two years previously. Her husband was in jail, she had 6 children, the youngest one had Down's syndrome, and she had recently discovered that her 15-year-old daughter was pregnant and that her 13-year-old daughter was addicted to crack. Jacinta felt that no one but the Pentecostal church had offered any form of support, signifying that the church offers services that the State should be providing.

Eight out of the 15 women cited their father or husband's alcoholism and domestic violence as the major contributing factors to their conversion. This suggests that the women were all looking for solace from a problem in their lives. Several of the women reported being on the verge of leaving their husbands due to the alcoholism and related problems, and were therefore looking for a practical answer to a serious problem in their lives. In addition, 5 out of the 15 women cited their own, their husband's or family member's drug addiction as a major contributor to their conversion. Drug addiction led to conversion for several of the women, whether it was the women's own drug addiction, or a close family member's. Therefore, domestic abuse in the form of physical violence from a spouse or family member often due to his alcoholism or drug addiction proved to be the dominant form of violence affecting the women.

Stories of urban violence were noticeably lacking. However, during an interview with Laura, she mentioned that her brother-in-law had been shot in front of her house. When questioned about whether she found the area dangerous to live in, Laura replied:

> I don't think it's very violent around here. We've been burgled but it wasn't by people from around here. The same with my brother-in-law, he was just unlucky, that bullet wasn't meant for him, it was meant for someone else.

Her narrative suggests that there is a high tolerance to urban violence which has become normalised due to its pervasiveness in their everyday lives. According to Valentine (1989) this 'othering' of violence – locating it in 'other' people and 'other' places – is a coping strategy which allows women to operate in a climate of fear and maintain some level of control over their lives by attempting to minimise risk, because they cannot be afraid of all men all of the time (1989: 171).

What are the life-cycle consequences of the women's involvement in a Pentecostal church and how do they vary?

Violence, intimate power relations and indeed, religious or spiritual practice themselves do not necessarily remain at the same level or intensity over time. Gooren's (2007) theory of conversion careers and a life-cycle approach to women's conversion stories is useful in analysing the temporal outcomes of women's use of religion and spirituality throughout their lives, particularly in relation to the violence they experienced. The following section analyses women's life-cycle conversion stories with levels of spirituality described by Gooren (2007) above.

Maintaining the confession level of religious conversion

Some women consciously married a Pentecostal and brought their children up within the church in order to protect themselves and their family from urban and domestic violence. Sara and Flora, currently aged 39 and 28, were the women who had converted youngest in this study, aged 6 and 17 respectively. They had both converted due to their father's alcoholism, which had negative impacts on the family. Now married with children, they both remained very active within the Assembly of God church, attending at least two or three times a week with their families. Both reported a conscious awareness of wanting to marry a man with the same religious beliefs they had, and a strong determination to bring up their children within the church.

> There is danger everywhere in Brazil, but here it's getting worse. There are so many drugs. Here we raise our children inside the home and don't let them play outside, because we're scared they'll fall into it [drug taking]. In the job we do [health assistants] we know about a lot of abuse, especially child abuse. The Evangelical church gives you a structure and that's how we want to raise our children, so they can learn to make the right choices and be as safe as possible.
>
> (Sara)

This narrative demonstrates that in some cases, Pentecostal women within the church are keen to marry men who are Pentecostal themselves, in order to maintain the tight, family structure the church emphasises so heavily. It also reveals an awareness of violence in the form of child abuse in Mauá. These women show that core member identity is seen as a way of protecting themselves from both urban and domestic violence. This also demonstrates that conversion is spatialized, as the women felt more protected through conversion, both in the private space of the home and in the street.

Outside of the church, several women developed a new role for themselves through evangelising and proselytising in their neighbourhood (Drogus, 1997; Mariz and Machado, 1997). For all the women interviewed, conversion to Pentecostalism gave them a renewed sense of self-esteem from which they drew strength and happiness. Camila, now 47, converted aged 20 after her daughter's tragic death and developed a new role for herself evangelising in the local area. She claimed that other people sought her out for advice, which was an obvious source of pride. Similarly, Julia, 47, who converted aged 40, took on a leading role evangelising in prisons and orphanages which she found very fulfilling.

Ana's husband passed away but she felt moved to rent out a space in front of her home and open a small branch of her preferred church, Renewed Presbyterian. The small, but growing group of members hired a pastor and evangelised in the local area to bring in more members. These examples suggest that conversion allowed the women a voice in the public sphere they hadn't had before (Birman, 2007), therefore pointing to social empowerment (Stoll, 1990, Smilde, 2007), and challenging traditional gender roles with the women finding roles for themselves outside the home.

Roberta, 53, who converted at the age of 40 because three successive husbands had left her, found that marriage counselling from the pastor helped her and her husband so the whole family converted. She saw it as her responsibility to pray for her neighbours and evangelise in the local area.

> You have to be careful around here at night, there are a lot of drugs and people are dangerous. The problem is crack, there's a lot of it here, I pray for my neighbours and their children who aren't Evangelical, spread The Word, and thank God that my son isn't doing drugs.

Roberta felt it was her duty to evangelise since it had given her a role in life, as it had to a greater or lesser extent in many of the women's lives. However, her testimony also highlights awareness of local urban violence and suggests that she and her family felt more protected from that violence due to their conversion. Marital counselling had solved her immediate problem of domestic strife, but the sense of safety derived from the whole family's conversion partly explains the continued attraction for remaining within the church. These examples show women who have remained at the highest *confession* level of religious affiliation, demonstrating that this intense level of religious participation, taking place in the church, in the home and in the street, is necessary for them to maintain the sense of safety from everyday violence that they have gained.

Empowerment

Many women felt empowered through their conversion and that translated into empowerment within their relationships, leading to elements of change in socially constructed gender roles. Eight out of the 15 women who converted to Pentecostalism reported improvements in their marriages and in all but two of these situations, the husbands also converted. Varlene, who converted with her husband due to his alcoholism and abuse, demonstrated the ability to speak in tongue. An obvious source of pride and admiration within the family, this could be seen as a form of empowerment through religion, elevating her importance and respect within the family and in church. At the same time, her husband's focus returned to the family, stopping the highly negative characteristics of machismo he had previously displayed through drinking and womanising. This is what Brusco (1995) called the domestication of men, which occurred concurrently with Varlene's own growth in self-esteem through her role in the church. Varlene's conversion changed the balance of power within the home and she no longer suffered from domestic violence. In this case, conversion and its effects were played out in the church and in the home.

It is important to note that the Pentecostal church does not set out to empower women or change socially constructed gender roles, but its family focus and asceticism create the realignment of a family's goals (Brusco, 1995). As these goals are family-orientated, they can be considered more feminine goals, making Pentecostalism a feminine, although not feminist, religion (ibid). In a similar way, Maria-Claudia's conversion and the important leadership role she developed within the Assembly of God church gave her strength after her alcoholic and abusive husband left her. She opened a little shop to sustain the family and was supported emotionally by the church and visions from members that he would return. Eventually, her husband did indeed return and seeing the whole family had converted, he did too. According to Maria-Claudia, conversion therefore reunited the family, whose economic situation improved due to her entrepreneurialism and the fact that the husband was no longer drinking away the family's income. The couple were both highly active within the church and Maria-Claudia was particularly respected for her gift of visions and premonitions.

Forgiveness of sins is a strong theme in Pentecostalism, as is the sacredness of marriage. Together with the social status many women feel they gain once married, Maria-Claudia was keen for her husband to return, despite the domestic violence she had suffered. In this case, however, the return and conversion of a wayward husband, her role as the main breadwinner and the maintenance of her conversion at the highest confession level, allowed for a change in power relations within the relationship which protected her from further violence, especially within the home.

Similarly, Yolanda's conversion 15 years ago, aged 20, led to her husband's conversion and the creation of new roles for the couple as leading religious figures in the community. He became a pastor of 5 Assembly of God churches in Mauá. By encouraging her husband's conversion, Yolanda's conversion allowed her to escape the everyday violence she had been experiencing, in this case high levels of domestic violence while her husband had been an alcoholic. Although the socially constructed roles of patriarchal/pastor husband and submissive/helper wife were still present, Yolanda was empowered by her husband's domestication through his rejection of his previous life of drink and drugs. There was a growth in equality between the sexes within the relationship. If the empowerment was not the kind expected by Western feminist standards, within the context of economically poor and socially disenfranchised women, this level of empowerment represents a significant, positive change.

Conversion and problem solving

Conversion to Pentecostalism can also be used as a problem-solving technique (Rostas and Droogers, 1993). When Laura's womanising husband began drinking and staying out late with his friends, instead of arguing with him, she prayed as she had learned to do at church and said that God told her she needed to find a job, having previously been a housewife.

> I used to be so jealous, my God, to an extreme! It was ruining our marriage, I was even jealous of his mother and sisters! The church cured me of my jealousy, and when I got the job, that made him start to think more about me. . . . I'm still an attractive woman you know, and suddenly I had all these male colleagues and I was in the street, not just at home. . . . It made him a bit insecure and he began to desire me again!

Her new job's working hours meant that her husband had to be at home in the evenings to look after the children, but the extra money relieved his burden as the sole breadwinner and Laura's confidence grew because she was more occupied and earning money independently. The result was that her husband stopped his late-night drinking and staying out with friends, and the family's focus was realigned to achieving common goals. The variation in this situation was that her husband did not convert, but the balance of power within the family became more equal, improving family life. Laura displayed signs of conversion which were all played out in church, in the home and in public spaces, but did not display the missionary zeal signs of confession the other women had attained.

This suggests that the level of confession may be more common in women whose husbands have also converted, where family life revolves solely around church life. It also suggests that conversion to Pentecostalism at 'conversion' level solved Laura's problem, so she had no need to go up to confession level. This is unlike the women in the previous examples who gained and maintained their growth in female power by reaching confession level.

Conversion and separation

In some cases, the women's sense of empowerment through conversion led to separation. Marcela had lived most of her life in favela Pedreirinha. She had been evangelised in her 30s through the radio – media has been proven to be an important vehicle in attracting converts (Oosterbaan, 2006). Marcela admitted that before converting, she would give most of her monthly salary to a local shaman, who cast spells on her husband, to make him stop drinking and beating her.

Marcela initially turned to Afro-Brazilian spirit religions, and then Pentecostalism, as a potential solution to domestic violence. Pentecostal churches strongly denounce the 'black magic' of Afro-Spiritist religions, which they see as the Devil's work (Birman, 2007). In addition, the church emphasises female submission to husbands and the importance of prayer in cases of marital conflict. This would allegedly help their partner be released from the devil by whom he was possessed:

> I realised that I hadn't been a good wife and that I must be more obedient. I went back home and my husband couldn't understand what had happened to me, I sat on his knee and wept for forgiveness.
>
> (Marcela)

Told, essentially, that the domestic violence Marcela suffered was her own fault due to her use of shamans and lack of consistent prayer for her husband, Marcela claimed that their domestic situation improved because she learned to be less argumentative with her husband. Marcela learned that changing her own behaviour in the home could improve her husband's behaviour, although the change entailed a greater level of submissiveness.

Over time though, the situation worsened, as Marcela's husband refused to convert and continued drinking, which meant that the beatings continued. However, Marcela claims that the teachings of the church made her a calmer person, allowing her to finally see the need for separation. She firmly stated, 'if I hadn't converted, I think we would have killed each other', admitting that during the violent outbreaks, she too fought back as hard as she could. Marcela maintained the 'conversion' level in Gooren's (2007) conversion careers. Maintaining conversion level ultimately resolved the domestic violence she was suffering and helped her through life emotionally and spiritually. Now, living alone and unable to attend church services due to her health and fear of street crime, Marcela admitted, 'I invite Jesus to come and lie down next me in bed, that way I am never alone'. Marcela's religion was obviously a source of comfort throughout her life which brought her solace in different ways and in different moments.

Disaffiliation

The following examples demonstrate disaffiliation with the church, which happened over time once the main motive for conversion had been resolved. These examples also show that the high levels of discipline, moral asceticism and time dedicated to the church, which led to the women's conversion, also led to their disaffiliation.

Carla, now 38, had converted at the age of 26 because of her and her husband's addiction to crack, and found the courage to separate from her husband, a year or so after converting. As with Marcela, Carla's husband's refusal to convert and change his negative habits after her conversion led to their separation and her escape from violence. Carla quickly attained confession level of religious affiliation, evangelising in the local neighbourhood. However, she later met another man who moved into her family home. As Carla still wasn't officially divorced from her husband, in the eyes of the church she was living in sin with another man. This highlights the fact that even though empowerment is sometimes gained from the teachings of the Pentecostal church, it does not aim to change traditional social roles and holds very conservative views on marriage.

Carla was still allowed to attend services, but she was banned from performing any leadership duties due to her family situation. The church therefore demoted Carla from confession to affiliation or even disaffiliation level. Carla still identified as Pentecostal but did not go to church very often as a result. This demonstrates religious intolerance for a family that no longer fitted the married husband and wife mould, despite Carla overcoming her addiction and finding a more suitable man. This situation is unlikely to be unique to this case study in Mauá, and could therefore indicate one of the reasons for the Pentecostal church's equally high drop-out rates.

Teresa, now 29, chose to leave the church of her own accord. She had converted aged 22, in order to remove herself from an abusive relationship and an addiction to cocaine. Over the years, Teresa worked as a leader in the church and as a missionary in favelas around São Paulo, trying to convert other addicts. But a year ago she suddenly left the church:

> It was due to problems at work, stress and too much pressure. I regret leaving because of that, it's not that I couldn't return, I could, but it's up to me and I want to be selfish, I want to do my own thing, I want to make the most of things and have fun. I wanted to live something new and threw everything up in the air.

Teresa had also reached confession level, but this time the high levels of discipline and morality as well as the personal time she sacrificed made her decide to drop out. The fact that Teresa was young, single and had friends who were not Pentecostal obviously influenced her desire to change from such an abstemious lifestyle, demonstrating how social factors are important in conversion and continued religious participation.

Conclusion

The study finds that the women interviewed in Mauá were using religious conversion and different levels of religious adherence and spiritual practice in order to negotiate everyday violence and in particular, domestic violence. It extends current theories on evangelical Protestant conversion in the Americas by highlighting a clear link between conversion and domestic violence. This study demonstrates that conversion can help women escape violence, and shows how women do so by employing Gooren's (2007) concept of conversion careers, examining violence and conversion from a long-term, life-cycle perspective. Each woman's conversion is a highly complex and heterogeneous process, although the link between conversion and domestic violence is unlikely to be unique, given the high levels of interpersonal violence in Brazil and growth of Pentecostal churches throughout the country.

Data from this study found that several of the women interviewed felt more protected from urban violence having converted to Pentecostalism *when* that conversion entailed the conversion of their husbands and children. They believed that it protected them and their families from drug-taking. More importantly, the subsequent involvement of the women in the Pentecostal church allowed them to negotiate different forms of domestic violence in the home. Some women found jobs outside the home or developed leadership roles within the church, while others found great satisfaction evangelising and proselytising non-members. There were visible signs of female empowerment leading to greater equality between husband and wife and to changes to their socially constructed gender roles.

The conversion or non-conversion of their spouses proved important as it was closely linked to the women's own subsequent levels of religious adherence. Also, the resolution or non-resolution of the problem or problems that they had been facing affected their level of continued religious adherence. Disaffiliation occurred due to the church's strict ascetic doctrine on marriage and non-attendance at social events such as parties outside of the Pentecostal group, which contributed to the loss of some of its adherents. This strict code of conduct suggests that while conversion to Pentecostalism may 'protect' converts from urban and domestic violence, it also alienates converts from society.

However, it is evident that while violence is a push factor for female conversion, conversion by itself does not save women. In fact, Pentecostal focus on female submission and placing the blame for violence on spiritual entities or even women's failure to pray allows violent men off the hook and could place women in even greater danger. This study highlighted how women used the teachings of the church, as well as different levels of religious adherence, in order to find alternative ways of addressing the forms of everyday violence they experienced. The religious effects of conversion were played out in various spaces, especially the church, the street and the home. This creative negotiation of violence and the use of faith had positive, practical outcomes in their lives and empowered the women to negotiate and counteract domestic violence.

Notes

1 Pentecostalism is a form of evangelical Protestantism and the majority of converts to Protestantism in Brazil are Pentecostal. Therefore, for this study, I use evangelical Protestantism and Pentecostalism interchangeably.
2 Participants' names have been changed in order to protect their identities.
3 The WHO estimates 35 per cent of all women around the world have experienced either physical and/or sexual violence from an intimate partner and/or sexual violence from a non-intimate partner. In addition, 38 per cent of femicides are committed by the woman's intimate partner (WHO, 2013).

References

Abi-Eçab, A., 2011, Religião e Violência na Periferia de São Paulo, *Revista Anagrama: Revista Científica Interdisciplinar da Graduação*, Ano 5 – Edição 1 Setembro–Novembro de 2011, Sao Paulo

Agencia Patricia Galvão, 2016, *Dossiê Violencia Contra as Mulheres*, Agencia Patricia Galvão, iniciativa do Instituto Patricia Galvão

Birman, P., 2007, Conversion From Afro-Brazilian Religions to Neo-Pentecostalism: Opening New Horizons of the Possible, in: *Conversion of a Continent: Contemporary Religious Change in Latin America*, edited by T. Steigenga and E.L. Cleary, New Brunswick, NJ and London, Rutgers University Press, pp. 115–132

Brenneman, R., 2012, *Homies and Hermanos, God and Gangs in Central America*, New York, Oxford University Press

Brusco, E., 1995, *The Reformation of Machismo: Evangelical Conversion and Gender in Colombia*, Austin, TX, University of Texas Press

Burdick, J., 1996, *Looking for God in Brazil: The Progressive Catholic Church in Urban Brazil's Religious Arena*, Berkeley, CA, University of California Press

Caldeira, T. and Holston, J., 2000, Democracy and Violence in Brazil, *Comparative Studies in Society and History*, 41(4), 691–729

Chesnut, A., 2007, Specialised Spirits, Conversion and the Products of Pneumacentric Religion in Latin America's Free Market of Faith, In: *Conversion of a Continent: Contemporary Religious Change in Latin America*, edited by T. Steigenga and E.L. Cleary, New Brunswick, New Jersey and London, Rutgers University Press, pp. 72–91

Drogus, C. A., 1997, Private Power or Public Power: Pentecostalism, Base Communities, and Gender, in: *Power, Politics, and Pentecostals in Latin America*, edited by E. Cleary and H. Steward-Gambino, Boulder, CO, Westview Press

Freston, P., ed. 2008, *Evangelical Christianity and Democracy in Latin America*, Oxford, Oxford University Press

Garmany, J., 2013, Slums, Space and Spirituality: Religious Diversity in Contemporary Brazil, *Area*, 45(1), 47–55 (2012)

Goldstein, D., 2003, *Laughter Out of Place, Race, Class, Violence and Sexuality in a Rio Shantytown*, Berkeley, Los Angeles, CA and London, University of California Press

Gooren, H., 2007, Conversion Careers in Latin America: Entering and Leaving Church Among Pentecostals, Catholics and Mormons, in: *Conversion of a Continent: Contemporary Religious Change in Latin America*, edited by T. Steigenga and E.L. Cleary, New Brunswick, NJ and London, Rutgers University Press, pp. 52–90

Kruijt, D., 2001, Low Intensity Democracies: Latin America in the Post-Dictatorial Era, *Bulletin of Latin American Research*, 20(4)

Lehman, D., 1996, *Struggle for the Spirit, Religious Transformation and Popular Culture in Brazil and Latin America*, Cambridge, Blackwell Publishers Ltd.

Mariz, C. L. and Machado, M., 1997, Pentecostalism and Women in Brazil, in *Power, Politics, and Pentecostals in Latin America*, edited by E. Cleary and H. Steward-Gambino, Boulder, CO, Westview Press

Martin, D., 1993, *Tongues of Fire: The Explosion of Protestantism in Latin America*, London, Blackwell Books

Neild, R., 1999, Democratic Police Reforms in War-Torn States, *Conflict, Security and Development*, 1(1)

Oosterbaan, M., 2006. *Divine Mediation: Pentecostalism, Politics and Mass Media in a favela in Rio De Janeiro*, Unpublished doctoral thesis, University of Amsterdam

Rostas, S. and Droogers, A., 1993. The Popular Use of Popular Religion in Latin America: Introduction, in *The Popular Use of Popular Religion in Latin America*, edited by S. Rostas and Droogers, A., Amsterdam, CEDLA Latin American Studies

Scheper-Hughes, N., 1993, *Death Without Weeping, The Violence of Everyday Life in Brazil*, Berkeley, Los Angeles, CA and London, University of California Press

Schraiber, L., *et al.*, 2003, Violência Vivida: A Dor que Não Tem Nome – Violence Experienced: The Nameless Pain, *Interface: Comunicacão, Saúde, Educacão*, 7(12), 41–54

Smilde, D., 2007, *Reason to Believe: Cultural Agency in Latin American Evangelicalism*, Berkeley, Los Angeles, CA and London, University of California Press

Stoll, D., 1990, *Is Latin America Turning Protestant? The Politics of Evangelical Growth*, Berkeley, Los Angeles, CA and Oxford, The University of California Press

Valentine, G., 1989, The Geography of Women's Fear, *Area* 21(4), 385–390

Waiselfisz, J.J., 2012a, *Mapa da Violência 2012: Os Novos Padrões da Violência Homicida no Brasil*, pp. 1–245

Waiselfisz, J.J., 2012b, *Mapa da Violencia 2012 – Caderno Complementar 1: Homicidio de Mulheres no Brasil*, São Paulo, Instituto Sangari

Waiselfisz, J. J., 2015, *Mapa Da Violência 2015, Homicídio de Mulheres no Brasil*, 1ª Edição Brasília – DF – 2015, Brasil, Flacso

WHO, World Health Organisation 2013, *Global and Regional Estimates of Violence Against Women: Prevalence and Health Effects of Intimate Partner Violence and non-Partner Sexual Violence*, Geneva, available at: http://apps.who.int/iris/bitstream/10665/85239/1/9789241564625_eng.pdf (accessed June 2017)

Wilding, P., 2011, "New Violence": Silencing Women's Experiences in the Favelas of Brazil, *Journal of Latin American Studies*, 42(4), 719–747

Section 3
Spiritual transformations

Sara MacKian

> If we look back into the past history of mankind, we find, among many other religious convictions, a universal belief in the existence of phantoms or ethereal beings who dwell in the neighbourhood of men and who exercise invisible yet powerful influence upon them.
>
> (Jung, 2008, 128)

The very nature of spirituality requires us to acknowledge and confront that which we cannot easily grasp or sense; the invisible yet powerful otherworldly influences which lie at the heart of religious conviction. Attempting to engage with that empirically challenges us to find new ways of knowing and to push at the habitually imposed boundaries of our epistemological inquiries. For there are aspects of almost all spiritual ontologies which seem intuitively impossible to the outside observer; Jung's invisible, unknowable, unfathomable beings and forces which lie hidden to our regular senses.

This aspect of spirituality is therefore difficult to comprehend as rational, reasonable, intellectual beings who rely on things we can touch, see, categorise or at least measure with some degree of certainty. However, it is to this seemingly impossible task which the authors in this final section turn. Like the White Queen in *Alice's Adventures in Wonderland* who claimed sometimes to believe 'as many as six impossible things before breakfast', you are being challenged now to see the value of engaging with the world empirically from a position of believing what you might actually feel is logically impossible, or at the very least improbable.

Throughout this volume, we have seen that the spiritual, in all its manifest forms, appears to share some common attributes. Whether in strict affiliation to a particular religious doctrine or in relation to a more fluid and individually carved path, cutting across these experiences we witness some common threads: a spirit of the political; a sense of the everydayness of something which for so long has been conceptualised as existing somehow *beyond* the mundanities of the everyday; and the transformative power of engaging with the spiritual. Yet for all their apparent mundaneness and ultimate predictability, the spaces of spirituality are also inherently otherworldly and *un*predictable. By their very nature, spaces of spirituality also open up complex, unwieldy landscapes which are difficult to apprehend and tame with the academic and intellectual tools at our disposal. The phantoms and ethereal beings dwelling in the neighbourhood of men [sic] are not limited to Gods and prayer sanitised by familiar religious doctrines. They consist also of dead people, nature spirits, goddesses and magic, and a host of other eldritch energies which are less familiar to our academic narratives of contemporary religion. Although geographers have historically been reluctant to engage with this side of religious and spiritual practice, recently the discipline has become ever more open to engaging with the occult, the otherworldly and the impossible things which lurk in the unmapped territories of these outer spiritual realms. You are invited in this final section, therefore, to step with us into a space of academic liminality to reflect on how attending to the spiritual might also transform our practice as researchers and commentators. Whilst we may not be asking you to personally believe in the power of witchcraft, mediumship or mother goddess, we *are* asking you to recognise the value of responding to these forces faithfully as a part of your enquiries, and to consider seriously the possibility that picking up a pack of Tarot cards or engaging in ritual may open new spiritual spaces and possibilities for academic exploration.

Imagining impossible spiritualities

You may or may not be reading this before breakfast, but what six impossible things are you being asked to believe as you work through this final section? First off, Julian Holloway (Chapter 13) begins by reminding us of the sheer diversity of religious and spiritual traditions which might be shaping contemporary geopolitical spaces. Suggesting that certain spiritual geographies can manifest as occult geopolitics, he argues the implications of this should be in our sight-line for study. Holloway asks us to consider the role of witches in the mechanical and technological atrocities of Western warfare. Rather than focusing on the familiar and recognised geopolitical maps of the Second World War, we are invited behind the scenes of a counter-geopolitical reality; the occult shadow of Britain's war effort. One must always, argues Holloway, seek to make sense of and take seriously the spiritual geographies which are enacted and produced in such battles. As I write, covens are convening around the world to cast binding spells on President Donald Trump in an attempt to limit the geopolitical damage his presidency may unleash. Holloway's example may be historical, therefore, but such ongoing attempts at using witchcraft to influence global geopolitics suggest this is not something we can easily dismiss or confine to the great dust-heap of history.

Secondly, James Thurgill (Chapter 14) turns to explore not how the occult might have influenced the making of place, but rather how the spectral might aid in the *recovery* of hidden historical spaces. Looking at the work of Frederick Bligh Bond in 1918, he explores the role of psychic archaeology in uncovering the memories, people and landscape of Glastonbury Abbey. Though sceptical readers may consider the use of automatic writing and Ouija boards to ascertain information to guide archaeological digs quite impossible, Bond's findings through such methods were not without subsequently uncovered material archaeological evidence, in the form of lost foundations and skeletal remains. Nearly one hundred years later, Philippa Langley – President of the Scottish Branch of the Richard III Society – stood in a Leicester carpark and experienced the 'strangest feeling' (Kennedy, 2013). In that exact place beneath her feet, the remains of the last Plantagenet King of England were subsequently retrieved, perhaps asking us again to reflect on Bond's psychic excavations and the palimpsestic nature of place. Not as impossible, but as simply another layer we can choose to dig into for clues as we reimagine our connections to the narratives of place and the implications for the geographical imagination.

In Chapter 15, David Wilson reminds us that making the familiar strange and the strange familiar is good anthropological (and geographical) practice. Examining material with fresh eyes (in this case as an apprentice medium at a Scottish Spiritualist church), Wilson destabilises the popularly assumed distinction between shamanism (as 'othered' and 'elsewhere') and British mediumship. Instead, he suggests, there is much to be gained from considering each as culturally specific examples of essentially the same thing. However, he warns that the current research funding climate is not conducive to the sort of long-term extended commitment such participatory fieldwork engagement requires. Clearly the challenges to taking seriously the impossible in academic explorations are not just about the otherworldly, ontological and methodological, but are also about the real-world practicalities and mundane pragmatics of getting research done. This is our third impossible thing.

Patricia 'Iolana's Chapter 16 brings together the influences of Jung and second wave feminism on the development of the Western goddess movement. Once again, we see the political and cultural influences of the time making their mark on how people seek to align their innate spiritual drive in a meaningful relationship with the sociocultural circumstances they find themselves in. By reframing Jung's theory of individuation, a spirituality emerged which released women from the gender restrictions of a patriarchal worldview to provide them with a multidimensional, transformative, yet uniquely personal relationship with the god(dess) within. This takes us to the heart of the fourth impossible thing. Here, the challenge lies in accepting 'Iolana's proposition that such a spirituality offering adherents the tools and ability to 'heal themselves' can also, ultimately, heal the world.

As we turn to Alison Rockbrand's Chapter 17, the fifth impossible thing you are asked to believe is that you are now entering a ritual; a liminal space from which we will all emerge transformed. To study the esoteric is to participate in it on some level, she argues, as in the space of ritual there are no 'outsiders' and everybody becomes an 'insider'. You cannot feign abstinence on grounds of intellectual objectivity, as a blurring occurs between the public and the private with everybody in the ritual space taking part in the magickal change that occurs. In this way, she presents esoteric theatre as a means of re-enchantment in a society lacking in ritual. Rockbrand's invitation to join the ritual might be seen as a rehearsal for encountering the next chapter in this collection.

Finally, we come to professor dusky purples, who arguably presents the most impossible things to believe (Chapter 18). For a start, professor purples is destabilising the very act you are now engaged in, for she implores: 'We must continually read differently, read more, and read better' (page 293). She provocatively prods the ritualised nature of reading as a means of knowledge advancement and reveals to us the pitfalls of 'reading only one way'. She invites us instead to contemplate (and participate) in a *new* way of engaging with the geographies of spirituality and their reading in the academy; to loosen the epistemological binds that tie us. Indeed, she argues, survival can only be achieved by maintaining a foot in other worlds . . .

Stepping into other worlds

With this final section then, we ask you to question *how* you know. We are challenging you, admittedly, when we encourage an embrace of the otherworldly, the unknown and the seemingly impossible. However, there are truly incredible things to be gained from different ways of knowing; from daring to even imagine the impossible. It may not necessarily manifest the impossible as a reality, and it may at times feel unsettling; nonetheless, it can release us from intellectual battles, providing ontological relief and epistemological rebirth. In order to achieve this, as the authors in this final section show, it is time to abandon the strictures of the carefully considered, emotionally barren ways of knowing which have previously defined our safe empirical journeys through well-rehearsed religious landscapes, and to embrace instead the art of the impossible, the creativity of uncertainty, and the multiple spatialities of the spiritual. Reading through these chapters reveals the world (and our place in it) to us in new and unfamiliar ways. As you start to embrace the possibility of the impossible, it opens up a new way of approaching the world and our place within it. Gaining novel knowledges through the magical art of reading *differently* thereby prompts us to question what spirit(s) guides our own investigations and experiences of the world? And whether there is space to expand that spirit, in the spirit of advancing academic enquiry.

I opened this introduction with a quote from Jung, and I turn once more to him to close with the reminder that we should 'never lose sight of the limitations of our knowledge' (2008: 126) but constantly strive to push beyond them. Ask yourself, dear reader, if you are asking the right questions, and reading the right way. Believe the impossible and you might see something worth exploring further. Rather than dismissing ways of reading, we need to bring them together, to break the chains which separate the ideal (pretence) of rigorous intellectual pursuit from more intuitive, creative and exploratory ways of knowing. Attending to the spiritual in its more eldritch and occult manifestations is shown as a fertile ground for generating new alternative knowledges which can contribute to our understanding of our place in the world. Attending to the spiritual then can – and perhaps *should* – also transform our practice as researchers. It is in that spirit which we invite you to immerse yourself in this the final section of the book.

References

Jung, C.G. (2008) *Psychology and the Occult*. London: Routledge.

Kennedy, M. (2013) 'It's like Richard III wanted to be found', 5 February, *The Guardian*. Available at www.theguardian.com/uk/2013/feb/05/king-richard-iii-found (accessed 4.4.17)

13 The magical battle of Britain

The spatialities of occult geopolitics

Julian Holloway

Introduction: religion and geopolitics

Where the geography of religion and political geography meet, a productive range of writings have emerged around the theme of religious geopolitics (see Dijkink 2006; Dittmer 2007, 2009; Dittmer and Sturm 2010; Megoran 2010). In his agenda setting piece on this conjunction, Sturm (2013) sets out a distinction between 'the geopolitics of religion' and 'religious geopolitics'. The former refers to actors who view the geopolitical map of the world through theological spatial divisions and religious discourses of valued significance; here one may cite the contrasting contemporary examples of Daesh's self-declared caliphate or the Dali Lama's vision of Tibet as a regional 'zone of peace' facilitating a 'scalar jump' to world peace (McConnell 2013: 164). On the other hand, 'religious geopolitics' demarcates how manifestly secular geopolitical discourse often deploys and is organised through religiously inspired ideas, languages and practices; here one might mention the Christian Right in the USA whose policies, both foreign and domestic, are often framed or underpinned by Evangelical readings of Scripture. Sitting somewhere between these demarcations, one could also look to debates over postsecularism and its (re)configuration of state and religious relations and practices, to see further intersections between religion and (geo)politics (Cloke 2010, 2011).

This chapter develops Sturm's (2013) notion of geopolitical religion, and hence religious and spiritual discourses and practices which give meaning to, predict or seek to intervene in the relations between nations and the map of political spaces. Thus, it aims to explore how religious and spiritual actors give meaning to, practice and create visions of geopolitical space, and how they seek to enact religious and political power both within and beyond the nation (Dittmer 2007). However, as McConnell (2013: 162) points out, critical analyses of the geopolitical-religious nexus have 'been dominated by the study of two faiths: Anglophone Christianity and, post 9/11, Islam'. In light of this, this chapter seeks to supplement the scope of these analyses by taking as its primary focus a spiritual movement with both a long and varied history and a diverse constitution: the Western Occult tradition and movement.

For the purposes of the analysis presented here, the Occult tradition is a broad philosophical practice that concerns itself with unseen or hidden forces or aspects of cosmic reality which can, in various ways, be contacted, drawn upon or manipulated in the service of particular goals. These concealed 'realities' pervade the universe and Occultists often argue that knowledge of these forces has been lost or obscured. Therefore, uniting occult thinking from Renaissance Hermeticism, through to the nineteenth century occult revival and more contemporary visions of the contemporary New Age and magick, is the assurance that there is a Universal, underlying hidden structure to the world and cosmos, that has been forgotten or deliberately masked. Via practice, training and knowledge the occultist thus seeks an awareness of the secrets of nature and the cosmos and hence seeks some form of union or realisation with them.

Interest in the occult and the supernatural more broadly has grown in human geography in recent years (see Bartolini *et al.* 2013; Dixon 2007; Holloway 2000, 2003a, 2003b, 2006; MacKian 2012; Pile 2005a, 2005b, 2006, 2012; Thurgill 2015). However, there has been little work in human geography that seeks to explore the connection between the occult and geopolitics despite the numerous examples wherein the occult has intersected with, arguably influenced, contested and ran as a counter-geopolitical reality to more mainstream geopolitics. Acting as a hidden shadow to these explicit geopolitical strategies, imaginaries and practices, the geopolitics of the occult has drawn and practiced its own spatialities of the map of power. Thus, occult geometries of power often reproduce, but also skew and refract, national, militarist, statist and popular geopolitics. In particular, the occult simultaneously resonates with and differs from explicit meanings and narratives of national identity, history and destiny. Furthermore, how the geopolitical map is shaped, the processes that configure it and how it might be re-drawn, are re-imagined by occult movements which develop spatialities driven and organised by irrational forces, other cosmic realms and topologies of unseen power. Simultaneously sharing commonalities with mainstream geopolitical strategies and discourse whilst running counter to them, occult geopolitics seek to understand or shape national and world affairs according to their own visions, esoteric imaginaries and cosmic goals.

In this chapter, I will seek to analyse a particular example of occult geopolitics – Dion Fortune's Fraternity of Inner Light and what has become known as the 'Magical Battle of Britain' – which sought to direct the course of World War Two through connecting to and manipulating an esoteric topology of unseen forces. This occult geopolitics both drew upon and differed from more secular geopolitics whilst being formed through a series of material, immaterial and affective spatialities that were patterned through a judgemental and essentialist cosmology of difference and divine destiny.

Intersections of the occult and geopolitics

In addition to the case study that forms the majority of this chapter, examples of the intersections of the occult and geopolitics are manifold. Arguably, the strengths of these links range from the sturdy to the more tenuous, and as such can be given to considerable contest, both historically and evidentially. Here I will discuss briefly and sequentially some of the more well-known examples of these intersections. For example, in early seventeenth century Europe, the three infamous books associated with Rosicrucianism (*Fama Fraternitas*, *Confessio Fraternitatis* and *The Chymical Wedding of Christian Rosencreutz*), filled with alchemical, hermetic and astrological knowledge, appeared at a time of Catholic and Protestant enmity across Europe. For some these books have been deemed as both occult treatise and geopolitical propaganda in their call for religio-political reformation in Europe, and as supporting the hermetically inclined Fredrick V (1596–1632) to lead the effort to undermine Catholic European supremacy (what Reformation Protestants deemed heresy). With the start of the Thirty Years War and the defeat of Fredrick V at the Battle of White Mountain in 1620 by Emperor Ferdinand II, this secret brotherhood supposedly fled east taking their occult knowledge and geopolitical discourses with them (directing the occult gaze of later nineteenth century groups, such as Theosophists, Eastwards). As such, the occult-geopolitical import of these works faded, but arguably was present and influential at their inception (Lachman 2008).

The philosopher, mathematician, astrologer, alchemist and geographer Dr John Dee (1527–c.1608) is a widely acknowledged thinker whose magical and occult work entwines in many ways with geopolitics. Not only did Dee astrologically calculate the date for Elizabeth I's coronation, he was asked by political advisers to the Queen to predict Spanish naval invasions which were expected as part of a more general Catholic conspiracy both within and beyond the borders of the realm (Parry 2012). Moreover, Dee is widely acknowledged with inventing the phrase 'The British Empire' in his writings on Elizabethan geography. Dee's mythological belief in the empire of the Welsh sixth century King Arthur (and before him Brutus) allowed him to assert that Elizabeth had a historical and genealogical right to the possession of certain territories, including Greenland, Iceland and into North America (MacMillan 2001; Parry 2006). This justification for the geopolitical expansion of the realm intersected with Dee's reading of Johannes Trithemius' philosophy of angelic spirits controlling different ages of history. He thus saw Elizabeth as having an eschatological and apocalyptic role and destiny to 'civilise' and restore these lands of the Empire in order to prepare the world for the second coming of Christ. Indeed, for the most part these geopolitical occult prophecies garnered their detail through the practice of ritual 'scrying', wherein his (ultimately treacherous) companion Edward Kelley summoned and received messages from angelic presences (Wooley 2001).

England in the seventeenth and eighteenth centuries was another key period where links and intersections between occult knowledge and groups, and geopolitical events and discourses flourished. For example, the development and popularity of Speculative Freemasonry amongst liberal Whigs, including Prime Minister Robert Walpole (1676–1745), meant a belief in a Hanoverian monarchy and Protestantism mixed with esoteric lore of divine rationality and sacred geometries. Furthermore, the popularity of the fraternal Masonic lodges in pre-revolutionary France has been argued to have been a key aspect of the revolution itself, through the lodges facilitating a mixing of social strata and acting as forums for radical anti-Christian thought (Boyet 2014; McIntosh 2011). Indeed, in post-Revolutionary France occultism thrived particularly in the form of explicitly geopolitical prophecy wherein predictions for a return of the 'Great King', and reestablishment of a (reformulated) Ancien Régime, abounded: for example, the Prophecy of Orval of the 1830s predicted the overthrow of Louis-Philippe and the 'arrival of a Great King, who would establish the European peace under French hegemony' (Harvey 2003: 688; see also Holloway 2015).

In the same century, the Theosophical leader Helena Petrovna Blavastky (1831–1891) not only promoted a racialised view of the Aryan 'root' race with Altantean origins, but upon moving to India became a strong proponent of Home Rule. Here she laid the foundations for the social reformer, suffragette and theosopher Annie Besant (1847–1933) who 'was to be interned in India for activities relating to her support of Indian Home Rule, and in 1917 was elected president of the Indian National Congress' which itself was founded by another theosophist Octavian Hume (1829–1912) (Owen 2004: 31). Indeed, there is a need to explore the links between late nineteenth and early twentieth century occultism and geopolitics more widely. One possible avenue for such research are links and influences between the former Society for Psychical Research president and British Prime Minister Arthur James Balfour's (1848–1930) spiritualist experiences (known as 'The Palm Sunday Case'), The Balfour Declaration and Zionist messianic prophecies. Or one could pursue a critical analysis of the British Israelite Movement whose religious nationalism (active from the 1870s onwards) was worked through a reading of the Great Pyramid of Khufu as the 'Bible in Stone' and hence a material manifestation of the Egyptian possession of arcane and lost spiritual knowledge and truths. Through multiple geometrical calculations of the Pyramid's dimensions and structure, not only was the Movement's belief in the British as descendants of the Biblical Lost Tribes confirmed, but also the reading of geopolitical events of the two World Wars were interpreted in eschatological terms (see Moshenska 2008).

The relationship between occult practice, knowledge and the geopolitics of war is where this chapter sits. As such, one might look to the work of anthropologists on war magic and warrior religion which 'demonstrate that warrior religion and war magic are ubiquitous social phenomena that have arisen across the globe in a diverse range of cultures' (Farrer 2014: 3). Here ritual practices are embodied and performed in order to harness occult and magical forces to inflict harm on enemies, or sometimes heal (in various ways) the victims of war. In the extreme one might discuss war sorcery as a form of occult geopolitics whereby ritual practices are deployed to 'harness magical, spiritual, and social-psychological forces that result in an opponent's misfortune, disease, destruction and death' (Farrer 2014: 4). Yet this more strategically offensive practice is often only part of the broader war magic which also seeks to counter harmful forces and offer safeguards in times and spaces of conflict. One example is that of Javanese *kanuragan*, a ritual process wherein cosmological knowledge and entities named *aji* are used to foster invulnerability and power, which was practiced amongst fighters in the Indonesian War of Independence (1945–1949) against the Dutch colonial powers (de Grave 2014).

Of course in discussing the occult in relation to war and geopolitics it is impossible not to mention the links between Nazism and the occult tradition. A source of considerable debate, contest and popular myth (propagated by many non-academic books, TV shows and, of course, films), this intersection is one of some legitimacy and definite complexity. Colouring rather than directly causing Nazism, some of the occult intersections with the Third Reich include: the influence of Guido von List's völkisch and esoteric ideology of Ariosophy, and the supposedly ancient Aryan Teutonic Gnostic religion of Wotanism; the anti-Semitic and proto-Nazi Thule Society and its links with the DAP of whom Hitler joined; the self-proclaimed mystical teacher Karl Maria Wiligut who was made head of the SS Pre-and Early History department; Hess and Himmler, who 'trafficked themselves in a range of esoteric beliefs, from astrology and pendulum dowsing to natural healing' (Kurlander 2015: 504); and the supposedly magical space for the SS 'brotherhood' at Wewelsburg which Himmler allegedly believed would be a stronghold for a millennial conflict between a Germanic Europe and Asia, and where the room of the 'Obergruppenführersaal' was purportedly designed for twelve officers of the SS, akin to the Knights of the Round Table, to 'commune' with the 'Aryan Race Soul'. Add to this list the 'acceptable border sciences including World Ice Theory, cosmobiology . . . "scientific" astrology, and biodynamic agriculture' and the mix of racism, occultism and geopolitical thought under National Socialism is a heady, disputed and often conflated one (Kurlander 2015: 521; see also Black and Kurlander 2015; Goodrick-Clarke 2003; Trietel 2004).

Therefore, the geopolitical hostilities of World War Two had an occult shadow that accompanied and coloured some of its discourses and events. To this end the so-called Nazi Occult is well known. What is less well known are the occult activities leading up to and during the war that occurred in Allied nations, particularly Britain. Indeed, subject to much speculation and a degree of conjecture, the Great Beast himself, Aleister Crowley (1875–1947) was said to have links to British Intelligence in the 1930s, and in the 1920s he and his followers were expelled by Mussolini's authorities from Sicily and the 'Abbey of Thelema' allegedly due to the Italians believing him to be a British spy. One of Crowley's friends, whom he dubbed the magical 'Adept of Adepts', was the bohemian and notoriously eccentric Evan Morgan, 2nd Viscount Tredegar (1893–1949), who was both a member of the occult group The Black Circle in London and was appointed in 1939 head of MI8, the Radio Security Service. Admiral John Godfrey, who himself seemingly recruited astrologers in the war effort, was boss to Ian Fleming at British Naval Intelligence and it was the latter who exploited (through producing fake astrological charts) Rudolf Hess' pre-war British high society links and occult interests to persuade him to fly to Scotland (to seek a meeting with the Duke of Hamilton – a member of the Hermetic Order of the Golden Dawn) to seek peace with Britain. This, in part, was due to Hess' own astrologically influenced concerns over Hitler's second front with the Soviet Union (see Howard n.d; Spence 2008). Finally, the 'Father of Wicca' Gerald Gardner (1884–1964), by his own account, performed 'Operation Cone of Power' in 1940 wherein 'a Great Circle was cast at night in the New Forest and a cone of magical energy raised and directed against Hitler' (Hutton 1999: 208). It is to a similar attempt to thwart and undermine Nazi war efforts through magickal means that this chapter turns now, namely Dion Fortune and her Fraternity of Inner Light.

Dion Fortune and the 'Magical Battle of Britain'

Described by Hutton (1999: 188) as a 'complex thinker, whose career defies any simple formulations', Dion Fortune (1890–1946) was one esoteric thinker and practitioner whose rituals and actions during World War Two allow us to explore the spatial intersections of the occult and geopolitics more closely. Born Violet Mary Firth in Llandudno on December 6th 1890, she was brought up in a household who followed the teaching of Christian Science. With the outbreak of World War One, Fortune joined the Land Army along with many other women and was given a laboratory job (where she discovered the possibility of making cheese from soya bean milk). After the war she had set herself up as psychotherapist in what was then still a nascent area of inquiry. Fortune then moved from her study of psychology to occultism; for her the two were not mutually exclusive as they both involved 'hidden forces'. As Benham (1993: 253) explains:

> Although both defer to hidden forces that inform the appearances of life, the phenomena of psychology emerge from the inner depths of the individual psyche; the phenomena of the occult are understood as invisible forms and powers *outside* and *beyond* the psyche – although they may primarily engage it at the unconscious level. For Violet the two were quite compatible.

Post World War One, Fortune joined the Theosophy movement. Leaving in 1927 she became President of the Christian Mystic Lodge and a member of the London 'Alpha et Omega' temple of the (by then much divided) Hermetic Order of the Golden Dawn wherein she changed her name to Dion Fortune (taken from her family's motto '*Deo non Fortuna*', approximately 'By God, not Fate'). By 1929 she had left the Golden Dawn and the Lodge and had formed her own Community of the Inner Light, later to become The Fraternity of Inner Light and today The Society of Inner Light.

Broadly one can divide Fortune's esoteric and occult biography into three periods: unorthodox and mystical Christianity (1914–30); more pagan and Goddess inspired thinking (1930s); and a return to esoteric Christianity after 1939 until her death in 1946 (see Hutton 1999; also Chapman 1993; Fielding and Collins 1998; Johnston Graf 2007; Richardson 2007). The events that concern us here occurred in this last period. However, it is important to consider her earlier thinking and writing in order that her actions and ideas during World War Two are suitably contextualised. In short, Dion Fortune's thinking was defined by a geopolitical discourse of race and nationhood wherein the two were often conflated – a mode of thought not uncommon at the time. Indeed, she believed that each race/nation had a 'Group Soul' based on racial differences (or subconscious) which gave rise to and could be influenced by the 'Group Mind' (or consciousness) of the nation:

> The innumerable individuals who make up a nation share a common subconsciousness [sic] below the personal subconsciousness of each one; this is called the racial or collective subconsciousness, and it plays a very important

> part in both individual and collective life. It is this level of the subconsciousness that is appealed to by national heroes; it is this level that is manipulated by spellbinding demagogues. It is here that the trend and limitations of the national character are determined, and from here that its inspiration is drawn.
> (Fortune 2012: 72)

Fortune's was very much a cosmological and esoteric form of nationalism and racism. She saw these Group Souls and races as ultimately descended from the oneness of the 'Divine Mind'. From there 'cosmic rays' descend to organise the 'group soul' of a race which, 'being rooted in the Earth, acquires a Place aspect with the passage of time', equating to both the 'Group Soul' and 'Group Mind' of the nation (Fortune 2012: 155). For Britain, Fortune portrays the three races of 'the Kelts, the Norse, and what can only be described as the Conglomerate, being that which is composed of all the different elements that have ever struck roots in British soil' and hence a 'unified Group Mind has grown up on the basis of a diversified Group Soul' (Fortune 2012: 155). Furthermore, for each Group Soul of nation/race there existed an ascended cosmic Master: 'racial types are guided in their destiny by Racial Angels and initiated by racial Masters' (Fortune 2012: 154, 158). However, and telling in the quoted passage above is the use of the word 'limitations', implying a hierarchical vision of the advancement of certain nations/races above and beyond others. As Hutton (1999: 182) reveals, in Fortune's writings, and particularly her novels, a racial hierarchy was presented through descriptions 'of the "wily Teuton" and the "savage races" of the Balkans' and a 'general fear of contamination by other nations, races or classes which runs through her books at this time'. Indeed, she stated 'the instinct for racial purity is a sound one' (Fortune 2012: 154).

In this context of a cosmological geopolitics of race and nation the 'Magical Battle of Britain' occurred (as it has become known). As such, between October 1939 and July 1942, Fortune and her Fraternity of Inner Light sent out 134 weekly letters that amounted to an occult and spiritual geopolitical strategy to magically defend Great Britain and counter the 'brute force' of Nazism, initiated in the belief that 'the knowledge of the Secret Wisdom is going to play an important part in what has to be done for the winning of the war and the building of the stable peace' (Fortune 2012: 40). This belief was driven by Fortune's understanding that the Nazis themselves were waging war on the astral or spiritual plane: Hitler, who surrounded himself by a 'relatively small and apparently obscure group of those who realise that there are subtle forces that can be enlisted to serve their ends', was according to Fortune 'a natural occultist and highly developed medium' (Fortune 2012: 81). Yet, the cosmological belief that necessitated this occult geopolitical strategy was wedded to a more material and mundane necessity: with travel restrictions in place during the war, and paper rationing meaning the printing of the Fraternity's magazine ceased in 1940, letters were a means by which the spiritual work of the group could be continued despite the geographical dispersal of its members. As such, the letters were sent out every Wednesday, to be read and consumed by the Sunday in order that the occult diaspora of the

Fraternity could join in the 'united meditation' scheduled for that day at 12:15, whose 'nucleus of trained minds' would be based at the Fraternity's headquarters at Queensborough Terrace in London (Fortune 2012: 15).

Immediately we see that this occult geopolitics was played out across two interwoven and mutually dependent spatialities: one spiritual and imagined; the other material and this-worldly. Indeed, as I have argued elsewhere, what we might call a network spirituality is both mundane and cosmologically patterned (Holloway 2000). Here the material (the letters) and immaterial (the magical warfare) are necessary to the performance and perceived success and development of the spiritual geopolitics through which they take shape: the diasporic occult geopolitics of the Fraternity thus took form and coalesced through this material-immaterial spiritual network. Moreover, the materiality of the body and embodiment – its positioning, training and deportment – was central to the realisation of the Fraternity's spiritual geography and its occult geopolitics. As such, a preliminary document was sent out to members containing a series of instructions, composed of seven stages, on how to perform the 'work' of the esoteric warfare and hence generate this spiritually networked geopolitics. These included a series of directives on how to hold the body and its geographical positioning:

> [. . .] take your seat if possible in a quiet, dimly lit room, secure from disturbance; face towards London; sit in such an attitude that your feet are together and your hands clasped, thus making a closed circuit of yourself. Your hands should rest on the weekly letter lying on your lap, for these letters will be consecrated before they are sent out in order that they may form a link. Breathe as slowly as you can without strain, making a slight pause at the beginning and ending of each breath.
>
> (Fortune 2012: 16)

The occult geopolitical strategy sought here thus began with a body composed and configured in immediate space. Yet the immediate was married to, supported and generated a scaling of the spiritual geography to an extended space, in both physical and cosmological senses. Moreover, this was a composed and trained body that would inform and configure an occult geopolitical sensibility of judgement and salience towards space and geopolitical events (Holloway 2012). In other words, this was a somatic composition which would seek to realise a psychic defence from the evil forces of (occult) German Nazism and its ultimate defeat, both in the seen and unseen world. This spiritual geography was one composed of an assemblage of geopolitical scales and spaces of practiced embodied cosmological evaluation, which were simultaneously immediate, distant, national and international, material and immaterial, ordinary and extraordinary, and patterned as good versus evil.

In order to understand this patterned occult geopolitical strategy and scaling more closely, we need to attend to the immaterial and spiritual geographies produced by the Fraternity's warfare. Therefore, in order to explore the geopolitics of occult and indeed wider spiritual movements, one must always seek to make sense and take seriously the spiritual geographies enacted and produced. These spiritual mappings often centre upon key places of divine and spiritual intervention, cosmological significance or mysterious events. Mostly due to her unorthodox Christian beliefs, Fortune's spiritual geography coalesced around the early medieval monasteries of Iona, Lindisfarne and Glastonbury (Hutton 1999: 184). This was further realised and strengthened during the Fraternity's occult warfare: early on in the geopolitical work conducted by the group, visions and symbols emerged through meditation which centred on Glastonbury.

> Starting from the symbol of the Rose upon the Cross, we immediately found it surrounded by golden light of great brilliance. [. . .] It was then perceived that the golden light and the Cross were formulated inside a cavern. [. . .] This cavern is known to the initiates as the cavern beneath Mount Abiegnus, the Hill of Vision, of which the earthly symbol is Glastonbury Tor. [. . .] [I]n future those who join with us in the meditation exercises should visualise the Rose Cross as standing in the cave under the Hill of Vision, for this is now our meeting place.
>
> (Fortune 2012: 23–24)

Glastonbury, its earthly and material environs, is confirmed and made apparent here as a place of spiritual energy and insight. Through the embodied action of meditation, a space of spiritual centrality is performed and perceived. Patterning the network, this action simultaneously unites the dispersed Fraternity, as they supernaturally travel to and reside in the 'Hill of Vision', whilst enacting a mode of communal spiritual subjectification composed in and through this immaterial geography, and providing a place of identification for a scattered spiritual community. Moreover, the patterning, scaling and differentiation of this spiritual geography and the subjectivities produced, become central to the performance of the group's occult geopolitical action: for it is under the Hill of Vision where the group will meet and be contacted on the 'Inner Planes' by those Masters who will 'bring to the race mind a realisation of the support afforded to it by cosmic law' (Fortune 2012: 23). In other words, this (im)material space became the key geopolitical arena in which different geographies – astral, material, disaporic, conflicted – were performed and intersected. Hence, here the Fraternity were given the esoteric knowledge, sourced from otherworldly spatialities, of how to wage occult war and protect the nation.

At this juncture we see the Fraternity's spiritual geography finding some coincidence and overlap with more widely held nationalist geographies. Fortune and the Fraternity sought to 'evoke primordial energies from the primitive levels of the national group soul and harness them to archetypal ideas in the group mind of the race' (Fortune 2012: 84), and given that it was Glastonbury where their occult geopolitical strategies would be learnt and performed, it was the 'Watchers of Avalon' which would pursue war on the astral-spatial plane: 'Let us wake from their long sleep the primordial images of our race, King Arthur and his knights, with the wisdom of Merlin to guide them. These shall keep the soul of England against the invisible influences being brought to bear upon it for its undoing' (Fortune 2012: 85). Whilst these 'ideals' figure strongly in other British nationalist and populist geopolitical imaginaries, both in the past and in the present, it must be noted that they are deployed here in a manner which runs somewhat counter to their role as symbolic emblems of national identity. For here, the Fraternity evoked Arthur and Merlin as doing *actual* work on the soul and spirit of the nation, and hence were very much a real part of the conflict and war. Far from being just symbols to rally round and identify with, Fortune believed warfare was taking place in this other-worldly space and hence these figures were acting to shore up Britain's defences as the war ensued:

> In order to guard against any such subtle influencing, let us meditate upon angelic Presences, red-robed and armed, patrolling the length and breadth of our land. Visualise a map of Great Britain, and picture these great Presences moving as a vast shadowy form along the coasts, and backwards and forwards from north to south and east to west, keeping watch and ward so that nothing alien can be observed.
>
> (Fortune 2012: 34)

In making sense of spiritual geographies, here manifest as occult geopolitics, we must therefore take seriously the reality of the formations, ideas and cosmologies generated by these groups in their discourses and practices.

One way in which we might pursue a more sympathetic understanding of these geographies, yet one which, importantly, allows for a critique of such groups to emerge, is through an attention to the forces that bind and pattern them, and allow them to work. More precisely, one might examine how occult geopolitics is formed and enacted through series of transfers that amount to an esoteric spatiality of diasporic affect. Therefore, the transfers that formed and produced this occult geopolitical network are very much akin to Pile's (2011) study of telepathy as a form of affect over distance. Pile (2011: 4) informs that telepathy, as multi-faceted supernormal phenomena, 'refers to the ability to sense (to be affected) and also to perception beyond immediate cognition, such as intuitions, premonitions, or inklings'. As such, the Fraternity's occult warfare was composed and achieved through a series of transfers occurring at distance that welded imagined spaces, embodied acts and judgemental differentiations, such that a community of affected sensation emerged and solidified. Here 'the experience of shared affects between people at distance' was produced through an occult cosmology of racial-national differences and geopolitical esoteric conflict (Pile 2011: 7). These circulations allowed for an occult geopolitics to occur across distance and through im/material spaces and borders, melding the embodied and the imagined, the sensed and the spiritual. Essential to this network of geopolitical defence and combat was a series of binding occult transferrals. Here affect at a distance is envisaged as occult energies performed through supernormal flows across space. These transfers are understood through cosmologies of spiritual energy, thought-forms, visions and sensations circulating amongst the network. As this spiritual affect circulated, as the occult transference affected at a distance and as these energies bound the network, the geopolitical forces of the 'Watchers of Avalon' and their attendant geographies were generated.

This form of spiritual geopolitics thus amounts to an occult topology of affect spatially manifest across different and dispersed geographies: a geopolitical spatiality and strategy was built and cemented with a 'glue' of occult affect transference binding the network:

> From our Inner Plane contacts we draw strength and inspiration [. . .] It is not enough to make contact and receive inspiration. The inspiration will soon dry up unless it flows through us, ever renewing itself in flowing. For those who have the deeper knowledge, participation in the national war effort is a sacramental act whereby the power that has been drawn down is put in circuit. Break the circuit, and the power ceases to flow.
>
> (Fortune 2012: 53)

During the war years, this affective topology, composed of energetic circuits of 'sacramental' and 'inner plane' inspiration, produced and strengthened judgemental dispositions with regard to the events happening, how to affect these events and processes and, more significantly, why they were happening. Yet Fortune's letters rarely state, and at best only hint at, possible direct material impacts on the events of the war. For the most part, where a link is made it takes the form of prophecy: for example, on June 23rd 1940 she notes how the 'change of feeling' in the USA towards the war was seen in their meditation three weeks previous and 'how it will be recalled that the entry of Italy into the war was announced a fortnight before it occurred' (Fortune 2012: 54). However, Fortune warned her followers from dealing directly in this-worldly geopolitics stating:

> . . . our teaching concern[s] principles, not politics. [. . .] This is the way in which, as initiates, we work. We outline nothing; we meditate upon cosmic principles till these take intellectual form. [. . .] There is, in consequence, a gap between the initiates who bring through the archetypal ideas and the statesmen and economists who give them practical form. [. . .] The thought-forms that have developed as a result of group meditation work have to cross the gap by means of their own inherent energy.
>
> (Fortune 2012: 103)

For Fortune and the Fraternity, their occult warfare was happening through an other-worldly spiritual geography, wherein the nation's group soul and, literally, spirit was being fortified and advanced.

At the heart of this occult geopolitical strategy was belief in cosmic destiny whereby the Fraternity 'have to simply pull the lever, and the Machinery of the Universe does the rest. Our work is to formulate and reformulate day by day the mental link between the spiritual influences and the group mind of the race' (Fortune 2012: 25). The notion that there is a 'cosmic plan' to the universe that each of us is living out (whether we realise it or not) and that we need to learn our spiritual and cosmological destiny is something common to occult, esoteric and spiritual groups (Holloway 2000). Consequently the more spiritually advanced and developed a seeker is, the more they are aware of their part in the divine plan and eschatological affairs: 'We believe that there is a cosmic plan being worked out, of which the present conditions form a phase, and that we can consciously co-operate with the working of that plan' (Fortune 2012: 20).

Fortune, given the spiritual messages she and others in the group received, was thus able to state as early as 1941 that 'the question of the ultimate outcome of the war and the form of the final peace was never considered a matter for speculation because it was taken for granted' (Fortune 2012; 90). This assumption was based on the very appearance and intervention of the Masters or Elder Brethren that waged astral combat, protected the nation and allowed the Fraternity to do their spiritual geopolitical work during the war. Indeed, Fortune argued that the 'opportunity to establish contacts with the Masters' was ripe during the war: 'for it does not often happen that the veil is as thin as it is at the moment' (Fortune 2012: 32). Moreover, this appearance signalled a proto-New Age version of history which envisioned the war as the movement from the Piscean Age 'as the pure Aquarian types made their appearance among us' (Fortune 2012: 148).

With the spiritual assurance that a New Age was dawning and the nation would be protected through the astral combat of the Masters and the work done by the Fraternity, in her later weekly letters Fortune began to spell out her geopolitical spiritual vision for the future. This proposal took the form of a post-national cosmopolitanism wherein the spiritual would supersede the material geopolitical conflict experienced and suffered:

> Nations must not be looked upon, nor think of themselves, as self-contained units; they are simply sub-sections of human society thus divided up for convenience. [. . .] It is the men and women of the New Age in all nations who must take control across the national barriers as soon as the fighting is over, and they must meet as Aquarians, not as English, French or Germans.
>
> (Fortune 2012: 100–101)

Whilst this occult geopolitical future is shaped here by a doctrine of spiritual-global post-nationalism, Fortune's vision of the divine plan must still be read as one where racism and nationalism are driving forces. Therefore, generated through affective relation and communities of esoteric sensation, this was singularly *not* an ethics of open becoming, but one where differences were absolute and destinies preordained. Here a network of affective relations did not present a future open to becoming (contra Anderson and Harrison 2010; Connolly 2011), but a spiritual geography where the future is pre-given and closed down: as Fortune (2012: 158) opined, 'each race has its own destiny to pursue under its own leaders, and change can only come from within, if change be needed'. This destiny was a teleological path of spiritual enlightenment, yet one where history and fundamental differences in national Group Souls were paramount and conditional:

> When the Germans open up the primordial levels of their racial mind they release the elemental energies of the old gods – the bloodstained, mindless

> images of the heroes of Norse myth. [. . .] A good thousand years intervenes between the [Christian] conversion of Britain and the conversion of Germany; consequently the influences of Christianity reaches to a far deeper level of racial consciousness with us than with them, and when the surface consciousness of the British group soul peels off we find, not the mindless heroes of Valhalla, but the chivalry of the Table Round; Excalibur instead of Nothung; and the Quest of the Grail instead of the looting of Rhinegold.
>
> (Fortune 2012: 85)

As such, pre-destined spiritual enlightenment is made provisional across space due to evolutionary and essential cosmological differences. Indeed, Fortune (2012: 157) expands these national differences to a global scale in her discussion of the Western and Eastern mystery traditions when she states: '[t]o the East belongs the glory of a great past from which we may learn, but to which we may not return'. This occult geopolitics is thus one formed through a doctrine, embodiment and practice of essentialism, spiritual evolutionism and ultimately, esoteric supremacy at a national and global scaling, despite (or because) of her claims to a cosmopolitan New Age dawning.

Conclusion

Understanding how spiritual geographies are produced and sustained through communities of sensation and affective topologies, and how they give rise to and pattern geopolitical discourses of essential division and hierarchical evolution, is one way a critical approach to such movements might be developed. Here I have traced how a spiritual geography is formed that seeks to intercede in national conflict and foresee and configure the future map of political spaces, both nationally and internationally. I have examined how occult geopolitics is played out and performed across two interwoven spatialities – the immaterial and the material. Spiritual geographies are manufactured through the real and the imagined, and are stitched together with the thread of spiritual energies and occult affective transfers. Dion Fortune and her Fraternity of Inner light sought to wage war on an immaterial plane against hostile forces, and in-so-doing plaited together an occult geopolitical imaginary and practice that simultaneously drew upon and differed from national symbolism organised around iconic material spaces. This was a spiritual geography performed through a network glued and bound by supernormal affects at a distance and embodied action that affected geopolitical judgement in and towards other nations and identities. This judgement was manifest as a spiritual evolutionism, with some identities and nations closer and more inherently capable of divine realisation than others.

This chapter has explored in detail an historical example of where occult discourses and practices coincide with and shadow geopolitical events. Yet to believe that the practices and discourses of occult geopolitics are a thing of the past would be a mistake. For example, Pop (2014) has examined how during the Romanian Presidential election campaign of 2009, Aliodor Manolea, a staff member of the incumbent president Traian Băsescu, used occult powers (specifically the energy of the 'violet flame', a source of mystical power) during a live television debate to 'negatively influence' the counter-candidate Mircea Geoană. Furthermore, there is a contemporary coincidence between the occult, geopolitical imaginaries and conspiracy theory.

For example, David Icke, whose popularity is widespread amongst New Age countercultures, has produced a series of publications and video podcasts detailing his discourse of geopolitical institutions, such as the European Union. Icke believes the world is being run by a 'hidden kabbal', deemed the Illuminati, whose goal is a 'Global centralised society, based on a world government, world central bank controlling all finance, and a world army imposing the will of the world government' (Icke 2016a). The EU is a 'Super State' within this 'world government': 'The plan within those [super states] is to destroy all countries, to end all sovereignty, to end all nations, and break these nations into regions. [. . .] The idea is to have a world government dictating to these union super states and the union super states dictating to the regions of the super state'. To this end, Icke has spent much time and effort revealing the 'evidence' that proves this occluded geopolitical agenda. For example, he notes how the 'Twelve stars of the union is the symbol of the Babylonian goddess. [. . .] The European Union is not a political union, it is the union of the Illuminati Goddess which they wish to enslave the whole of Europe within' (Icke 2016b).

Given the popularity of Icke – someone who can sell out the 6000 capacity Wembley Arena in 2012 – and the circulation of other occult inspired conspiracy theories both today and in the past, it seems appropriate and indeed crucial that geographers of religion and spirituality seek to analyse and critique the spatialities of occult geopolitics, and their significance and ramifications. Indeed, in a world where political power is increasingly practiced through statements of 'alternative facts', it seems imperative geographers of religion and spirituality seek to critically investigate and unpack the consequences and implications of the intersection of geopolitics and religion, especially in the 'post-truth' world of occult conspiracy theory wherein Icke and his ilk reside.

References

Anderson, B. and Harrison, P. (eds.) 2010. *Taking Place: Non-Representational Theories and Geography*. Ashgate, Aldershot.

Bartolini, N., Chris, R., MacKian, S. and Pile, S. 2013. Psychics, crystals, candles and cauldrons: alternative spiritualities and the question of their esoteric economies. *Social and Cultural Geography* 14, 367–388.

Benham, P. 1993. *The Avalonians*. Gothic Image, Glastonbury.

Black, M. and Kurlander, E. (eds.) 2015. *Revisiting the Nazi Occult: Histories, Realities, Legacies*. Camden House, Rochester, NY.

Boyet, J. 2014. *Old Gods in new clothes: the French revolutionary cults and the "Rebirth of the Golden Age"*, Master's Thesis, Liberty University. http://digitalcommons.liberty.edu/cgi/viewcontent.cgi?article=1331&context=masters

Chapman, J. 1993. *The Quest for Dion Fortune*. Red Wheel/Weiser, Newburyport, MA.

Cloke, P. 2010. Theo-ethics and radical faith-based praxis in the postsecular city. In Molendijk, A., Beaumont, J. and Jedan, C. (eds.) *Exploring the Postsecular: The Religious, the Political, and the Urban*. Brill, Leiden, pp. 223–241.

Cloke, P. 2011. Emerging geographies of evil? Theo-ethics and postsecular possibilities. *Cultural Geographies* 18, 475–493.
Connolly, W. E. 2011. *A World of Becoming*. Duke University Press, Durham, NC.
de Grave, J-M. 2014. Javanese Kanuragan ritual initiation: A means to socialize by acquiring invulnerability, authority, and spiritual improvement. *Social Analysis* 58(1), 47–66.
Dijkink, G. 2006. When geopolitics and religion fuse: a historical perspective. *Geopolitics* 11, 192–208.
Dittmer, J. 2007. Intervention: Religious geopolitics. *Political Geography* 26, 737–739.
Dittmer, J. 2009. Maranatha! Premillennial dispensationalism and the counter-intuitive geopolitics of (in)security. In K. Dodds and A. Ingram (eds.) *Spaces of Security and Insecurity: New Geographies of the War on Terror*. Ashgate, Aldershot, pp. 221–238.
Dittmer, J. and Sturm, T. (eds.) 2010. *Mapping the End Times: American Evangelical Geopolitics and Apocalyptic Visions*. Ashgate, Aldershot.
Dixon, D. 2007. A benevolent and sceptical inquiry: exploring 'Fortean Geographies' with the Mothman. *Cultural Geographies* 14, 189–210.
Farrer, D. S. 2014. Introduction: Cross-cultural articulations of war magic and warrior religion. *Social Analysis* 58, 1–24.
Fielding, C. and Collins, C. 1998. *The Story of Dion Fortune: As Told to Charles Fielding and Carr Collins*. Thoth Publications, Loughborough.
Fortune, D. 2012. *The Magical Battle of Britain*. Skylight Press, Cheltenham.
Goodrick-Clarke, N. 2003. *The Occult Roots of Nazism*. I.B. Tauris, London.
Harvey, D. A. 2003. Beyond enlightenment: occultism, politics, and culture in France: from the old regime to the fin-de-siècle. *Historian* 65(3), 665–694.
Holloway, J. 2000. Institutional geographies of the New Age movement. *Geoforum* 31, 553–565.
Holloway, J. 2003a. Spiritual embodiment and sacred rural landscapes. In P. Cloke (ed.) *Country Visions*. Pearsons, Harlow, pp. 158–175.
Holloway, J. 2003b. Make-believe: Spiritual practice, embodiment and sacred space. *Environment and Planning A* 35, 1961–1974.
Holloway, J. 2006. Enchanted spaces: The séance, affect, and geographies of religion. *Annals of the Association of American Geographers* 96, 182–187.
Holloway, J. 2012. The space that faith makes: Towards a (hopeful) ethos of engagement. In P. Hopkins, L. Kong and E. Olson (eds.) *Religion and Place: Landscape, Politics and Piety*. Springer Publishing, Dordrecht, The Netherlands, pp. 203–218.
Holloway, J. 2015. Sealing future geographies: religious prophecy and the case of Joanna Southcott. *Transactions of the Institute of British* Geographers 40(2), 180–191.
Howard, M. n.d. The occult war: Secret agents, magicians and Hitler. *The Cauldron: Witchcraft, Paganism and Folklore*. ONLINE: www.the-cauldron.org.uk/Resources/Occult%20Wara.pdf (Last accessed 21st June 2016).
Hutton, R. 1999. *The Triumph of the Moon: A History of Modern Pagan Witchcraft*. Oxford, Oxford University Press.
Icke, D. 2016a. *OUT! David Icke talks Brexit*. ONLINE: www.youtube.com/watch?v=uDiGONVJqcU (Last accessed 30th June 2016).
Icke, D. 2016b. *European Union of evil*. ONLINE: www.youtube.com/watch?v=EQ3m_vWXaZc (Last accessed 30th June 2016).
Johnston Graf, S. 2007. The occult novels of Dion Fortune. *Journal of Gender Studies* 16, 47–56.
Kurlander, E. 2015. The Nazi magicians' controversy: enlightenment, "Border Science," and occultism in the Third Reich. *Central European History* 48, 498–522.

Lachman, G. 2008. *Politics and the Occult: The Left, the Right, and the Radically Unseen.* Quest Books, Wheaton, IL.

MacKian, S. 2012. *Everyday Spirituality: Social and Spatial Worlds of Enchantment.* Palgrave Macmillan, Basingstoke.

MacMillan, K. 2001. Discourse on history, geography, and law: John Dee and the limits of the British Empire, 1576–80. *Canadian Journal of History* April, 1–25.

McConnell, F. 2013. The geopolitics of Buddhist reincarnation: contested Futures of Tibetan leadership. *Area* 45(2), 162–169.

McIntosh, C. 2011. *Eliphas Lévi and the French Occult Revival.* State University of New York Press, Albany.

Megoran, N. 2010. Towards a geography of peace: pacific geopolitics and evangelical Christian Crusade apologies. *Transactions of the Institute of British Geographers* 35(3), 382–398.

Moshenska, G. 2008. 'The Bible in Stone': pyramids, lost tribes and alternative archaeologies. *Public Archaeology* 7(1), 5–16.

Owen, A. 2004. *The Place of Enchantment: British Occultism and the Culture of the Modern.* The University of Chicago Press, Chicago, IL.

Parry, G. 2006. John Dee and the Elizabethan British empire in its European context. *The Historical Journal* 49(3), 643–675.

Parry, G. 2012. Occult philosophy and politics: Why John Dee wrote his *Compendious Rehearsal* in November 1592. *Studies in History and Philosophy of Science* 43, 480–488.

Pile, S. 2005a. *Real Cities.* SAGE, London.

Pile, S. 2005b. Spectral cities: Where the repressed returns and other short stories. In J Hillier and E. Rooksby (eds.) *Habitus: A Sense of Place.* Ashgate, Aldershot, pp. 235–257.

Pile, S. 2006. The strange case of Western cities: occult globalisations and the making of urban modernity. *Urban Studies* 43, 305–318.

Pile, S. 2011. Distant feelings: Telepathy and problem of affect transfer over distance. *Transactions of the Institute of British Geographers* 37, 44–59.

Pop, D. 2014. The wizards of the violet flame: A magical mystery tour of Romanian politics. *Journal for the Study of Religions and Ideologies* 13(38), 155–171.

Richardson, A. 2007. *Priestess: The Life and Magic of Dion Fortune.* Thoth Publications, Loughborough.

Spence, R. 2008. *Secret Agent 666: Aleister Crowley, British Intelligence and the Occult.* Feral House, Port Townsend, WA.

Sturm, T. 2013. The future of religious geopolitics: Towards a research and theory agenda. *Area* 45, 134–140.

Thurgill, J. 2015. *Enchanted geographies: Experiences of place in contemporary British landscape mysticism.* Ph.D Thesis, University College London.

Trietel, C. 2004. *A Science for the Soul: Occultism and the Genesis of the German Modern.* Johns Hopkins University Press, Baltimore.

Wooley, B. 2001. *The Queen's Conjuror: The Science and Magic of Dr. Dee.* HarperCollins, London.

14 'Where should we commence to dig?'

Spectral narratives and the biography of place in F. B. Bond's psychic archaeology of Glastonbury Abbey

James Thurgill

Introduction

In 1918 Frederick Bligh Bond published his report on the archaeological excavations he and his team had conducted at the ruins of Glastonbury Abbey in the West of England. Bond, a practicing architect at the time of his appointment to excavate the site, possessed a deep interest in medieval ecclesiastic architecture and design, publishing a number of architectural reports and essays on case studies in the West of England from 1902 onward, as well as his first major work on the subject, *Roodscreens and Roodlofts*, in 1909. Bond's existing notoriety as an architectural historian, particularly of pre-reformation churches, had led to him being appointed Director of Excavations at Glastonbury Abbey in 1908. Though merely an amateur in the field of archaeology, Bond made significant progress in uncovering sections of the abbey's foundations hitherto believed to have been lost for centuries. The success of Bond as an archaeologist is, however, not quite as straightforward as it might first appear; rather the case of Glastonbury Abbey remains one of Britain's most perplexing unsolved mysteries.

As with Glastonbury itself, the town from which the abbey takes its name, the ruins are shrouded in religious myth, with Bond's contribution serving as only one of myriad mythical and mystical narratives that are rooted in the area (Hopkinson-Ball, 2007, 2012; Michell, 1989). Marion Bowman (2009) speaks of the complexity of Glastonbury, positioning the site as a multitude of place-based encounters. To be sure, Glastonbury is not a single space; it has a variety of natural features and constructed sites that are imbued with different resonances, attractions and meanings (Bowman, 2009, 167). Through such a lens, Glastonbury becomes represented by a fluid identity; one that is both essentialized and yet capable of shifting the conceptualization of its 'rootedness' in and between myriad mythical readings, from biblical to spectral (Bowman, 1993, 2004; Holloway, 2000, 2003a, 2003b; Prince and Riches, 2000; Wylie, 2002). What sets Bond's story apart is perhaps the evidential nature of his findings, that his claims are (somewhat) supported by material archaeological evidence, not that this has deterred skepticism. In his *The Gate of Remembrance: The story of the psychological experiment which resulted in the discovery of the Edgar Chapel at Glastonbury* (1918), Bond describes how he and his co-investigator, John Alleyne, made use of automatic writing and necromancy in order to communicate with the otherworldly characters of the monks who had lived in and built the abbey. Moreover, Bond makes use of the experiment to test his theory of 'Greater Memory', a collective (perhaps cosmic) historical memory that transcends and interpenetrates our own (1918, vii). Bond's is perhaps the first example of what we might term psychic archaeology (Williams, 1991) – a form of excavation that initially relied on Spiritualist practices in order to uncover the material past hidden beneath us.

Archaeology and Spiritualism might well seem like unlikely bedfellows, particularly when it is so often the case that the former is considered to challenge, 'sanitize' or undermine the complex spiritual topology of the latter. Yet both fields are inextricably linked through the mechanism(s) that each employs in order to galvanize an understanding of place via the chronicling of history and the human experience. Both archaeology and Spiritualism make strategic use of narrative insofar as they seek to explore and recount events, encounters and practices that are rooted in memory, people(s) and landscape. Furthermore, both the archaeologist and the spiritualist privilege an embodied experience of time and space: The archaeologist quite literally places their body within the physical space in which they seek to excavate – the earth itself – whilst the spiritual is encountered as a psychophysical experience, with the body being utilized as a physically located material receptacle for interaction with the immaterial divine. From the outset, then, archaeology is a spectrally inflected practice; it seeks to uncover traces of the past, physical memories that are concealed by time and earth, uncovered through intuition, a mental projection of where material history might lay, as well as making use of both recorded history and professional expertise. As such we might view the archaeological discipline as engendering a sort of spectral materialism through which a greater understanding of the past-world is generated in the bringing forth of objects consigned to an otherwise forgotten history.

For both archaeologist and spiritualist, then, the world evolves around them, with relationships to place becoming of greater importance where affective interactions with the unseen are made manifest. It goes without saying that there are, of course, obvious differences between the *telos* of archaeology and that of spirituality (if the latter can even be said to possess one) and I do not seek to make a case for reconciling the ultimate aims of the two here. However, I wish to make use of both practices so as to demonstrate narrative and experience as central to an understanding of place and our connection to it, as illustrated through the example of Bond and his psychical experiments.

Using Bond's excavations of Glastonbury Abbey as a case study, this chapter aims to demonstrate the integral role spectrality plays in the building and rebuilding of spiritual-spatial narratives, positioning place (and its biography) at the heart of Bond's ghostly encounters. However, the case of Bond's Glastonbury excavations expresses a much wider relevance for readers than of the importance of spectrality alone; rather, Bond's work exemplifies a direct challenging of the affective qualities of place – one that sees the workings of the immaterial revealed to us as an autonomous, communicative force worthy of our consideration. Put another way, in thinking beyond the supernatural, Bond's report provides us with an opportunity to rethink how we relate to places and their histories, showing us through overtly practical (and material) means the innovative ways in which we might reconnect and engage with the world around us.

Spectral geographies: haunting, narrative and memory

Before entering into an account of Bond's excavations at Glastonbury it is worth considering how and in what ways spectrality itself functions as a component of spatial experience. Spectral and spectro-geographies have, in recent years, emerged to become a foundation for discussions of the unseen, affective qualities of place (Thurgill, 2014; Trigg, 2012). Undoubtedly influenced by Derrida's *Spectres of Marx* (1994), readings of spectral geographies can be seen as responding to the sense of loss or haunting that came with the *fin-de-siècle* of the 20th century (Maddern and Adey, 2007). Increasing developments in the use of the revenant as a mode of cultural (and spatial) enquiry has led to what Luckhurst terms the 'spectral turn', an analytical framework negotiated through 'a language of ghosts and the uncanny . . . of anachronic spectrality and hauntology' (2002).

Geographers in particular have placed much emphasis on the roles of haunting, spectrality and Forteana when assessing our engagement with place and space. Though not an exhaustive list, a number of notable spectral tropes have developed in light of contemporary geographic research: affect (Pile, 2012); death and dying (Maddrell, 2013; Maddrell and Sidaway, 2010); phantasmagoria (Hetherington, 2001; Pile, 2005a, 2005b); supernatural agency (Dixon, 2007; Holloway, 2006, 2016); literary hauntings (Magner, 2015; Matless, 2008; Wylie, 2008); spiritual practice (MacKian, 2011, 2012); ruination (Edensor, 2005, 2008; Trigg, 2012) and tourism (Holloway, 2010; Inglis and Holmes, 2003). Such a wide range of geographic interest in haunting serves to demonstrate the relevance of spectral geographies to scholars of the spatial humanities, galvanizing the need to analyze and interpret the workings of memory, history and absence in our experiences of place. This occulted vision of place is dependent upon a spatial memory that we are, for the most part, unable to tap into. Haunting has a way of readdressing the balance between the future and the past, allowing us to move beyond the confines of the present and affording us a temporal experience that oscillates between things both present and absent. Haunting provides an antidote to the anxiety of spatial and spiritual kenophobia that pervades our daily lives – the fear that all which lay before us is but nothingness, the abyss. Spectrality allows for a return, moreover, it necessitates one.

Like all forms of narration, spectro-spatial narratives are place-based stories, accounts expressed as a 'complex network of relations' (Cobley, 2001). It has been argued elsewhere that alternative spiritual practices add to the creation and narration of a biography of place through their apparent willingness to deal with a place's immaterial qualities (Thurgill, 2015a, 2015b): Bond's psychic archaeology is surely an example of this process. How a place comes to be known to us is largely determined through a combining of spatial experience and narrative.

Places are necessarily rooted within a socio-historical context and encountering them comes about through a negotiated enquiry whereby one seeks to become informed of previous uses, inhabitants and events that have occurred there. As such, we might well consider place to be palimpsestic, inscribed with the writing and rewriting of its past, laden with memory.

The philosopher Edward Casey (2000) describes place as a situated repository for memory, 'a mise-en-scene for remembered events precisely to the extent that it guards and keeps these events within its self-delimiting perimeters'. In this sense one might argue that any encounter with place is an archaeology of sorts. In attending to place we are forced to unearth the memories stored within it, deciphering particularities, histories and our own experience in order to construct a biographical understanding of the space we (temporarily) occupy. In doing so, place comes to be understood as existing within a continuously unfolding storyline, one that is always both historical and contemporary.

The haunting qualities of place, the manner in which the past permeates and lingers within the present, is a much documented interpretation of the affective mechanisms at play in spatial encounters (Benjamin, 1999; de Certeau, 1988; Pile, 2005a, 2005b, 2012; Till, 2005). Historical narratives, both officially recorded and those presented in oral traditions, feed the idea that place exists within a progressing narrative, that there is some sense of a (linear) plot which can be traced or followed in order to *know* about where we are. This spatial story is made up both of tales told and re-told but also from the memories of a place that are left to be uncovered. This narrative or biography of place informs our spatial experience, setting to work a psychical engagement with the affectual qualities of a site. The psychic archaeology of F. B. Bond works, as we shall see, to speak to the haunting of place, to allow its history to unfold and its story to be uncovered.

The quest at Glastonbury

In 1908 Bristol-based architect and archaeology enthusiast, Frederick Bligh Bond, was appointed Director of Excavations at the ruined ecclesiastical site of Glastonbury Abbey. Prior to the start of the excavations, Bond had spent much of the previous year studying manuscripts and plans of the abbey as well as making a number of field visits to the ruins themselves. Bond was later commissioned to dig the site in search of 'lost' foundations, particularly those belonging to an apse of the so-called Edgar Chapel. It was during these initial archaeological investigations that Bond first began to experiment with psychic enquiry, seizing the opportunity to 'test' his theory of 'Greater Memory' (Bond, 1920, vi). Rejecting conventional belief that memory is confined to the corporeal life (and death) of the individual, Bond conjectured that postmortem memory could be consigned to a greater, infinite source of memorial record and that, through certain psychological training, this cosmic repository of memory could be accessed remotely by the living. Bond viewed this 'Greater Memory' as 'transcending the ordinary limits of time, space and personality' (1920, 20), an ever-growing, omnipresent resource at the disposal of man's mental faculties. Furthermore, Bond sought to test his hypothesis through the (then future) excavations of Glastonbury Abbey, developing a working methodology for psychic archaeology. Bond framed his unusual archaeological approach as a 'psychological experiment', a term he went on to place in the subtitle to his accounts of the Glastonbury dig.

Bond was not alone in his belief that psychical methods could be of use in the excavation of historical sites and, together with like-minded former naval officer, close friend and Spiritualist, Captain John Allen Bartlett, who worked under the pseudonym of John Alleyne, Bond began experimenting with the use of automatic writing and Ouija boards. The pair began preliminary investigations of the ruins in 1907 when Bond was anticipated to become Director of Excavations at Glastonbury Abbey on behalf of the Somerset Archaeological Society (1920, 7). Bond describes in the preface to his account how he and Alleyne developed their practice so as to successfully excavate the abbey site 'in the absence of physical remains and [a] lack of trustworthy evidence from documents' (1920, 8).

In setting out the principles and methods of psychology applied to his archaeological research, Bond criticized the over-rationalized thinking of Western society, claiming that the West looked only to measurable phenomena for explanations of the world, whilst the East turned inwards, to imagination: 'Western folks' Bond suggested, 'think it unpractical to cultivate this gift (of imagination) . . . we have no system of training it' (1920, 18). From the outset, Bond's thinking appears to have been influenced by Eastern spiritual and philosophical traditions, namely Buddhism and Hinduism as well as Kabbalistic mysticism. His intention to qualify the uses of his theory through a blending of these spiritualities and their associated mystic practices can be seen at least two years prior to the physical excavations taking place. The excavations held at Glastonbury can be viewed in part, then, as Bond's attempt to 'train' the mind, or his mind at least, reconciling this difference in thinking so that the truth of the (spatial-) past might be revealed: 'a more contemplative element in the mind would seek to revive from the half-obliterated traces below' (1920, 19). Bond outlines his thinking as follows:

> The germination of new and profitable ideas in the mind may in this respect be brought about, firstly, by a suitable system of mental exercise and culture; secondly, by a willingness to hold back all mental preferences and preconceptions, and to restrain also the surface activities of the brain, so that the channel of pure 'idea' which resides in the subconscious mind may be maintained, and the finer activities allowed to percolate.
>
> (1920, 24)

The experiments were themselves predominantly practiced remotely from Bond's architectural practice's offices in the Alliance Chambers, 36 Corn Street, Bristol. Bond describes how he would hold a piece of foolscap paper in place with his left hand whilst gently resting his right hand on top of Alleyne's, which in turn would be lightly clasping a pencil. Bond would start by asking a question 'as though addressed to some other person' (1920, 32). Alleyne's fingers would start to move and the pencil would begin to make small, irregular lines. Bond notes that '[t]he agreed method was to remain passive, avoid concentration of the mind on the subject of the writing, and to talk casually of other and indifferent matters' (1920, 32). Indeed, Bond writes in a footnote found in the Preface to the Second Edition (of *The Gate of Remembrance*) how he would often read aloud from a novel to Alleyne throughout the process so as to draw the medium's mind away from consciously answering the questions posed and allow for the unadulterated recording of 'small voices from a distant time' (1920, viii). Neither man would attempt to read or decipher the writing until after the communication had ended, thus removing any additional element of intention from the automatist process.

Bond and Alleyne's experiments with automatic writing produced a seemingly 'queer' and 'curious patchwork of Low Latin, Middle English of mixed periods, and Modern English of varied style and diction' (1920, v) with responses to Bond's questions growing of greater accuracy and detailed instruction 'as though in obedience to some preordained intention and settled plan' (1920, v). Naturally, Bond's account appeared all the more genuine because of the 'details of its structure and history, written in the archaic language of the old monks' themselves (Michell, 1989). However, it should be noted that the languages Bond claimed to be in use throughout the sessions of automatism were rife with error and inconsistency. Knowing this to be a point of contention, Bond accounted for any flaws or errors in the transcribed Latin by conjecturing that the communications merely reflected the pervasive illiteracy that would have been present during the lives of the communicators, therefore, erroneous spelling was to be expected (Bond, 1920, 31). Defensive of his technique, Bond sought to instate automatism as a 'gift of tongues' parallel to those present in the biblical imagination, citing Saint Peter so as to argue for a 'great revival' of these powers – a revival that Bond himself undoubtedly hoped to initiate (1920, 22–23). Both Bond and Alleyne considered these experiments to engender a 'fruition of memories and experiences long dormant and inaccessible to us' (1920, 25).

Though both men had, in one way or another, been involved in the Spiritualist practices that led them to pursue this line of investigation, Bond notes the automatism used to investigate Glastonbury Abbey as a marked departure from Spiritualism per se: 'neither F.B.B. nor J.A. favoured the ordinary spiritualistic hypothesis which would see in these phenomena *the action of discarnate intelligences from the outside upon the physical or nervous organisation of the sitters*' (1920, 19). Such a view is important here in that it moves away from the idea that Bond and Alleyne were under the influence of spirits, as such, and toward the notion of an engaged interaction with a greater source of shared memory, similar to that which was later alluded to by Jung in his writings on psychology and the occult as the 'collective unconscious' (2008). If what Bond describes is to be believed, then both he and Alleyne were novices in spirit communication. Indeed, Bond makes the claim that although Alleyne had previously experienced a talent for divination twice before, it had been without intention or a willing of the gift and so should not be considered connected to the Glastonbury Abbey experiments.

Bond conducted his initial archaeological surveys of the site during the months of May through August 1907, his seasonal approach being somewhat enforced by insufficient funding and a dwindling team of labourers (Kenawell, 1965). Initial experiments with psychical phenomena produced intriguing results for Bond. The first experiment with automatism took place in Bond's offices on 7th November 1907; Bond posed the question '*Can you tell us anything about Glastonbury?*' to which he received the answer: '*All knowledge is eternal and available to mental sympathy*' (1918, 32). After a short pause in communication a second message was transcribed: 'I was not in sympathy with monks – I cannot find a monk yet'. As the experiment continued information on the history of the abbey started to be received as well as what appeared to be a hand drawn plan of the abbey, including a sketch of the hitherto lost Edgar Chapel. Glastonbury Abbey was dissolved in 1539 AD during Henry VIII's Dissolution of the Monasteries act, at which point, according to Bond's research, the Edgar Chapel disappeared from historical record (1918, 4). Much like the presence of the chapel itself, Bond noted how communicative narratives would terminate without warning, only to be resumed with the flow of information re-established in the following séance (1918, 30).

In order to qualify the 'scientific' nature of Bond's experiments, he enlisted Everard Feilding, Secretary of the Society for Psychical Research, as an independent adjudicator so as to provide an outside testimony to the truth of the account he would later give on the excavations and the psychical communications that led to his success (1918, 6). Bond included an extract of a letter from Feilding written in March 1917 within the body of the excavation script. In the letter, Feilding encourages Bond to publish his findings and reiterates his belief that no knowledge of the Edgar Chapel's location existed prior to Bond and Alleyne beginning automatism.

By the sixth sitting Bond had been instructed on where to begin his excavations:

> Sitting VI, 26th November 1907
>
> Question: 'Where should we commence to dig?' Answer: 'The east end. Seek for the pillars and the wall(s) at an angle. The foundations are deep'.
>
> (1920, 41)

Further sittings continued to provide information on the whereabouts of the Edgar Chapel's remains, as well as detailed descriptions of the personalities that communicated through Alleyne. Amongst the most frequent and detailed communications derived were those emanating from three particular spirits: Gulielmus (William the Monk), Johannes (truant monk and nature-lover) and Abbot Beere (received as Bere in the communications) (1918, 45). These three spirit characters, Bond claims, provided substantial accounts and descriptions of daily life at the abbey; of feuds and battles; of leisure pursuits and the integral role of nature; historical accounts of the construction of the chapel; of traders, visitors and founders of the abbey (including its connection to Joseph of Arimathaea); as well as of rites and rituals conducted within the abbey walls. These spectral figures formed part of a collection of spirits who termed themselves simply 'The Watchers' and gave Bond the exact locations of those parts of the abbey he was seeking to excavate, in turn demonstrating a bond (to Bond) between Bond's work and the historical guardianship of the Abbey (1918, 93). Sitting XXVII conducted on 17th March 1908 at Bond's Bristol offices gave the amateur archaeologist the final sign he was waiting for: 'The time is ripe for the stones to be studied', and in May of the same year Bond finally acquired a licence to excavate the ruins, acting as a representative of the Somerset Archaeological Society (1918, 45).

Bond and Alleyne continued their own psychical experiments in private and on 16th June 1908 the final detailed description of the Edgar Chapel was recorded, providing Bond with measurements of its foundations along with hand drawn plans of the abbey buildings (1918, 28–29). When Bond finally began to excavate the remains of the abbey, following the instructions afforded him via the automatism, he successfully uncovered the foundations of a large rectangular chapel east of the abbey's retro-quire. Bond recalls in his text that many of the measurements received in the sittings appeared conflicting and as such he was required to repeat and probe in his questioning in order to make sense of the dimensions. The information recorded in Bond and Alleyne's transcription of the sittings provides sixteen references to the Edgar Chapel and East End of Quire (1918, 70), in addition to supplementary information on King Arthur ('The tombe of Arthur in shining black stone was in fronte ye altare' (Bond, 1918, 65) and skeletal remains found outside the church walls (named in automatic writing as Eawulf, Earl of Somerton: Bond, 1918, 106) which were found to have been buried along with a second human skull.

By the time the third edition of his account was published in 1920, Bond had been provided with yet further archaeological data through his psychical workings and had moved on to excavating the remains of a second formerly 'lost' structure, the Loretto Chapel. On the discovery of the Loretta Chapel Bond wrote:

> [T]his discovery sets the seal upon the veridical nature of the writings, and emphasizes the importance of the method employed by the author for the recovery of latent knowledge.
>
> (1920, 2)

This second major discovery was, for Bond, the point at which the theory of 'Greater Memory' was ratified by reality, a demonstration to his critics that otherwise (seemingly) impossible material finds were made possible through 'kindred knowledge from the great reservoir of the memory of nature' (1920, 112).

Bond's psychical and physical excavations were doubtless a continuation of his existing interest in both Spiritualism as well as geomantic and Eastern traditions. In addition to the publication of his script as a monograph, Bond published further reports as articles in *Psychic Research Quarterly*[1] and the *Journal of the American Society for Psychical Research*, as well as a collection of nine pamphlets entitled *The Glastonbury Scripts*.[2] His meditations on The Chapel of Our Lady (A.D. 1184) reportedly built on the site of the Church of Joseph of Arimathaea (confirmed in Bond's spirit communications) demonstrate an architecture that follows the spiritual instruction of the Gematria in the Greek scriptures, a practice that utilizes geometry and numerology for the purposes of engendering sacrality.[3]

Furthermore, as a trained architect Bond was a talented illustrator and used this skill to produce conjectural reconstructions of the abbey and its associated structures; this he did so as to augment the descriptive accounts arrived at through spirit communication and automatism. In doing so Bond provided yet another demonstration of his unique insight into the place and biography of the abbey.

For more than a decade Alleyne and Bond conducted séances and varying forms of divination in order to commune with the spirit-memories of the monks who had frequented the abbey all those centuries ago. Over fifty communications were made between 7th November 1907 and 30th November 1911, with a number of additional spirit writings being gathered from 1912 and later. The script generated by the pair was in fact so vast that Bond published an additional text from the experiments in 1919.[4] This second text Bond titled *The Hill of Vision*, and which saw a gathering together of the messages received from initial sittings that did not directly correspond to Glastonbury Abbey or its locale. The most significant claim of this secondary publication was that the messages received from Bond and Alleyne's sittings had prophesied the First World War. Together with *The Gate of Remembrance*, Bond had succeeded in bringing the entirety of his psychological experiments to the public's attention.

The experiments went on in private, many sessions taking place remotely whilst at Bond's Bristol based architectural office, behind the backs of the conservative church authorities who Bond had suspected would be less than enthused by the prospect of his psychical methods. Bond was right to be concerned. Following the publication of his findings in 1918, word of Bond's occult practices spread. Bishop Armitage Robinson eventually dismissed Bond from his role in 1922 on grounds of using necromancy at a consecrated site. Following his rejection by the Church, skepticism from the archeological community and a damaged reputation, Bond left the UK to continue his psychical research in America. Treatment of Bond's claims was (and remains) in many ways far less fair then it ought to have been. Accusations of fraud undermined the very concrete findings (seen in the form of archaeological evidence) that Bond presented his readers with. Furthermore, from the very beginning of his account, Bond set the tone as one of both enquiry and experimentation; he defined his work as suggestive, imploring the reader to treat the text 'with an open mind' (Bond, 1920, 112) and concluded by stating that the account was a demonstration of his working method and the results presented were 'not to be accepted with credulity, but are subjects for critical analysis' (1920, 155). To follow Bond's conjecturing on the nature of memory is to surrender to the spectral, and in doing so, he suggests 'we should stand at the threshold of the Gate of Remembrance' (1920, 144).

Coda

Notwithstanding the obvious criticism that such an account would face, Bond's description of his psychic archaeology, and the successes he had in discovering both the Edgar and Loretta chapels at Glastonbury Abbey, remains one of the most fascinating examples of Forteana to date. The series of questions and responses that were recorded by Bond highlight the specificity of the information that was 'coming through' during the automatic writing sessions and support Bond's theory that his apparent success at locating remains at the site was partially down to inexplicable occurrences.

The methodologies employed by Bond are the first documented instances of psychic archaeology; a fringe discipline that remains practiced today. Bond's techniques for uncovering the material past through spiritually inflected archaeological practices provided the foundations for later spiritual enquiries into landscape such as those seen in the works of Broadhurst and Miller (1990); Devereux (1991, 1994, 2010); Foster-Forbes (Foster-Forbes and Campbell, 1973); Lethbridge (1957, 1963); Underwood (1968) and Watkins (1922, 1925). My interest in this case is not so much to validate claims of the existence of the paranormal, nor to prove or disprove the information gathered by Bond and his team to have emanated from spectral sources. Rather, the subject of interest here is one of re-imagining our connections to the narrative, history and sacrality of place and moreover, what this might mean for the development of a biography of place, an unending spatial-story in which we play an active role in shaping, challenging and re-writing place. Of wider interest still is Bond's unique way of challenging the tension between matter and immateriality. The ghosts given voice through Bond's experiments call for us to pay attention to place, to listen to its stories. In analyzing the affective nature of place(s), we often overlook any sense of purpose to being moved or disturbed by our surroundings, but Bond teaches us otherwise. By returning spirits to place, Bond and Alleyne's experiments work to demonstrate that affect teaches us about place, that through being affected by a site's physical and psychical offerings we can gain a greater understanding of our role in its history. It is not so much spectrality, then, that determines the significance of Bond's work, rather, it is the innovative way in which he works with materiality to 'show' us the value of the immaterial.

Furthermore, it is worth considering what implications a case such as Bond's might have for the geographic imagination. How might the spectres of the abbey change the way we view the relations between people, place and affect? It appears that whether real or figurative, Glastonbury Abbey retains its own unique set of spirits. Bond made a connection to these through a multiplicity of materialities; the abbey site, masonry, soil, grass, shovel, paper, the very lead of John Alleyne's pencil that permeated the foolscap and recorded the communications. Through each of these objects, Bond was able to ascertain the location of the hidden chapel(s). He did so through an engagement with the immaterial via uncontestably material means: the site spoke to Bond, spectrally narrating its own story through unquestionably corporeal processes. The tension between these two, the material and the immaterial, is what forces open a space for the engendering of the spectral, and moreover is the site at which we engage with a biography of place. The mobilization of a mysticism that shrouds the abbey, as well as the wider setting of the Somerset town of Glastonbury (Cope, 1998; Michell, 1989), makes for a deeply affective environment that is situated within a landscape saturated by its past and associated narrative(s) of myth. Such a landscape is where new meanings and re-imaginings can take place. We might refer to such spatialities as being the place where the (im)material comes to exist; where the immaterial inflects and works upon materiality. The absence of the definite here gives way to a place of exploration; the chapel remains were simultaneously lost and present, both there (existing) and not there (unseen). Bond exploited the tension between this absent-presence further through the replication of the process; he engaged with unseen subjects, using material *things* to bring their voices to life.

It strikes me that the interaction between the historical figures of the abbey (the architect and the monks) has further developed the hauntological existence of the site. If, as Tim Edensor (2005) posits, ruins act as spaces for rethinking history of their own accord, then they possess the affectual qualities that lead one to contemplate and imagine previous acts of habitus that would have occurred within them. In this sense structural remains haunt us by continually allowing the past to permeate the present. In bringing the 'spirits' of the Glastonbury ruins to life, Bond created a duel haunting whereby ghosts appeared both as absences made present (the ruins) and as unseen agents (the spirits). The spectralization of such a site is amplified further through the retelling of Bond's discoveries here: the abbey, its ghostly narrative and indeed Bond himself will continue to occupy the space; the (after)lives of all three caught up in a continuous cycle of haunting.

Notes

1 Bond, F. B. (1920–1) 'The discoveries at Glastonbury', *Psychic Research Quarterly*, 1, pp. 302–312. Online at www.iapsop.com/archive/materials/psychic_research_quarterly/psychic_research_quarterly_v1_1920-1921.pdf. Last accessed 7 July 2016.
2 Coates, R. (2015) *Frederick Bligh Bond (1864–1945): a bibliography of his writings and a list of his buildings*. Working Paper, University of the West of England (Research Repository), Bristol. Online at http://eprints.uwe.ac.uk/25679.
3 Bond, F. B. and Simcox Lea, T. (1917) *A preliminary investigation of the cabala contained in the Coptic Gnostic books and of a similar gematria in the Greek text of the New Testament, shewing the presence of a system of teaching by means of the doctrinal significance of numbers, by which the holy names are clearly seen to represent aeonial relationships which can be conceived in a geometric sense and are capable of a typical expression of that order*. Oxford, UK: B. H. Blackwell.
4 Bond, F. B. (1919) *The Hill of Vision*, a forecast of the Great War and of social revolution with the coming of the new race, gathered from automatic writings obtained between 1909 and 1912, and also, in 1918, through the hand of John Alleyne under the supervision of the author. Boston, MA: Marshall Jones Co.

References

Benjamin, W. (1999) *The Arcades Projects (1927–40)*. Cambridge, MA: Harvard University Press.
Bond, F. B. (1918) *The Gates of Remembrance*. First Edition. Oxford: Blackwell Publishing.
Bond, F. B. (1920) *The Gates of Remembrance*. Third Edition. Oxford: Blackwell Publishing.
Bond, F. B. and Bede Camm, D. (1909) *Roodscreens and Roodlofts*. London: Isaac Pitman. Online at https://archive.org/stream/roodscreensroodl01bond. Last accessed 4 July 2016.

Bowman, M. (1993) *The Avalonians*. Glastonbury: Gothic Image Publications.
Bowman, M. (2004) 'Procession and possession in Glastonbury: continuity, change and the manipulation of tradition', *Folklore*, 115(3), pp. 273–285.
Bowman, M. (2009) 'Learning from experience: the value of analysing Avalon', *Religious Studies*, 39(2), pp. 161–168.
Broadhurst, P. and Miller, H. (1990) *The Sun and the Serpent: A Journey of Discovery Through the British Landscape, Its Mythology, Ancient Sites and Mysteries*. Launceston, Cornwell: Mythos Press.
Casey, E. (2000) *Remembering: A Phenomenological Study*. Second Edition. Bloomington and Indianapolis, IN: Indiana University Press.
Cobley, P. (2001) *Narrative: The New Critical Idiom*. London and New York: Routledge.
Cope, J. (1998) *The Modern Antiquarian: A Pre-Millennial Odyssey Through Megalithic Britain*. London: Thorsons.
de Certeau, M. (1988) *The Practice of Everyday Life*. London: University of California Press.
Derrida, J. (1994) *Spectres of Marx*. Oxford: Routledge.
Devereaux, P. (1991) *Earth Memory*. London: Quantum.
Devereux, P. (1994) *The New Ley Hunter's Guide*. Glastonbury: Gothic Image Publications.
Devereux, P. (2010) *Sacred Geography: Deciphering Hidden Codes in the Landscape*. London: Octopus Publishing Group.
Dixon, D. (2007) 'A benevolent and sceptical inquiry: exploring 'Fortean Geographies' with the Mothman', *Cultural Geographies*, 14(2), pp. 189–210.
Edensor, T. (2005) *Industrial Ruins*. Oxford: Berg.
Edensor, T. (2008) 'Mundane hauntings: commuting through the phantasmagoric working-class spaces of Manchester, England', *Cultural Geographies*, 15(3), pp. 313–333.
Foster-Forbes, J. and Campbell, I. (1973) *Giants, Myths and Megaliths*. Self Published.
Hetherington, K. (2001) 'Phantasmagoria/Phantasm Agora: materialities, spatialities and ghosts', *Space and Culture*, 11(12), pp. 24–41.
Holloway, J. (2000) 'Institutional geographies of the New Age movement', *Geoforum*, 31(4), pp. 553–565.
Holloway J. (2003a) 'Make-believe: spiritual practice, embodiment and sacred space', *Environment and Planning A*, 35(11), pp. 1961–1974.
Holloway, J. (2003b) 'Spiritual embodiment and sacred rural landscapes', in P. Cloke (ed.) *Country Visions*. Harlow, Essex: Pearson Education Limited, pp. 158–175.
Holloway, J. (2006) 'Enchanted spaces: the séance, affect, and geographies of religion', *Annals of the Association of American Geographers*, 96(1), pp. 182–187.
Holloway, J. (2010) 'Legend-tripping in spooky spaces: ghost tourism and infrastructures of enchantment', *Environment and Planning D: Society and Space*, 28(4), pp. 618–637.
Holloway, J. (2016) 'On the spaces and movements of monsters: the itinerant crossings of Gef the Talking Mongoose', *Cultural Geographies*, 24(1), pp. 21–41.
Hopkinson-Ball, T. (2007) *The Rediscovery of Glastonbury: Frederick Bligh Bond, Architect of the New Age*. Stroud: Alan Sutton.
Hopkinson-Ball, T. (2012) *Glastonbury: Origins of the Sacred*. Bristol: Antioch Papers.
Inglis, D. and Holmes, M. (2003) 'Highland and other haunts: ghosts in Scottish tourism', *Annals of Tourism Research*, 30(1), pp. 50–63.
Jung, C. (2008) *Psychology of the Occult*. London: Routledge.
Kenawell, W. W. (1965) *The Quest at Glastonbury: A Biographical Study of Frederick Bligh Bond*. New York: Helix Press and Garrett Publications.
Lethbridge, T. C. (1957) *Gogmagog: The Buried Gods*. London: Routledge and Keegan Paul.

Lethbridge, T. C. (1963) *Ghost and Diving Rod*. London: Routledge and Keegan Paul.

Luckhurst, R. (2002) 'The contemporary London gothic and the limits of the "spectral turn"', *Textual Practice*, 16(3), pp. 527–546.

MacKian, S. (2011) 'Crossing spiritual boundaries: encountering, articulating and representing otherworlds', *Methodological Innovations Online*, 6(3), pp. 61–74.

MacKian, S. (2012) *Everyday Spirituality: Social and Spatial Worlds of Enchantment*. Basingstoke, Hampshire: Palgrave Macmillan.

Maddern, J. F. and Adey, P. (2007) 'Editorial: spectro-geographies', *Cultural Geographies*, 15(3), pp. 291–295.

Maddrell, A. (2013) 'Living with the deceased: absence, presence and absence-presence', *Cultural Geographies*, 20(4), pp. 501–522.

Maddrell, A. and Sidaway, J. (eds.) (2010) *Deathscapes: Spaces for Death, Dying, Mourning and Remembrance*. Farnham: Ashgate Press.

Magner, J. L. (2015) 'Looking through time itself: Henry Handel Richardson and the haunting of lake view', *Literary Geographies*, 1(2), pp. 195–212.

Matless, D. (2008) 'A geography of ghosts: the spectral landscapes of Mary Butts', *Cultural Geographies*, 15(3), pp. 335–357.

Michell, J. (1989) *The New View Over Atlantis*. London: Thames and Hudson.

Pile, S. (2005a) *Real Cities*. London: SAGE Publications Ltd.

Pile, S. (2005b) 'Spectral cities: where the repressed returns and other short stories', in J. Hillier and E. Rooksby (eds.) *Habitus: A Sense of Place*. Aldershot: Ashgate Publishing Group, pp. 235–257.

Pile, S. (2012) 'Distant feelings: telepathy and the problem of affect transfer over distance', *Transactions of the Institute of British Geographers*, 37(1), pp. 44–59.

Prince, R. and Riches, D. (2000) *The New Age in Glastonbury: The Construction of Religious Movements*. New York and Oxford: Berghahn Books.

Thurgill, J. (2014) *Enchanted geographies: Experiences of place in contemporary British landscape mysticism*, unpublished manuscript, last modified November 2014.

Thurgill, J. (2015a) 'Mobilities of magick', in T. Gale, A. Maddrell and A. Terry (eds.) *Sacred Mobilities: Journeys of Belief and Belonging*. London: Ashgate Press, pp. 53–70.

Thurgill, J. (2015b) 'A strange cartography: leylines, landscape and "Deep Mapping" in the works of Alfred Watkins', *Humanities*, 4, pp. 637–652.

Till, K. (2005) *The New Berlin: Memory, Politics, Place*. Minneapolis, MN: University of Minnesota Press.

Trigg, D. (2012) *The Memory of Place: A Phenomenology of the Uncanny*. Athens: Ohio University Press.

Underwood, G. (1968) *The Pattern of the Past*. London: Museum Press.

Watkins, A. (1922) *Early British Trackways, Moats, Mounds, Camps and Sites*. London: Simpkin, Marshall, Hamilton, Kent and Co.

Watkins, A. (1925) *The Old Straight Track*. Third Edition. London: Methuen.

Williams, S. (1991) *Fantastic Archaeology: The Wild Side of North American Prehistory*. Philadelphia, PA: University of Pennsylvania Press.

Wylie, J. (2002) 'An essay on ascending Glastonbury Tor', *Geoforum*, 33(4), pp. 441–454.

Wylie, J. (2008) 'The spectral geographies of W.G. Sebald', *Cultural Geographies*, 14(2), pp. 171–188.

15 Categorizing Spiritualism as a shamanism

Lessons in mapping

David Gordon Wilson

Introduction

This contribution explains the background to 'Redefining Shamanisms' (Wilson 2013), to the doctoral thesis upon which it was based, and to the underlying research, with the intention of summarizing some of the insights gained. The methodologies I adopted drew in part upon geography at its most traditional, and were a potent reminder that comparative and other established methodologies, which may seem to be exhausted, or perhaps just overly familiar, can continue to offer new insights.

I will explain my interest in the modern Anglo-American tradition of Spiritualism, by which I mean the religio-philosophical movement that makes use of human mediums to communicate with the spirits or souls of the deceased (perceived as people who are alive but no longer incarnate), and which has found its particular home in Anglo-American culture of the mid-nineteenth century to date (Nelson 1969). I also detail my approach to studying contemporary Spiritualist practices, and to categorizing Spiritualism as a distinct religious tradition. I show how my interest in shamanic traditions alerted me to the useful comparisons they offer in the endeavour to categorize Spiritualism. In this I follow in the footsteps of those who maintain that categorization is a core activity in human understanding, something that was first brought home to me by a particularly able Professor of Roman law during my undergraduate legal studies (Birks 1997).

It is in light of these concerns, and particularly the question of how to categorize Spiritualism, that I came to see my efforts as an exercise in mapping. I mean this in two distinct senses, which I term geographical and mathematical. A basic geographical approach might be to take a map of the world and proceed to plot areas where this or that religious tradition is dominant, or has been identified. At any given time there will be overlaps, porous boundaries, co-existence, local blendings, or simply gaps in scholarly knowledge. In plotting those parts of the world where scholars have identified shamanic traditions, it became apparent that there was an obvious lacuna, namely Europe (especially northern Europe), and, from the early to mid-nineteenth century, North America; to that list might be added recent Australia and New Zealand. This indicated that where Protestant Christianity is the normative tradition, shamanic practices tend not to be identified; to the extent that shamanism is identified, it tends to be perceived in non-Protestant, often indigenous, marginal sub-cultures.

It struck me that Protestant Europe and modern North America represent an obvious missing piece of the puzzle when assembling a map of shamanic traditions around the world, while also being the classic locus when mapping the presence of Spiritualism as a distinct tradition. Might these be pieces of the same jigsaw puzzle?

This question led me to the more mathematical sense in which we use the term mapping. In mathematics, if a shape or area can be transposed from its existing position onto a shape or area in a different location, and match, one can be said to map onto the other. This is a particular example of enquiry into the extent to which each of the elements of a set can be associated with those of another set, into the extent to which they can be said to be the same. For the purposes of the current exercise, the question becomes, 'Can the essential or characteristic elements of Spiritualism be identified and, if so, can they be associated with the essential or characteristic elements of shamanic traditions?' Neither of these tasks is easy.

This is a question that takes us back to the old practice of comparing things as a way of comprehending them, but the process of answering it, while revalidating comparative approaches, highlighted the importance of attending to our choices as to which characteristics to compare and contrast. At its broadest, this is the question of which traditions to compare; interestingly, this question is answered in part by ignoring some of the characteristics often assumed to be of the essence of religion, and it is certain characteristics of Spiritualism and shamanisms that alerted me to the need to do this. It is also important to point out that some previous scholarly attempts to locate Spiritualism suggested the shaman as a role with which the Spiritualist medium might bear comparison (Nelson 1969: 246), although, to my knowledge, that suggestion had not been explored in any detail.

A particular challenge presented by Spiritualism is that it is a singularly undogmatic religious tradition; it has no single founder, it has competing narratives, it has many written texts but no one set regarded as authoritative (Nelson 1969: 238–246). Ask ten Spiritualists what they believe and they will probably provide ten different answers, possibly even the outright response that it is not a matter of belief, although it is reasonable to expect them to centralize mediumistic spirit communication in some way. The point to highlight is that the element of consistency in the answers is not a belief but a practice; this is true regardless of one's own views as to whether any actual spirit communication is or can be present. Prioritizing belief misses the point of Spiritualism, as it often does with other traditions; at the very least, it can lead us to overlook those practices that actually maintain the tradition. This well illustrates why scholars of religion often prefer to focus upon practices, upon reading behavioural texts rather than written ones, or upon interpreting physical artefacts (including texts) as giving access to the ways in which people relate to each other so as to maintain a tradition (Jordan 2001). Which of these approaches is useful depends upon the tradition; traditions that are not significantly characterized by written records, or where scripture is not central to maintenance of the tradition, can be difficult to access absent personal involvement on the part of the researcher.

This is important as it indicates the methodology that is needed, something that is not always obvious. Methodologies embody (often implicit) assumptions, not only as to how to study but also as to what is worth studying. For this reason, methodologies can be powerful political tools: used foolishly, they can obstruct learning; used wisely, they can illuminate ideas and objects of study hitherto marginalized but which risk challenge.

In order to compare two things, one requires knowledge of both. In choosing two things to compare, there should be elements of commonality that offer a preliminary indication that the comparison might prove useful; an important favourable indication here is if it is apparent that an appropriate methodology is the same for both objects of study. Given an obvious need for analysis of practices rather than beliefs, some element of ethnography is indicated, especially given the paucity of existing examples. The need for this approach was strengthened by my awareness that shamanic studies as an academic discipline has relied heavily upon anthropologists, and therefore upon ethnography as a methodology. The classic example is the work of Russian ethnographer Sergei Mikhailovich Shirokogoroff, who worked among the Evenki (Tungus) tribes or clans of Siberia and northern China from 1913 to 1918 (Shirokogoroff 1929, 1935). It is from the Evenki, and as a result of Shirokogoroff's work, that we have the term šamān or shaman, which roughly translates as 'knower' ('gnostic' would be a good transliteration, were it not already employed elsewhere). Shirokogoroff felt that the description shaman should not be applied to comparable practitioners in other cultures, demonstrating a sensitivity to context that needs to be accommodated in making instructive comparisons; both commonalities and differences matter.

Despite influential efforts to do so, scholars of shamanism(s) have struggled to establish a definition of shamanism such as might allow the field to be clearly distinguished; indeed, discussion has been so prolonged, and the word now applied, academically and popularly, to so many particular examples that many, if not most, scholars have given up the attempt and maintain that the effort is naïve. I adhere, however, to the view that a category that does not work indicates that some things habitually included in it should not be there or have been inadequately comprehended; either way, it is likely that the question underpinning the category description has not yet been adequately formulated.

My interest in Spiritualism and (other) shamanisms is not only based upon my recognition of those traditions as fascinating in their own right, and for what they reveal about the societies in which they are bounded (both enabled and constrained), but is also prompted by the analytical challenges they present to scholars. Some of those difficulties arise simply because we bring ourselves to whatever we study, and it can take us time to realize that sometimes we must not only observe and analyze so as to learn but change in order to comprehend, perhaps especially so when engaged in ethnography.

Exploring Spiritualism and shamanism: mapping the globe and other initial comparisons

How then to show that Spiritualism is worth studying, how to identify that which is essential or strongly characteristic of Spiritualism so as to enable comparison, and how to choose an appropriate comparator?

It is difficult entirely to remove my personal experience from answering these questions. In September 2001, I began reading for an undergraduate degree in Divinity at the University of Edinburgh; with no intention of going into ministry, I was free to include courses in shamanism and African religious traditions. From personal curiosity, I also began attending services at Portobello Spiritualist Church in Edinburgh. Initially, I was curious as to the possibility that there might be elements of Spiritualist mediumistic practice that might parallel practices in the early Christian churches of the first century CE. That interest gradually took a back seat as I became a member of Portobello Spiritualist Church's development circle (the church's teaching forum), and increasingly began to draw comparisons between my experiences and observations there with the shamanic traditions I was learning about at Edinburgh.

From joining Portobello Spiritualist Church's development circle in January 2003, to leaving it in June 2007, I underwent a fascinating series of learning experiences; these are detailed more fully in the extended ethnography that forms Chapter 4 of 'Redefining Shamanisms' but, in brief, I gradually became a demonstrating medium at Portobello Spiritualist Church and across Scotland, with occasional forays into England and across to Canada. This extended involvement was recorded and reflected upon at length, initially for personal benefit, and from September 2005, as research towards my doctoral thesis at Edinburgh.

I began by making a simple comparison of mediumship within the Spiritualist tradition with shamanism, with the intention of testing whether Spiritualism could be categorized as a shamanism, the thesis being that it could indeed be so categorized. Initially, I had no particular expectations as to what might come from making that categorization, and was ill-placed even to say why it might be worth doing.

In its origins, the identification of shamanism is closely bound up with the traditions of the Evenki (Tungus) and other nomadic reindeer-herding peoples of Siberia, whose tribes or clans tended to maintain shamans, whose social role was to act as carriers of clan lore (creation myths, history, knowledge of animal behaviour and uses, plants for healing), and to communicate with spirits (human, animal, elemental/nature spirits) on behalf of the clan. Communication is usually undertaken with practical outcomes in mind, such as knowledge of the whereabouts of game for hunting, healing of illnesses not susceptible to the usual physical remedies, or personal or collective advice from the ancestors. A sometimes overlooked aspect of a shaman's role is to act as psychopomp, conductor of a soul, whether into this world by ensuring a safe birth (midwifery) or, upon death, by escorting the person to their due place in the spirit world, the afterlife. The healing aspect of a shaman's work can also include the recovery of those who become lost souls during their embodied life; many traditions entertain the teaching that a soul can become fractured or fragmented, or that the boundary between the living and the dead can be inadvertently crossed at risk of loss of life, the shaman's responsibility being to correct or repair the situation. There are many potential activities that might be called healing in a shamanic context.

My early studies of shamanism took me to examples identified in Siberia, and among indigenous North American peoples, particularly Alaska and Greenland (Jakobsen 1999). This was partly down to availability of secondary material, as shamanic models have been heavily utilized in English-speaking North American scholarship in comprehending the traditions of indigenous North American, or 'First Nation', peoples (Jones 2006, 2008). In large part, this was due to the work of Mircea Eliade, who was based at Chicago and whose cross-cultural model of shamanism was very influential, both as to particular practices identified as shamanic, and as to the idea that there could be a cross-cultural model (Eliade 1964). Specifically on Spiritualism, there are some intriguing hints in early Shaker practice as to the tradition of Spiritualist mediums with native North American spirit guides (Bennet 2005), and some early to mid-twentieth century descriptions of native North American spiritual traditions made very heavy use of Spiritualist terminology and perspectives (Spence 1914; Seton 1939).

Further enquiry soon reveals that scholars have identified shamanic traditions in societies across the globe (Atkinson 1992). The claim that shamanic traditions have been identified across the world has proved slightly controversial as regards countries where scholarship has been dominated by British scholarly traditions employing models of possession, with its traditionally negative connotations. Broadly speaking, scholars have often fallen into habits of thought that characterize shamanic practitioners as active, masculine, travelling or journeying, employing techniques that have been proactively learned and which require the practitioner to practise a degree of self-control; by contrast, similar traditions in countries where British scholarship was enabled by British colonial authority (especially Africa, India, and parts of south-east Asia) have often been characterized as involving some form of possession, generally characterized as passive, feminine, spontaneous, and uncontrolled or otherwise lacking in expertise. These preconceptions have often been evident in preliminary attempts to understand Anglo-American mediumship, and have in some degree shaped that tradition by being characteristic of the society within which is functions.

Increasingly, however, modern scholarship concludes that traditions of spirit communication have both active and passive aspects, much like any conversation; the work of Smith (2006) in examining south-east Asian traditions has been especially useful here, showing that it is often simply a matter of scholarly habit and semantics as to whether a tradition is labelled shamanic or possessory (Smith 2006: 60–66). The use of the word shamanism in relation to Indian or African traditions can seem unfamiliar, as can the use of the word séance in relation to shamanic demonstrations, but this strangeness has little to do with the traditions and practices being examined, and can have its uses if seeking to examine the material with fresh eyes. Making the familiar strange and the strange familiar is good anthropological practice.

Looking more closely at European history, it becomes possible to identify some practices that can be labelled shamanic elsewhere, but they have largely been forced underground as folk religion or, more traditionally, witchcraft. This is especially true of those areas of Europe where Protestant Christianity has been dominant in recent centuries, Protestant Christianity having been especially keen to exclude experiential knowledge of spirits, or spirit, something Roman Catholic and Orthodox Christian traditions have retained particular ways of accommodating (Sluhovsky 2007).

Modern European culture, particularly Protestant culture, is the part of the world where shamanism is most completely missing; it is also the part of the world where the mediumistic tradition of Spiritualism is most strongly and obviously present.

It can be objected that the modern Spiritualist tradition began in upstate New York in the 1840s, with the activities of Andrew Jackson Davis and the Fox sisters, among others. This is correct, but highlights the importance of identifying location in time and culture as well as in physical space when seeking to identify the places, or spaces, where human beings conduct religious or spiritual activities. Shamanism is identifiable in the traditions of North American First Nations and, following the transplantation of Protestant European culture, continues to be identifiable in that location in the form of Spiritualism, within the migrated culture. A similar point can be made as to the presence of Spiritualism in South Africa, Australia, and New Zealand; also parts of India and, to a lesser extent, Nigeria.

By mapping cultural presence over time, and by identifying Spiritualism as the form of shamanism found in traditionally Protestant Anglo-American culture, it is immediately obvious that we plug the one significant gap in the shamanic mapping of our cultural world over time.

Initial conclusions and observations: the varieties of mediumistic and other shamanic experience

A major factor obstructing consensus on a definition of shamanism is the perceived variety of experiences attested to by practitioners, and observed by researchers. There is also an academic history of resistance to the idea that there is a robust cross-cultural model that can be identified or developed. Although Eliade's model was influential, it was quickly apparent that it had its problems, to the extent that it, and even the possibility of a useful cross-cultural model, became deeply unfashionable.

Eliade maintained that, historically, the Evenki were the people among whom shamanism was to be found in its purest form. Like many other early attempts to determine the origin of a religion (or even of religion itself), this followed early linguistic studies of cultural migration, in this case of the word shaman itself. In this, Eliade drew heavily upon Shirokogoroff, whose work is excellent but now little used.

Shirokogoroff was a careful anthropologist and ethnographer, reluctant to use the term shaman (Evenki šamān) to describe practitioners beyond the Siberian Tungus culture he was examining; he was perfectly open to the possibility that similar practitioners might be found in other cultures but strongly argued against employing the term shaman to indicate them. Despite Eliade's own position that the Evenki were the host people to the archetypal shaman, the weight of his argument was that a cross-cultural model is indeed possible, and for a time scholars almost universally spoke of shamanism, rather than shamanisms. As the deficiencies of Eliade's model became apparent, including his Christian preconceptions and evidential selectivity, it was only a question of time before the scholarly pendulum swung again in favour of appreciating, indeed emphasizing, the cultural context of particular shamanic traditions; my own attempt to develop a useful cross-cultural model has had to contend with this contemporary bias.

If we treat shamanism as an academic field of enquiry, rather than something that is out there in the field, we can usefully designate experiences and practices as shamanic, while appropriately acknowledging their cultural settings, and therefore their variety. On this basis, the category of practices traditionally recognized as shamanic include: the ability to perceive spirits, and to discriminate between them, both as to different kinds of spirit, as to individual spirits, and as to their motivation or purpose. These skills are generally honed through some communally recognized and validated process of training or apprenticeship involving an existing, recognized practitioner, and will include the ability not only to perceive spirits but to communicate with them purposefully in order to achieve some practical, useful objective.

The spirits perceived will generally include human ancestor spirits, but may also include animal spirits, sometimes individual animals but possibly a kind of species spirit, and other nature or elemental spirits, including those we might term gods, namely spirits with some responsibility for creating or maintaining the existence of the world, or with responsibility for the care or wellbeing of a clan or people.

Methods of communication by a practitioner can involve some form of divination, overshadowing by a spirit influence (the degree of which can amount to full possession, where the personality of the practitioner is overshadowed almost to the point of absence), or journeying (astral) travelling, by which is meant that the practitioner's soul or consciousness travels out with her or his physical form so as to journey to the spirit world(s), where some form of conversation or other interaction takes place. As mentioned, healing is often noted as a regular element in a shaman's practice; many shamans employ a wide knowledge of herbs or other substances, and are to this extent recognizable as herbalists; many only employ their ability to communicate with the spirits if they diagnose an illness as having a spirit origin, such as the uncontrolled and unknowing overshadowing of the patient by a spirit, with unwelcome results.

The communal services expected of shamans inevitably vary according to culture, as do the detailed contents of the recognized training processes, the skills or methods employed to communicate, and, in some societies, the kinds and identities of spirits recognized. This might seem to make it self-servingly easy to draw parallels between recognized shamanic traditions and cultures on the one hand, and Spiritualist mediumistic practices in Anglo-American culture on the other, but it goes too far simply to reduce these various traditions to traditions of spirit communication: there are more particular and precise parallels that can be drawn.

The public settings in which Spiritualist demonstrations are given are generally modelled on the churches and chapels of nonconformist Christian denominations (Nelson 1969: 130–152); superficially, they look very different from the demonstrations of a Siberian shaman. This is misleading, in that one needs to look beyond language and clothing, the use of prayers rather than chants, hymns rather than songs, churches rather than tents, so as to identify the particular practices employed.

It is a popular, and often a convenient scholarly commonplace that shamans typically travel in spirit, whereas mediums are overshadowed or possessed; even scholars can be susceptible to the attractions of convenient typologies. This characterization owes much to Eliade's model, which emphasized journeying or travelling as typical of shamanic practice, together with an emphasis upon animal or other nature spirits. Re-examination of Shirokogoroff is instructive here: first, it transpires that travelling is not typical of shamanic practice but is instead better understood as its apogee, in the sense that it was regarded by the community as the mark of an experienced and proficient shaman; more commonly, especially in the earlier stages of a career, the shaman was much more likely to be used or overshadowed by the spirits; secondly, interaction with human or ancestor spirits is the norm in Shirokogoroff's descriptions, examples of which easily form the bulk of the spirit interactions documented by him.

As to Spiritualist practices, astral travel has long been recognized as well within the mainstream, with descriptions of this practice leaving little doubt that shamanic journeying and astral travel map very closely indeed (Sculthorp 1961). Although Spiritualist tradition is primarily focused on providing evidence of life after death through communication with human spirits, interaction with animal spirits is routine, if not heavily emphasized.

In 'Redefining Shamanisms', I undertook the exercise of detailing the correspondences between traditional Evenki practices and those to be found in Spiritualist tradition, mapping one to the other. Although there is little space to do so here, and little point in repeating the exercise, the core examples given help to illustrate why I consider a distinction between shamanism and mediumship (including possessory traditions) to be one that unhelpfully distinguishes between traditions that are more appropriately categorized together.

Testing the thesis against the varieties of shamanic experience: why does the comparison hold?

It was important to establish that various correspondences of practice can be observed between the Evenki shamanic and British Spiritualist traditions, if only because it challenges some of the categories, and therefore some of the language, that English-speaking Anglo-American scholars have habitually used; that said, the more interesting and revealing question is why do these correspondences arise? Is it possible to identify the underlying mechanisms or processes whereby variant but comparable traditions arise in apparently unrelated cultures across time and space? It is in answering this question that my extended participation in the development circle at Portobello Spiritualist Church proved invaluable.

Mediumship is often spoken of as being a 'gift' but it is a commonplace within the Spiritualist movement that this is only true in the sense that one might speak of a person as having a 'gift for music', or for playing the piano: the old saying that 'discipline is the means by which talent becomes ability' applies. The ability to work as a medium is widely understood as being the outcome of an extended training, undertaken over years rather than months, sometimes decades rather than years. The usual form of that training is apprenticeship to an existing medium, either on a one-to-one basis or through participation in a conducted group, usually referred to as a circle (from the traditional practice of sitting in a circle).

Reflecting on the process of developing as a medium at Portobello Spiritualist Church led me to perceive it as an apprenticeship in the management and development of my awareness so as to engage with additional realities, by which I mean independently existent places and persons beyond those familiar to us from interaction with our physical surroundings, including other physically embodied persons. I perceived a particular pattern and structure to this apprenticeship, which I summarized as follows:

> The apprenticeship comprises a process of learning enhanced cognitive abilities, achieved by communicating and developing relationships with spirit guides or helpers, facilitated by an existing practitioner, who usually also undertakes responsibility for passing on a body of accumulated traditional knowledge. The apprenticeship typically proceeds from an initial uncontrolled (often possessory and/or unwelcome) psychic experience or phase to a point where the apprentice is granted communal recognition as having the ability to manage her or his awareness so as no longer to be personally at risk from uncontrolled spiritual forces, and is able to use her or his spiritual skills to communicate, at will, in ways that are recognised as being beneficial to other members of the community.
>
> (Wilson 2013: 17)

Although this definition was developed in order to better understand the mediumistic tradition at Portobello Spiritualist Church, it became apparent, from my examination of shamanic traditions through secondary sources, that this definition also serves as an accurate and useful definition of shamanism more widely. Attending to mediumship as the outcome of a process of apprenticeship made me more sensitive to those aspects of other shamanic traditions that seemed puzzling or disparate until interpreted within the contact of similar apprenticeship processes found in those traditions. Upon rereading Shirokogoroff's work, it was a striking moment when I came across his comment that the essence or core of the tradition was probably to be found in the apprenticeship undertaken by potential shamans, even as he acknowledged that this was the part of the tradition most inaccessible to researchers.

Preliminary application of this definition, or model, to other traditions has been encouragingly productive, not least because it lends itself to a valuable focus on the social mechanisms whereby traditions are maintained, and bodies of lore transmitted across the generations.

Conclusion: the importance of processes and marginality

Much of what has been said is supportive of pleas for the importance of examining the processes whereby religious traditions arise, develop, and persist. Time is often the dimension that allows us to make sense of what exists in terms of the other three, whether comprehending the creation of sedimentary rocks or the ways in which human religious traditions arise. When analyzing human societies, and particular aspects thereof, it is important to develop an appropriate narrative, not least because this is often the means by which non-scholars create meaning and significance, the things that underpin social institutions. Narratives are often key to how we make our way in the world, how we navigate our encounters with others, and can offer rich examples of categorization in action.

My own research involved undergoing a process that lasted almost a decade, from beginning my involvement with Portobello Spiritualist Church to being awarded the PhD it led to. This is not the kind of extended and open-ended research that readily finds funding in today's world but it is what was required in order to categorize Spiritualist mediumship. Previously, it languished as some sui generis oddity, which was a clear indication that it was still to be comprehended; attempting to do so brought unexpected benefits in helping by bring a greater degree of order to a wider field.

This happened because I tested a definition by attempting to include within it something to which it had not previously been applied. Just as particular methodologies contain implicit assumptions as to how things should be examined, or even as to what should be examined, so too do the definitions we employ, simply because definitions tend to be answers to the questions we have asked. This is why it can be important to unpick and challenge our preferred definitions; unless we do so, we are not in charge of the questions we are asking, and risk repetition or error.

I felt it a mistake to give up on developing a new definition of shamanism; this was not because I think there is one correct definition of shamanism but because I think there are always new definitions that can be developed, the application of which can offer new insights and understandings. As a lawyer, I long since abandoned the notion that there is a correct, enduring, or essential definition of anything; we lawyers define our terms in all sorts of interesting and novel ways, to suit the needs of the moment, or the contract. The only worthwhile test of a definition is its utility, in terms of its ability to enable understanding, to facilitate clarity of expression. To find that one's preferred definition does not quite work is to have learned; to respond by developing and testing a new definition is to continue to enquire. To maintain that developing new definitions is futile is to admit to having run out of ideas, even if the form of one's admission is to maintain that there is nothing more to be learned.

Using definitions is to engage in a form of mapping. Definitions bound, exclude as well as include, meaning they require us to attend to what is excluded or marginalized; on occasion this can turn out to be central to an improved understanding. Testing the categorization of Spiritualism as a shamanism not only offered new insights into Spiritualism and mediumship, and as to why they arose and persist in modern western society, but also brought unanticipated new insights that have enabled a new definition of shamanism, which has since gone on to prove helpful to scholars whose expertise is in shamanic traditions I have little knowledge of; in other words, the definition I developed has proved to have a degree of utility. I very much hope that others will develop and apply additional definitions, thereby furthering our collective academic learning.

References

Atkinson, J. (1992), 'Shamanisms today', *Annual Review of Anthropology*, Vol. 21, pp. 307–330.

Bennet, B. (2005), 'Sacred theatres: Shakers, Spiritualist, theatricality, and the Indian in the 1830s and 1840s', *The Drama Review*, Vol. 49(3), pp. 113–134.

Birks, P. (1997), *The Classification of Obligations*. Oxford: Clarendon Press.

Eliade, M. (1989 [1951, 1964]), *Shamanism: Archaic Techniques of Ecstasy*. London: Arkana.

Jakobsen, M. (1999), *Shamanism: Traditional and Contemporary Approaches to the Mastery of Spirits and Healing*. New York: Berghahn.

Jones, P. N. (2006), 'Shamanism: an inquiry into the history of the scholarly use of the term in English-speaking North America', *Anthropology of Consciousness*, Vol. 17(2), pp. 4–32.

Jones, P. N. (2008), *Shamans and Shamanism: A Comprehensive Bibliography of the Terms Use in North America*. Boulder, CO: Bäuu Press.

Jordan, P. (2001), 'The materiality of shamanism as a world view: praxis, artefacts and landscape', in N. Price (ed.), *The Archeology of Shamanism*. London: Routledge, pp. 87–104.

Nelson, G. K. (1969), *Spiritualism and Society*. London: Routledge and Kegan Paul.

Ordan, P. (2001), 'The materiality of shamanism as a world-view: praxis, artefacts and landscape', in N. Price (ed.), *The Archeology of Shamanism*. London: Routledge, pp. 87–104.

Sculthorp, F. (1999 [1961]), *Excursions to the Spirit World*. London: Greater World Association.

Seton, E.T. (1970 [1939]), *The Gospel of the Redman: An Indian Bible*. London: Psychic Press.

Shirokogoroff, S. M. (1966 [1929]), *The Social Organization of the Northern Tungus*. Ooosterhout: Anthropological Publications.

Shirokogoroff, S. M. (1999 [1935]), *The Psychomental Complex of the Tungus*. Berlin: Reinhold Schletzer, Verlag.

Sluhovsky, M. (2007), *Believe Not Every Spirit: Possession, Mysticism, and Discernment in Early Modern Catholicism*. Chicago, IL: University of Chicago Press.

Smith, F. M. (2006), *The Self Possessed: Deity and Spirit Possession in South Asian Literature and Civilization*. New York: Columbia University Press.

Spence, L. (1914), *Myths and Legends of the North-American Indians*. London: George G. Harrap & Co.

Wilson, D. G. (2013), *Redefining Shamanisms: Spiritualist Mediums and Other Traditional Shamans as Apprenticeship Outcomes*. London: Bloomsbury.

16 Jung's legacy

The Western Goddess Movement[1]

Rev. Patricia 'Iolana

Introduction

Jungian and post-Jungian theory have contributed significantly to the development of the Western Goddess Movement and the focus on spiritual and psychological well-being that currently permeates Western Culture. My research revealed not only the historical development of several important threads of accepted wisdom pertinent to the birth and development of the Western Goddess Movement in the United States and beyond but also concretised the expansion of the inherently religious attitude of Jungian and post-Jungian thought. This essay demonstrates the influence of the substantial contributions of seven individuals, who, along with the second wave of feminism in the United States, significantly enhanced the development of the Western Goddess Movement and focused on the spirituality at the heart of Jungian analytical psychology. Chronologically, these revolutionaries include: (1) Carl Jung's analytical psychology (1912–1961); (2) Dr Mary Esther Harding's feminist revision of Jung's theories and the birth of women's analytical psychology in America (1935); (3) Jung's heir apparent and grand theorist, Erich Neumann's *The Great Mother* (1955); (4) Naomi R Goldenberg's 1976 call for a feminist revision of Jung; (5) Christine Downing's ground-breaking memoir, *The Goddess* (1981); (6) post-Jungian E C Whitmont's *Return of the Goddess* (1982); leading to (7) Jean Shinoda Bolen's *Crossing to Avalon* (1994) which creates a bridge from Jung's analytical psychological theory to religious or spiritual praxis by including rituals for her readers to follow whilst revealing the extent to which Jung and post-Jungian spirituality have been integrated into diverse emergent paths to Goddess.

Carl Gustav Jung (1875–1971)

Carl Jung, a prominent Swiss psychiatrist, psychotherapist, and founder of analytical psychology, posited a vast array of theories and models over his career – the extent of which can be found in his massive *Collected Works*. Though not all of Jung's theories and models appear to be relevant or useful to the Western Goddess Movement, and only Jung's primary theories on the Collective Unconscious, Archetypes, the Shadow, Union, and his Path of Individuation are cited by the authors in my study.[2] However the two Jungian models that seem to permeate the Western Goddess Movement, and certainly the five memoirs that serve as my primary source material, are Jung's Individuation and the post-Jungian transformation of Jung's Anima and *Anima Mundi* archetypes.

Individuation

Individuation, as a psychological imperative, is the crux of Jung's analytical psychology; he writes: 'I use the term "individuation" to denote the process by which a person becomes a psychological "in-dividual," that is, a separate, indivisible unity or "whole"' (Jung, 1968: 275). According to Jung, this psychological wholeness is actuated through a conflict between the conscious and unconscious minds; Jung writes:

> Conscious and unconscious do not make a whole when one of them is suppressed and injured by the other. If they must contend, let it at least be a fair fight with equal rights on both sides. Both are aspects of life. Consciousness should defend its reason and protect itself, and the chaotic life of the unconscious should be given the chance of having its way too – as much of it as we can stand. This means open conflict and open collaboration at once. That, evidently, is the way human life should be. It is the old game of hammer and anvil: between them the patient iron is forged into an indestructible whole, an 'individual.' [. . .] This, roughly, is what I mean by the individuation process. As the name shows, it is a process or course of development arising out of the conflict between the two fundamental psychic facts.
>
> (Jung, 1968: 288)

Therefore *Individuation* is a process, based on opposing binaries, of attaining a harmonious balance or union between two psychic components (beginning with the union of conscious and unconscious and followed by a host of oppositional unions) into one cohesive and psychically-stable whole Self. This totality of Self, or psychological wholeness through union, was the aim of Jung's psychotherapy. Individuation is an extremely difficult long-term psychological process that requires special (Jungian and/or post-Jungian) knowledge or trained assistance to navigate successfully (avoiding such potential complications as substantial psychological breakdown or neuroses) (see Jung, 1968: 350).

In addition to having some special knowledge of Jungian and post-Jungian theory, it is also essential, according to Jung, to possess a 'religious attitude' (Jung, 2001: 68). For Jung, it is more important to have religious faith rather than a belief in religious dogma. So in order to effectively navigate Jung's psycho-religious Individuation, one needs to be equipped with both a basic understanding of Jung's theories as well as possess a religious attitude – which includes some belief in a Divine Source or Creator. Jung believes that a religious attitude and frame of mind are integral to Individuation: 'With us, religious thought still keeps alive the archaic state of mind, even though our time is bereft of gods' (Jung, 2001: 149).

It is also important to note that Individuation is understood as an inherent unconscious psychological drive for wholeness fuelled by the autonomy of the archetypes that will press themselves into consciousness despite any attempts at repression (McGuire and Hull, 1977: 294). The central element for Jung is that Individuation '[. . .] aims at a living co-operation [. . .]' (Jung, 1976: 123). The living co-operation is exemplified through the concepts of 'being in relation to', union, and the web of interconnectivity demonstrated in the source material. As the centrepiece to Jung's analytical psychology, he asserts that Individuation stands above all other of humanity's achievements in value and importance; he writes:

> The great events of world history are, at bottom, profoundly unimportant. In the last analysis the essential thing is the life of the individual. This alone makes history; here alone do the great transformations first take place, and the whole future, the whole history of the world, ultimately spring as a giant summation from these hidden sources in individuals. In our most private and most subjective lives we are not only the passive witnesses of our age, and its sufferers, but also its makers. We make our own epoch.
>
> (Jung, CW 10: para 315)

Therefore, according to Jung, Individuation is as important to the individual as it is to the Western world. As Jung understood Individuation, one of the positive psychological outcomes would be to discover one's reason for being (her or his affirmation of destiny). This acknowledgement of one's true path is an important element in Jung's process of Individuation and a required step before one can proceed further. However, the term Individuation is also used by Jung to describe the transformed *Self* after successful union and integration of opposing psychic forces. In Jung's analytical model one can only attain true 'Selfhood' through Individuation because, for Jung, the Self is also an archetype (Jung, 1976: 142). Confusion ensues because for Jung the term *Individuation* is used not only to indicate the process but also the psychological destination of wholeness. Having Individuation as the title for both the process and the end-goal can be confusing to readers unfamiliar with Jung's theories.

To avoid confusion, I have opted to incorporate a term used by Jung (1995: 328) and widely used by post-Jungians, *Path of Individuation*, to differentiate *the process* from the fully Individuated Self, but also to acknowledge the Eastern-influenced religious aspects of Jung's theories and models. Especially with Jung's model of Individuation, the influences of Eastern traditions such as Taoism, Hinduism, and Buddhism are evident. Jung perceived Individuation as the ultimate psychological state that each human being should strive for in order to be *psychologically healthy* and internally unified as an individual – and this elevated state of Self is impossible without some form of personal relationship with the Divine; according to Jung one who walks Jung's Path seeks self-enlightenment. Nonetheless Jung's focus on one's psychological health does not detract from the striking similarities between Jung's Individuation and the Buddhist Eightfold Path.

Damien Keown writes: 'The Eightfold Path is thus a path of self-transformation: an intellectual, emotional, and moral restructuring in which a person is reoriented from selfish, limited objectives towards a horizon of possibilities and opportunities for fulfilment' (2000: 56). This description of the Buddhist Path resembles Jung's Path of self-transformation on many levels including that both focus on replacing the ego as the centre of the Self and subordinating it to the fully realised Self. Based on the similarities of the Buddhist 'Path' in Jung's theory of Individuation, the fact that all five authors in my study offer their rebirth memoirs as Paths for others to follow, combined with the assertion that Jung's analytical psychology is arguably a religion (Dourley, 2006; Noll, 1994; Wehr, 1987), the use of the term *Path of Individuation* appears highly applicable to describe what post-Jungian theologian John Dourley deems a 'religious event' (2006: 43).

It is interesting to note that out of all of Jung's theories that continue to permeate both contemporary post-Jungian analytical psychology and the Western Goddess Movement, Jung's Path of Individuation, other than being more appropriately named as a Path and being reshaped for women, has endured relatively unchanged as a construct of healing and well-being. Post-Jungians have expanded the writing on the various components and shifts of consciousness that permeate Individuation as a psychological event, but Jung's original model remains, essentially, intact. This is, perhaps, due to the fact that this particular model of Jung's is not a gendered construct as are his other theories and models; however, Jung's construct of Individuation is limited by the gender essentialist nature of its author. The necessary feminist rephrasing would be completed by Jung's student, M Esther Harding (1888–1971), who created a Jungian Path of Individuation for women, and that model is accessible in the source material through the lens of five different analytic and religious foundations demonstrating the fluid nature of Jung's model of Individuation and its continuing importance and relevance as an accessible Path to Selfhood and Goddess in the Western Goddess Movement.

Jung's Anima

At the heart of Jung's Path of Individuation and the centre of his Collective Unconscious is Jung's Goddess, Anima. In its initial inception the Anima, as a Jungian archetype, is essentialist and gender-locked as the contrasexual feminine principle of man which Jung defines an 'an archetype that is found in men [. . .]' (Jung, 1976: 151). Jung further defines the Anima as '[. . .] a natural archetype that satisfactorily sums up all the statements of the unconscious, of the primitive mind, of the history of language and religion' (Jung, 1968: 27). However, Jung firmly believed that only men possessed an Anima archetype (a gender essentialist concept that will be revised by several women who follow Jung). Jung wrote at great length about the archetypes of the Collective Unconscious but gave 'special reference to the Anima concept' (Jung, 1968: 54). Jung wrote:

> With the archetype of the anima we enter the realm of the gods, or rather, the realm that metaphysics has reserved for itself. Everything the anima touches

> becomes numinous – unconditional, dangerous, taboo, magical. She is the serpent in the paradise of the harmless man with good resolutions and still better intentions.
>
> (1968: 28)

While Jung aligns the Anima with the 'realm of the gods', he also images her as the snake in the Garden of Eden who tempted Eve to challenge God with a subtle critique of Christianity evident in much of Jung's writing. What is important to take away, however, is that Jung envisions the Anima as the purveyor of self-knowledge. Jung writes: '[. . .] for the anima can appear also as an angel of light, a psychopomp who points the way to the highest meaning [. . .]' (1968: 29). Jung further characterises the Anima in an Alpha/Omega pairing with his model of the Shadow that signifies not only both of these archetypes' importance to one's growth as in individual (Individuation) but also their importance as the beginning and end of one's Path:

> If the encounter with the shadow is the 'apprentice-piece' of the individual's development [Individuation], then that with the anima is the 'master-piece.' The relation with the anima is again a test of courage, an ordeal by fire for the spiritual and moral forces of man.
>
> (Jung, 1968: 29)

In essence, what Jung is saying is that in his model of analytical psychology the growth of an individual begins with an encounter with the Shadow and ends with a 'relation with the anima' (often described by post-Jungian feminists such as Downing, Bolen, and Perera as attaining Union with Goddess). This effectively outlines Jung's Path of Individuation, and as Jung himself states, the Anima is 'numinous,' 'dangerous,' 'unconditional,' and 'magical' (Jung, 1968: 28). It is therefore not surprising that Jung has deified the Anima and refers to her as '[. . .] the latent primordial image of the goddess, i.e., the archetypal soul-image' (Jung, 1982: 10). Thus, Jung began with an archetype that was the essence of femininity within men, imbued her with divinity and referred to her as 'goddess.' This is not uncommon in Jung's models, as he believed '[. . .] the idea of a deity is not an intellectual idea, it is an archetypal idea' (McGuire and Hull, 1977: 346). However, in his writings, it is the Anima archetype as Goddess that takes centre stage of Jung's theories and models.

Jung makes special note of the 'timelessness' of the Collective Unconscious; he writes: 'The anima and the animus live in a world quite different from the world outside – where the pulse of time beats infinitely slowly, where the birth and deaths of individuals count for little' (Jung, 1968: 287). What Jung means, in an extension of Platonic thought (Jung, 1968: 4), is that the Collective Unconscious and the archetypes who inhabit it are all not bound by the constructs of linear time. If the Collective Unconscious can move outside of time, then it is not constrained to our linear concepts of the past, present, and future. By extension, if the archetypes inhabit the timeless Collective Unconscious, then they, too, are timeless. In essence, this timelessness construct makes the archetypes, and most importantly, the Anima eternal.

As timeless, eternal, *and* autonomous, the Anima has the power to emerge from the Collective Unconscious into an individual's conscious mind in any form or dress; Jung writes: 'The anima is conservative and clings in the most exasperating fashion to the ways of earlier humanity. She likes to appear in historic dress, with a predilection for Greece and Egypt' (Jung, 1968: 28). Christine Downing and Jean Shinoda Bolen, both trained Jungians and the first two authors examined in my study, present the Anima as Goddesses from the Ancient Greek pantheon. Their use of the Greek Goddesses offers a validity to Jung's theories about the Anima's (Goddesses') predilection to cling to earlier visions and forms of humanity. Jung however, does state that the Anima's form can vary by culture or individual and tends to change over historical periods as well; Jung writes: 'To the men of antiquity the anima appears as a goddess or a witch, while for medieval man the goddess was replaced by the Queen of Heaven and Mother Church' (Jung, 1968: 29). So, as an autonomous and eternal archetype, Jung has imbued the Anima with tremendous psychological and spiritual power. It would be Jung who would transform the Anima from lead archetype into the Divine Creatrix *Anima Mundi*.

Anima Mundi

Anima's final transformation would be the most important to the Western Goddess Movement. In 'The Difference Between Eastern and Western Thinking' Jung calls the Anima the '[. . .] spark of the *Anima Mundi*, the World Soul' (Jung, 1976: 481). Not a Jungian designed concept, the *Anima Mundi* is a vital force or principle which is conceived of as permeating the world. With the Anima already established as a numinous Goddess by Jung (1968: 28), the Anima (Goddess) is now aligned with the *Anima Mundi*. Both Jung's Anima and *Anima Mundi* would be further revised and *amplified*[3] by several influential and popular post-Jungians and become pivotal archetypes in the Western Goddess Movement.

While Jung's writings on the Anima were inspirational to his students and patients, they were problematic for many women who followed. There is no denying that Jung's original theories and models were sexist, gender-essentialist, and limiting. His writings are from the 'male' perspective and include only masculine pronouns which may be problematic to those who don't identify as male. He posits that women's psyches are less developed than men's (1968) and makes a number of detrimental statements about the analytical ability of women – especially prevalent in 'Aspects of the Feminine' (published posthumously as a collection of articles and extracts in 1982). However, this essentialist and gender-restricting bias eventually brought about feminist revision of Jung's theories beginning with his student Dr M Esther Harding in 1935. Post-Jungian revision could be considered a tremendous positive outcome of his original gender-essentialist theories, especially seeing that these feminist modifications and following amplifications would be critical to adherents in the Western Goddess Movement. Feminist revisions of Jungian thought, particularly the role of the Anima and *Anima Mundi* in Individuation, were necessary to contribute further to the development of the Western Goddess Movement.

Revising Jung: constructing Goddess[4]

In the post-Jungian years, there would be a number of individuals who would revise or amplify Jung's theories with a feminist twist. Susan Rowland's *Jung: A Feminist Revision* (2002) is integral to my understanding of the varieties of post-Jungian revisions. Not all post-Jungian revision would be useful and integrated by the Western Goddess Movement. The first major revision of Jung's Anima would occur in 1935 when Esther Harding published *Woman's Mysteries: Ancient and Modern* in the United States which would greatly expand Jung's geographical sphere of influence. Post-Jungian Susan Rowland cites Harding as a 'key author' in the Amplification of Jung's Eros and 'Feminine Principle' (Anima) (Rowland, 2002: 56). Harding both embraced and revised Jung. Rowland writes: 'Amplifying Eros as "the feminine principle" enables her to cover a far greater and more powerful range of qualities than those envisaged by her mentor' (Rowland, 2002: 56).

M Esther Harding (1888–1971)

In *Woman's Mysteries: Ancient and Modern*, which created the first women's post-Jungian analytical psychology written specifically for women by a woman, Harding continued Jung's essentialist dualities with the Anima/Animus, but revised Jung by writing specifically to women and extolling the power of the female psyche; Harding writes: 'The neglect of the inner or subject aspect of life has led, particularly for women, to a certain falsification of her living values' (Harding, 1971: 9). She presents an openly feminist model for the women who have recently been granted the right to vote and the introduction to Congress of the Equal Rights Amendment in 1923. Harding writes: 'Our civilisation has been patriarchal for so long, the masculine element predominating, that our conception of what feminine is, in itself, is likely to be prejudiced' (Harding, 1971: 30). Harding calls for women to find a way to connect with Jung's Anima as Goddess and offers a rich and varied cultural and anthropological history of numerous societies who worshipped Goddess in the embodiment of the Moon. Harding not only revises Jung's essentialist theory by expanding the need to women, but she is also the first to amplify Anima to cultural and historical Goddesses, providing her readers with a host of concrete manifestations of Jung's intangible Anima.

Harding's work would serve as a source of inspiration for those who followed and demonstrates the potential in Jungian amplification. Despite Harding's radical transformation of Jung's theories and being a student of Jung, Erich Neumann, who studied with both Freud and Jung, found himself 'Jung's anointed intellectual heir' (Paglia, 2006: 3). Neumann's *The Great Mother: An Analysis of the Archetype* (1955) transformed Harding's female psychology and mixed it with a form of Goddess-centred spirituality. Neumann amplified (projected) Jung's *Anima Mundi* archetype to that of Creatrix in *The Great Mother*, offering the reader a vast array of images of Jung's *Anima Mundi* as Magna Mater. Christine Downing and E C Whitmont, both theorists in the emerging Western Goddess Movement, follow suit and continue the amplification of Jung's Anima and his *Anima Mundi* (Great Mother) (Downing, 1981; Whitmont, 1982). All five of the authors in my study amplify Jung's Anima (Goddess) and his *Anima Mundi* (Great Mother) in their memoirs, thus continuing the post-Jungian amplification of Goddess and the Great Mother and expanding Her role through an analytical religious attitude indicating both their prominence and relevance to adherents in the Western Goddess Movement. However, revisions of Jung were not limited to the world of analytical psychology or mythology; theologians would also engage with post-Jungian theory on the Anima as Goddess.

Naomi R Goldenberg (1947–)

In 1976 Naomi R Goldenberg published an essay entitled 'A Feminist Critique of Jung.' In this essay, Goldenberg is openly critical of Jung's sexist language and models and calls for feminist scholars to 'confront the sexism of Jung's theories' which have been ignored by his followers (Goldenberg, 1976: 444). She writes: 'Beyond the overt sexism in Jung's concept of the feminine, a feminist critique must examine the inequity of the anima-animus model of the psyche, which is never challenged by any of his immediate circle of followers' (Goldenberg, 1976: 445–46). Maggy Anthony, who studied and authored *Jung's Circle of Women* (1999), agrees with Goldenberg's assessment but also comments that 'It was no accident that he chose a woman to accompany him into the depths of his own unconscious [. . .] for him women and the unconscious were synonymous' (Anthony, 1999: 103). A theory, however, can be inferred from Anthony's statement: the initial circle of women surrounding and collaborating with Jung (such as Toni Wolff, Marie-Louise von Franz, M Esther Harding, and Jolanda Jacobi) were part of his Individuation process, representing Jung's Anima archetype, and therefore unable to see the models and theories objectively. As ones so intricately connected both to Jung's own Individuation and to the development of the central models and theories he puts forth in his analytical psychology, perhaps Jung's circle of women were unable to see beyond their own contributions to the process itself as theory. It would be Harding who would revise Jung the most, but she never directly confronts his gender essentialism or sexism. Only the following generation was able to engage with Jung's theories critically: perhaps the reason for a lack of criticism from the first generation of Jungian women lies in their participation in Jung's development process whereas the second generation held an objective critical distance from Jung's theories and models.

Goldenberg wasn't only calling for a critical examination of Jung's sexist language and models, she also recommends fundamental changes be made to Jung's original gender-biased constructs; she states: 'I would argue that it makes far more sense to postulate a similar psychic force for *both sexes*' (Goldenberg, 1976: 447, emphasis is mine). Goldenberg suggests that the Anima-Animus archetypes be removed from biological gender and the fundamental restrictions these forms of gender essentialism contain. Goldenberg's call for a deconstruction of gender from Jung's theories and models would lead to important changes in post-Jungian theory that would be instrumental to the feminist adherents in the Western Goddess Movement. Rowland cites Goldenberg alongside Demaris S Wehr as 'key authors' in Feminist Theories examining Jung from other disciplines (Rowland, 2002: 84). She states that Goldenberg, while influenced by Hillman, differs from him in that she insists on the archetypal image 'as a vehicle of *cultural* expression' (Rowland, 2002: 84, emphasis in the original).

Goldenberg is critical at a time where change is prevalent on political, cultural, and academic fronts. During this historic period the Equal Rights Amendment is passed by the US Congress in 1972 (although never fully ratified by the US government), quickly followed by Title IX of the Education Amendments which bans sexual discrimination in the schools; Roe v. Wade in 1973 establishes a woman's right to safe and legal abortion; and a host of publications by feminist thinkers exemplify the drastic political and social upheaval occurring parallel to a new shift in feminist religious thought, the inclusion of women to the American Academy of Religion, and a new emerging concept of the post-Jungian Goddess. The Western Goddess Movement evolves at the peak of the second wave of feminism in the United States and the writings and revisions seem to follow Goldenberg in droves. The preponderance of significant literature being published at this time demonstrates how the writings of the second wave of feminism aided in the expansion and amplification of Jung's Anima (Goddess) and *Anima Mundi* (Great Goddess or Magna Mater) and how those models are carried forward into contemporary times.

Further feminist transformation of Jung's Anima would be explored by a number of scholars in the 1980s. Now openly critical of Jung's gender essentialism, some of these feminist revisions would build on Jung's perception of the *Anima Mundi* as the world's salvation from the failure of patriarchy and logos-centred thinking in a movement Rowland calls 'Goddess Feminism' which is a direct descendent of Jung. In *Modern Man in Search of a Soul* Jung writes: 'It is from the depths of our own psychic life that new spiritual forms will arise; they will be expressions of psychic forces which may help to subdue the boundless lust for prey of Aryan man' (Jung, 2001: 221).

Goddess Feminism, as a reformation of Jung, could not have come about without making all the critiques and changes recommended by Goldenberg the decade before:

> Feminist scholars must examine the very idea of archetype in Jungian thought if sexism is ever to be confronted at its base. Indeed, if feminists do not change the assumptions of archetype or redefine the concept, there are only two options: either (1) to accept the patriarchal ideas of the feminine as ultimate and unchanging and work within those or (2) to indulge in a rival search to find female archetypes, ones which can support feminist conclusions.
>
> (Goldenberg, 1976: 447–448)

Post-Jungian feminist revisions of Jung not only kept Jung and his theories topical during the second wave of feminism in America, but also provided a way for women to explore their own Selfhood, discover a form of spirituality that was Goddess-centred, and take control of their psychological health and well-being without the social and cultural limitations and gender restrictions of both Jung and the patriarchal West.

Christine Downing (1931–)

Many significant writers contributed to this transformation of Jungian Anima to post-Jungian Goddess, including Christine Downing's rebirth memoir *The Goddess* published at the same time as Sylvia Brinton Perera's *Descent to the Goddess* (1981). As a foremother to feminist theology, thealogy, and the Western Goddess Movement, Christine Downing's memoir and her engagement with both Jung and Harding are crucial; Downing constructed her memoir based on Jung's Path of Individuation by examining several Jungian concepts, including the Archetype, Shadow, Jung's affirmation of destiny, and the post-Jungian and feminist union of Anima/Animus or Eros/Logos. Downing is the first to provide a path of initiation into Women's Mysteries and follows Harding in her validation of women's search for meaning; she also provides the first bridge from analytical psychology to a post-Jungian religion by outlining a primer in thealogical polydoxy. Downing is the first to publish, and her memoir would be significantly influential to those who follow. Rowland cites both Downing and Perera as 'key authors' in post-Jungian Goddess Feminism alongside E C Whitmont (*Return of the Goddess* in 1982) and Jean Shinoda Bolen's *Goddesses in Everywoman* in 1984 (Rowland, 2002).

Whitmont's revisions of Jung had a tremendous influence on those who follow (especially Bolen), as he not only amplifies Jung's *Anima Mundi* archetype as the *Magna Mater* (Great Mother) Creatrix in a spiritual rather than analytical sphere, but also introduces Pagan pantheons and mythology interlinked with the Christian Grail myth. Whitmont would provide the beginning of the contemporary Western Goddess mythology; Bolen would later build on Whitmont's Grail myth and make it the centre of her argument of the return of Goddess to Western culture. Moreover, Whitmont makes a vital contribution to the revision of the highly-problematic gender essentialism of Jung's original theories by removing Jung's constructs of Eros and Logos from their gender-essentialist roles and offering Jung's 'ways of knowing' as an accessible path to Goddess for all genders. Through the projection ('amplification') of the Anima and the *Anima Mundi* onto existing cultural and mythological figures such as the Ancient Greek pantheon of gods and goddesses, Downing, Whitmont, and Bolen forever transformed Jung's Goddess from a purely analytical archetype to a spiritually-significant Goddess quest which helped spur the dramatic growth of the Western Goddess Movement in the past few decades.

Jean Shinoda Bolen (1936–)

Jungian analyst Jean Shinoda Bolen follows Whitmont's revisions to Jung's gender essentialist theories and rises to Goldenberg's 1976 critique in 1984 with her publication *Goddesses in Everywoman.* Bolen removes the anima/animus from bodily gender and gender essentialism. She presents anima/animus as elements of Goddesses for women, removing gender binaries as well as the inherent opposition in Jung's original anima/animus construct. In the foreword to *Goddesses in Everywoman* Gloria Steinem, a renowned and vocal feminist, writes: 'The author's sensitive analysis of archetypes takes them out of their patriarchal framework of simple exploits and gives them back to us as larger-than-life but believable, real women' (Bolen, 1984: xii). Bolen follows this work in 1994 with her memoir, *Crossing to Avalon,* in which Bolen combines thealogical experiences and enquiry alongside analytical psychological theories and models.

In her memoir, Bolen's Individuation pilgrimage is offered as a pilgrimage of 'rebirth,' and unlike Downing's memoir, Bolen's journey begins with her enrapt in a State of Grace. Bolen uses Jungian and post-Jungian threads to weave her memoir, and she relates additional, and in some ways, more complex, steps along the Path of Individuation. Bolen focuses on child-like wonder and being vulnerable; documents her psychological 'rebirth' as 'pilgrim' from the womb of Chartres; experiences various forms of Jungian Union, including an embodied Union of deities, of beliefs, and of imagery. Bolen's memoir complements Downing's and builds upon it in several important ways. The underlying theories behind Bolen's memoir are vast, varied, and complex. A long history of depth contemplation is shown to be a contributing factor in Bolen's beliefs and assertions in *Crossing to Avalon.* Jungian Feminism, Goddess Feminism, Joseph Campbell's post-Jungian Monomyth, and creation of a Monotheaism combined with Bolen's bridge from analytical theory to religious or spiritual praxis were all necessary modifications of Jung's original theories and models to create an almost perfect Goddess storm in the West.

Jung's legacy: the birth of a post-Jungian Goddess religion

Did Jung intentionally create a new religious movement? Richard Noll has been quite critical of Jung, calling him a prophet of his own religion. In *The Jung Cult*, Noll writes: 'Now the prophet of a new age, Jung promised a direct experience of God' (Noll, 1994: 240). Noll continues:

> Jung offers the promise of truly becoming an individual after becoming a god, or rather, after learning to directly experience the god within. This is a process of self-sacrifice and struggle during which one must give up one's former image of god [. . .].
>
> (Noll, 1994: 257)

Jung stands accused of creating a personal religion in analytical psychology. Noll's assessment is not altogether wrong. In his memoir, as in interviews, Jung does not deny his personal tumultuous relationship with organised religion. Jung explains in *Memories, Dreams, Reflections* (1995) how his upbringing in a Christ-centred home (his father was a pastor in the local church) served to first alienate Jung from organised religion. However, Jung's denial of the Church and organised religions in general does not imply that Jung does not believe in a Divine Creator/Creatrix, appreciate the importance of faith and belief, nor value the human soul. On the contrary, Jung speaks at length about the need for a personal relationship with the Numinous, and most especially the immanent Goddess. The boundaries between theology and psychology in contemporary times, however, are inherently blurred. Dourley explains:

> Jung's theoretical understanding of religion makes of the analytic process a religious event. It recalls the Gods to their psychic origin and encourages unmediated conversation with them within the containment of the psyche [. . .] to be valued for a number of reasons. The internalization of divinity curtails enmity between religious communities bonded by external Gods. More than this, Jung's total myth contends that divinity can become conscious only in humanity.
>
> (Dourley, 2006: 43)

Dourley makes a very important theological point that is worth further consideration. Stating that Jung's divinity is 'conscious only in humanity' is, by extension, to say that God can only be experienced within the confines of the human psyche which is *psychologically* verifiable by the individual rather than the Abrahamic doctrine of a creator God which is *theologically* substantiated yet requires a leap of faith. It is this proposition that puts Jungian theory at odds with the Church, but not necessarily with all theologians. However, seeing as any image of the Numinous is merely a projection of the human attempt to describe the ineffable (McFague, 1982), analytical psychology offers a safe location for one to contemplate the various images of the Numinous outside the confines of traditional theology. If the only means humanity possesses to connect with and relate to the Numinous is imaging; then the human psyche is responsible for creating and maintaining conscious paradigmatic constructs of the Numinous. According to Jung, archetypes of the Numinous emanate from the Collective Unconscious, which, in and of itself, contains the Numinous (*Anima Mundi*) at its centre. The Numinous creates and brings to consciousness representational archetypes of its own choosing. In other words, the Jungian Divine decides how it wants to be imaged in humanity.

Demaris Wehr agrees with Dourley and Noll; Wehr states: 'Jung's psychology [. . .] actually is a religion' (Wehr, 1987: 79). Wehr also writes:

> Besides being a religion, Jung's psychology is in some ways a theology and ontology. Since this is so, it can be addressed appropriately by feminist theologians, who, like Jung, explore the realm of images and symbols. Also like Jung, they cross the boundaries between the disciplines of religion and psychology [. . .].
>
> (1987: xi)

It is precisely this blurred boundary between theology and psychology that makes Jung's theories and models so accessible to theologians and especially fruitful territory for thealogical analyses. It is also in this rich, spiritually-charged environment that the Western Goddess Movement has been allowed to evolve and grow. While Jung denied that he was creating a religion that is Goddess-centred, post-Jungians are more eager to describe Jung's Path of Individuation as a 'spiritual quest' for wholeness that culminates with the individual's union with the Divine *Anima Mundi*.

In summation

From the publication of Jung's theories and models shortly after the turn of the twentieth century to contemporary forms of Goddess Feminism, Goddess Consciousness, and Goddess-centred faith traditions which are integral components of the Western Goddess Movement, Jung's legacy is both visible and viable. Taken from their original purely analytic and internal state, Jung's theories have been revised and transformed by those who follow to offer both women and men a Path to Goddess which includes rituals and praxis and strives for psycho-religious union and wholeness. In short, Jung's theories and models have been taken from the psychodynamic space to the liminal space of thealogy and religious praxis. In this realm, as in the original theories of Jung, one's psychological health and well-being is directly connected to one's faith. What many authors, such as those in my recent study, offer their readers is a psycho-religious or psycho-spiritual path to union with Goddess that is transformative, interconnecting, and multi-dimensional and transcends time and space; moreover they demonstrate the greatest legacy of Jung: Goddess as Great Mother/Creatrix who offers her adherents the tools and ability to heal one's Self, and, by extension, the world.

Notes

1 This paper is revised from a section of my unpublished 2016 doctoral thesis: 'Jung and Goddess: The Significance of Jungian and post-Jungian Theory to the Development of the Western Goddess Movement.'
2 My initial analysis was drawn from a close reading of five 'spiritual rebirth' memoirs: Christine Downing's *The Goddess: Mythological Images of the Feminine* (1981); Jean Shinoda Bolen's *Crossing to Avalon: A Woman's Midlife Pilgrimage* (1994); Sue Monk Kidd's *The Dance of the Dissident Daughter: A Woman's Journey from Christian Tradition to the Sacred Feminine* (1996); Margaret Starbird's *The Goddess in the Gospels: Reclaiming the Sacred Feminine* (1998); and Phyllis Curott's *Book of Shadows: A Modern Woman's Journey into the Wisdom of Witchcraft and the Magic of the Goddess* (1998).
3 A term from Susan Rowland which means the projection of the Anima and *Anima Mundi* archetypes onto existing religious or spiritual pantheons; see Rowland 2002.
4 While a number of post-Jungians contributed to the revisions of and continuation of Jung's theories and models into present day, and relying on classifications established by Susan Rowland's 2002 *Jung and Feminism*, this paper will cite specific individuals who I recognise as key contributors to the transformation of Jung's Anima to Goddess and Jung's *Anima Mundi* to the Great Mother worshiped in the Western Goddess Movement. It is in no way an inclusive or exhaustive list of contributors writing during this time period.

References

Anthony, M. (1999) *Jung's Circle of Women: The Valkyries*. York Beach, ME: Nicolas-Hays.
Bolen, J.S. (1984) *Goddesses in Everywoman: Powerful Archetypes in Women's Lives*. New York: One Spirit.
Bolen, J.S. (1994) *Crossing to Avalon: A Woman's Midlife Pilgrimage*. New York: Harper Collins.
Dourley, J.P. (2006) Jung and the Recall of the Gods. *Journal of Jungian Theory and Practice* 8(1), pp. 43–53.
Downing, C. (2007 [1981]) *The Goddess: Mythological Images of the Feminine*. Lincoln, NE: Authors Choice Press.
Goldenberg, N.R. (1976) A Feminist Critique of Jung. *Signs* 2(2), pp. 443–449.
Harding, M.E. (1971) *Woman's Mysteries: Ancient and Modern*. Boston, MA: Shambhala Publications.
Jung, C.G. (1968) *The Archetypes and the Collective Unconscious*, Second Edition. Hull, R.F.C. (Trans.); Read Sir, H., Fordham, M., Adler, G. and McGuire, W. (eds.). Princeton, NJ: Princeton University Press, Bollingen Series XX.
Jung, C.G. (1976) *The Portable Jung*. Campbell, J. (ed.); Hull, R.F.C. (Trans.). New York: Penguin Books, The Viking Portable Library.
Jung, C.G. (1982) *Aspects of the Feminine*. Hull, R.F.C. (Trans.). Princeton, NJ: Princeton University Press, Bollingen Series XX.
Jung, C.G. (1995) *Memories, Dreams, Reflections*. Winston, R. and C. (Trans.). London: Fontana Press.
Jung, C.G. (2001) *Modern Man in Search of a Soul*. Dell, W.S. and Baynes, C.F. (Trans.). London: Routledge Classics.
Keown, D. (2000) *Buddhism: A Very Short Introduction*. Oxford: Oxford University Press.
McFague, S. (1982) *Metaphorical Theology: Models of God in Religious Language*. Philadelphia, PA: Fortress Press.
McGuire, W. and Hull, R.F.C. (eds.) (1977) *C.G. Jung Speaking: Interviews and Encounters*. Princeton, NJ: Princeton University Press, Bollingen Series XCVII.
Neumann, E. (1955) *The Great Mother: An Analysis of the Archetype*. Manheim, R. (Trans.). Princeton, NJ: Princeton University Press, Bollingen Series XLVII.
Noll, R. (1994) *The Jung Cult: Origins of a Charismatic Movement*. Princeton, NJ: Princeton University Press.
Paglia, C. (2006) Erich Neumann: Theorist of the Great Mother. *Arion* 13(3): 14 pages. Boston University Press.
Rowland, S. (2002) *Jung: A Feminist Revision*. Cambridge: Polity Press.
Wehr, D.S. (1987) *Jung and Feminism*. Boston, MA: Beacon Press.
Whitmont, E.C. (1982) *Return of the Goddess*. New York: Crossroads Publishing.

Further reading

Adler, M. (2006 [1979]) *Drawing Down the Moon: Witches, Druids, Goddess-Worshippers, and Other Pagans in America*. New York and London: Penguin Books.
Armstrong, K. (2005) *A Short History of Myth*. Edinburgh: Canongate Books.
Campbell, J. (1949) *The Hero With A Thousand Faces*. New York: MJF Books.
Campbell, J. (1986) *The Inner Reaches of Outer Space: Metaphor as Myth and as Religion*. New York: Harper and Row.

Campbell, J. (2001) *Thou Art That: Transforming Religious Metaphor*. Kennedy, E. (ed.). Novato, CA: New World Library.

Campbell, J. (2013) *Goddesses: Mysteries of the Feminine Divine*. Rossi, S. (ed.). Novato, CA: New World Library.

Campbell, J. and Moyers, B. (1988) *The Power of Myth*. Flowers, B.S. (ed.). New York: Anchor Books.

Christ, C.P. and Plaskow, J. (eds.) (1979) *Womanspirit Rising: A Feminist Reader in Religion*. San Francisco, CA: Harper and Row.

Curott, P. (1998) *Book of Shadows: A Modern Woman's Journey Into the Wisdom of Witchcraft and the Magic of the Goddess*. New York: Broadway Books.

Dourley, J.P. (1981) *The Psyche as Sacrament: A Comparative Study of C.G. Jung and Paul Tillich*. Toronto: Inner City Books.

Goldenberg, N.R. (1979a) Dreams and Fantasies as Sources of Revelation: Feminist Appropriation of Jung. In: Christ, C.P. and Plaskow, J. (eds.) *Womanspirit Rising: A Feminist Reader in Religion*. San Francisco, CA: Harper and Row, pp. 219–227.

Goldenberg, N.R. (1979b) *Changing of the Gods: Feminism and the End of Traditional Religions*. Boston, MA: Beacon Press.

Keller, C. and Schneider, L.C. (eds.) (2011) *Polydoxy: Theology of Multiplicity and Relation*. London: Routledge.

Kidd, S.M. (2007 [1996]) *The Dance of the Dissident Daughter: A Woman's Journey From Christian Tradition to the Sacred Feminine a Reprint of the 10th Anniversary Edition*. San Francisco, CA: Harper One.

Perera, S.B. (1981) *Descent to the Goddess: A Way of Initiation for Women*. Toronto: Inner City Books.

Starbird, M. (1998) *The Goddess in the Gospels: Reclaiming the Sacred Feminine*. Rochester, VT: Bear and Company.

17 Boundaries of healing

Insider perspectives on ritual and transgression in contemporary esoteric theatre

Alison Rockbrand

This ritual is now open.
It is now a participation.
As it is read,
Let it be performed
In the body of the reader.

Introduction and preliminary issues

From a historical perspective esoteric theatre spans the plays of Florence Farr (*Beloved of Hathor* and *Shrine of the Golden Hawk*, 1900) or Rudolf Steiner (*The Portal of Initiation, The Soul's Probation, The Guardian of the Threshold*, 1913) and Aleister Crowley (*Rites of Eleusis*, 1910)[1] to the transgressive and gnostic theatre of Antonin Artaud (Sontag, 1976), Rosicrucian Order performances of the 1930s (Hutton, 2001) and the cabaret style public rituals of Satanist Anton Lavey in the late 1960s. These are plays and public performances of various kinds (here called theatre/performance) which are produced within the context of western esotericism and which adhere to its main cultural-spiritual tenets.[2] Many of these above are still being produced today, but there is also a new form of esoteric theatre emerging with its own contemporary style and practices; the performances of Angela Edwards and her boundary breaking psycho-sexual rituals, the Metamorphic Ritual Theatre of Orryelle Defenestrate and the healing theatre of the multimedia group Foolish People. The focus of this article is a study of these western esoteric ritual performances, their transgressive ritual components and how they can lead to transformational healing experiences.

As part of this research I will be discussing the experience of liminality or 'in-betweenness' in these performances and the healing aspects of ritual as a change in a performer's core identity.[3] A change of identity can be healing because it transforms a practitioner/participant from one state of being to a new desired state of being, taking with it old modalities or creating a new experience of 'initiation' as part of an expansive and individual spiritual journey. This is an idea explored for instance by Edith Turner (Turner, 1992), who has written about the healing aspects of ritual culture in tribal religions. To fully understand and explore this experience, this research is also being done within the context of insider research and active participation as part of a research methodology; interviews and performances mentioned below having been conducted from an insider's perspective both on performance and western esotericism.

Much of the research on esoteric ritual has focused mainly on non-western tribal ritual forms as explored through anthropological and ethnographic methodologies (Turner, 1974, 2009; Van Gennep, 1909; Bell, 1997; Douglas, 1984). While these are valuable, they are limited in terms of insider perspectives and present a possibly problematic western context onto non-western ritual and performance culture.[4] Grimes's study of the rituals of media and art (Grimes, 2006) and Greenwood and Pearson's much more insider-based research on contemporary Neo-Pagan rituals (Greenwood, 2000, 2009; Pearson, 2001) are part of a new direction in the approach to ritual studies. These approaches can further be directed inward to autoethnography (Ellis, 2004, 2007; Didion, 2005, Denzin, Lincoln and Smith, 2010) and discussed in terms of phenomenology (Husserl, 1964; Plummer, 1983; Garner, 1994) in an effort to discuss ritual in a way which is an experience or a participation.[5] All of these are part of the interdisciplinary methodology which has been applied here to the study of contemporary western esoteric theatre.

As with many practices in the western esoteric tradition, ritual is also the central praxis of western esoteric theatre; in this case it is a public, theatrical or performative ritual which is either shared with or includes the audience. This form of ritual being practiced by esoteric theatre practitioners (which will be called here 'esoteric ritual') is similar to non-western forms previously studied by Turner and Van Gennep as 'Rites of Passage' (Van Gennep, 1909). These are rituals comprised of three phases: 'The first phase (of separation) comprises symbolic behaviour signifying the detachment of the individual or group . . . from an earlier fixed point in the social structure' (Turner, 2009: 80). This is a detachment from a former self in preparation for the 'transmutation' of the self via the ritual.

In the second phase, 'the attributes of liminality or of liminal *personae* ("threshold people") are necessarily ambiguous since this condition and these people elude or slip through the network of classifications that normally locate states and positions in cultural space' (Turner, 2009: 95). This is the 'in-between' phase where the individual is neither who they were before nor who they will become; their identity is in flux, non-concrete, liminal.

The third phase comprises of a re-integration in which 'the passage is consummated [by] the ritual subject' (Turner, 2009: 80). This can happen both physically and symbolically through the idioms of re-integration such as the 'sacred bond, [or] the sacred cord' (Van Gennep, 2004: 166).

The practice of esoteric theatre is similar to these rites of passage in structure, but the defining aspect of esoteric theatre is the liminal phase of the ritual which leads not to a change in social standing or cultural space but to a type of experience which has been called by Greenwood 'magical consciousness' (Greenwood, 2015). This type of consciousness she defines as 'an affective awareness experienced through an alternative mode of mind' (Greenwood, 2009: 64).[6] It is within this magical consciousness that the esoteric practitioner is able to change identities and experience 'transmutation', which in this case is a change in inner meaning, identity, consciousness or awareness which is a permanent and significant change (Greenwood, 2005; Hanegraaff, 1998: 42–61).

In terms of practices leading to or creating 'liminality' and change in identity, often taboo breaking and transgression are part of contemporary esoteric theatre praxis.[7] This is true of some practices outlined in this paper, along with other integrated practices such as durational ritual, pain, danger, rituals dealing with ideas of purity and impurity, spirit possession and healing (Lingan, 2006, 2014).[8] Other practices include a tendency to site-specific theatre in which holy, haunted or sacred places are used for performance, mythological texts and collective creation in which plays are crafted out of several contributors in order for all performers to be able to integrate individually into a sense of initiation or identity transformation.

What follows below is an introduction to the healing and transgressive aspects of the esoteric theatre of Metamorphic Ritual Theatre, Foolish People and Angela Edwards. My own experiences as a performer will also be addressed in the article.[9]

Foolish People and the Theatre of Manifestation

Foolish People is a theatre and film making collective based in the UK, well known for their series of plays *Dark Nights of the Soul*, performed over a two-year period in London (2005–2007). These plays were created around the collective work of local occultists as performers and drew large crowds of esoteric practitioners as participating audience members. The work of Foolish People is founded on dramatherapy, western esotericism and shamanism in an overall practice they refer to as the 'Theatre of Manifestation'.[10] For his article I've researched the elements of site-specific practice in their work and how this relates to their transgressive, healing and mythic theatre rituals.

Many of Foolish People's ritual performances are similar to the performative healing rituals discussed by dramatherapist Sue Jennings, who defines dramatherapy as 'healing through drama allowing the client through the use of dramatic structures, to receive insights and explore emotions in a special place in real and imaginary time, within a social encounter' (Jennings, 1994: 19). Jennings has based some of her dramatherapy thesis on her research into the ritual séances of the Senoi Temiars (Jennings, 1995) with whom she lived in Malaysia. In a recent public lecture Jennings also referenced the healing elements of liminal trance states; trance states she claims are necessary to 'get in touch with healing energy' (Jennings, 2012: Lecture). As director John Harrigan has a background in dramatherapy, healing has become a focal point for the rituals of the Foolish People, and within the particular, special or liminal space of site-specific performances, Foolish People explore the healing crisis.

Foolish People are focused ultimately on creating a ritual space in which audience members are invited to become initiated into a personal and healing form of western esotericism.[11] To do this they produce plays designed to create healing and transformational experiences in the audience such as *Dark Nights of the Soul*, a six-chapter extended ritual performed at the Horse Hospital venue over two years (2005–2007) in London. In this production transgressive techniques such as ritual cutting, bleeding, nudity and violence were performed by the actors as part of various personal explorations of initiation. Harrigan explains this by citing pain as a part of the liminal experience of healing which is then passed on to the audience:

> Through each ritual, each project we undergo healing and that is a process of reducing and refining ourselves. We heal the audience. It is about healing them and others though our work so that they can see their own truth and reason for being here, and through this you can see ailments and physical problems get healed. Pain we also accept as a part of healing. That pain and suffering is perhaps necessary.
>
> (John Harrigan Interview 2: 2016)

They also lead performative workshops in which their healing practices are shared with a small group of people. These are participatory and initiatory workshops in which dramatherapy, liminal shamanic techniques (such as trance) and pagan mythology are used to increase a personal connection which director John Harrigan calls the 'numinous':

> Certain actions will increase your connection to the numinous. Yesterday we did a long ritual in which we went very deep [into a trance] and we did not know what was going to happen next. It was transcendental. I had an out of body experience. And this has an effect on the audience definitely.
>
> (John Harrigan Interview 2: 2016)

Part of the process by which Foolish People create liminal healing performance is also contemporaneous with ideas of the walking ritual performances of mythogeography as explored by Phil Smith, in which the space itself inspires stories, myths and workings which are brought about by individual interactions. The space of mythogeography is interacted with in a nonlinear and possibly also liminal way for 'the space of mythogeography is neither bounded nor sliced by time' and 'it is also a geography of the body' as well as 'a philosophy of perception' which 'is self-reflective in the sense that it regards the mythogeography, the performer and the activist as being just as much multiplicitous and questionable sites as the landscapes they move in' (Smith, 2010: 113–115).

In many of the performances of Foolish People, such as *Desecration* (Galleries of Justice, Nottingham, 2007) or *The Abattoir Pages* (The Old Abattoir, London, 2009), audiences interact with performers in and around specific places and the space itself serves as the main medium of ritual modality and creation. Harrigan speaks about the space in terms of this very personal and healing relationship, implying even that a space can 'suffer', which will inform an individual and shared ritual practice:

> We start with the geographic space and this is the framework for the whole ritual and performance. The story for the ritual comes from the physical

> space. We do rituals in the space and allowing ourselves with the space first and we interact and have a relationship with the space. If it is a space of suffering then we respect that and that informs our practice.
>
> (John Harrigan Interview 2: 2016)

In this way since rituals of Foolish People are based very specifically on the meaning and nature of places, there is an aspect of 'admitting the ghost', which Gordon writes is a 'special instance of the merging of the visible and the invisible, the dead and the living, the past and the present – in the making of worldly relations' in which the marginal and 'what we never even notice' becomes the starting point of an experience (Gordon, 2008: 25). There may be also something in the use of mythologies to evoke this crisis or healing moment. In writing of the self-defining of new age and Neo-Pagan spirituality, Bloch, for instance, writes of mythology not only as ideology but also of the trend to mythologise crisis situations and 'one's spiritual identity' (Bloch, 1998: 102).

It is these marginalities, ghosts of space and place, which are integrated into mythologies both personal and historical to create the rituals and healing performances of Foolish People. As well it is within these marginal or liminal spaces that esoteric theatre differs from other contemporary theatre: there is after all a history of the use of ritual forms in theatre practice which is generally understood as part of experimental theatre work.[12] However, they do not go far enough, or to put it another way, they go and yet they return, the ritual meaning is not changed. Ritual was done, but it was not liminal, magickal or esoteric; meaning and identity were not permanently transcended, destroyed or altered according to the generally accepted meanings of initiation or transmutation among contemporary occult practitioners (Greenwood, 2015; Schechner, 1993; Innes, 1981).

Angela Edwards and the performance of sacred pain

Angela Edwards is a London based visual and solo performance artist who specialises in themes of sacred prostitution, pain, taboo and endurance in public esoteric rituals. She performs these rituals in a wide variety of spaces including sex clubs, graveyards, galleries and traditional theatre spaces. Recent work includes a series of rituals called *Death Shrine* and *Death Shrine to the Holy Whore* (Performance Space 2013: London, various graveyard spaces, London, 2015) and *Holy Corpse Sculptural Shrine:* St. Steven's Church Yard, London, 2016, *The Celestial Shroud* (Figure 17.1), ritual skin-sewing performance at *Chronic Illness of Mysterious Origin III*, London 2016. I have worked with her on several projects as an assistant. Contrary to the work of Foolish People, for Angela Edwards the body is the only ritual space and her ritual performances are designed to push her body into trance states through a transgression of her own physical and emotional boundaries.

In anthropological studies of tribal ritual culture, there has been a focus on the issue of pain, taboo, danger and durational experiences (Turner, 1992; Whitehead, 2002; Walsh, 2007; Douglas, 1984). Mary Douglas has for instance written about ritual uncleanness and the dichotomy of taboo as something dangerous and yet

Figure 17.1 Angela Edwards in *The Celestial Shroud*, ritual skin-sewing performance at *Chronic Illness of Mysterious Origin III*, London 2016.

Source: Alison Rockbrand.

existing as part of an integrated ritual meaning in culture. In the context of how this differs from western thinking, she states that:

> For us sacred things and places are to be protected from defilement. Holiness and impurity are at opposite poles. . . . Yet it is supposed to be a mark of primitive religion to make no clear distinction between sanctity and uncleanness. If this is true it reveals a great gulf between ourselves and our forefathers.
>
> (Douglas, 1984: 8)

This echoes Eliade, who likewise interpreted the sacred as potentially ambivalent in that 'it attracts or repels. The sacred is at once "sacred and defiled" ' (Eliade, 1958: 14–15). In Douglas' analysis, she finally turns the looking glass back towards her own culture and possibly herself when she sums up her discussion by deciding that 'we shall not expect to understand other people's ideas of contagion, sacred or secular, until we have confronted our own' (Douglas, 1984: 29). It is a self-confrontation of the sacred and the profane that inspires the pain-centred durational performances of Edwards' healing ritual cycles.

There are instilled in many of her performances ideas of religious inversion healing by religious or spiritual ordeal, found also in the 'sacred pain' rituals which occur cross-culturally (Glucklich, 2001). In one instalment of *Death Shrine to the Holy Whore*, performed in a 'voyeuristic' experimental box space at the Edinburgh Fringe (Edinburgh 2013), Edwards crucified herself on a free standing cross while her labia was stapled by an assistant, after which she was ritually buried under a pile of dirt and could only breathe through a straw for a duration of 30 minutes. Edwards, who is a practitioner of a non-traditional form of Voodoo, also cites shamanism as a reference to pain in rituals of healing:

> Those rituals were healing in that they were transformative. Many things to do with the body are healing such as with acupuncture, so when I use needles I also think of that as a healing touch on my body. There is that whole shamanic tradition of cutting as healing or initiation. Also having scars on my body and healing for months and months with the scars is a constant reminder of the healing process. All my endurance rituals require me to heal physically and that is a reminder of the healing process. It's a dedication of the body to the practice.
>
> (Angela Edwards Interview 3: 2016)

Pain and durational experiences in the context of ritual healing are something which can been seen in all of the performers being written about in this paper. For Angela Edwards however, pain, durational experience as part of the experience of the sacred is the main motivator of her work. Like Mary Douglas' inward gaze, she is using her own body as an experience, to explore the meaning of the sacred.

She does this through the creation of the abject body as her own body in the uniting of the sacred with the profane. One of her ritual modalities is that of the physical embodiment of the 'Holy Whore'. Edwards, who has worked as a sexworker, re-interprets the abject through its integration as sacred, in a way which echoes Kristeva's psychoanalytical approach to healing in *Powers of Horror* (Kristeva,

1982). This is possible because 'through abjection, bodily processes become enmeshed bit by bit in significatory processes, in which images, perceptions and sensations become linked to and represented by "ideational representatives" or signifiers' (Grosz, 2001). In the performance *Death Shrine to the Holy Whore Part 7* (London 2015), Edwards lay crucified over a bladed cross in a local cemetery for a period of 5 hours. In *Death Shrine to the Holy Whore Part 8*, she lay within a specially created death shroud, completely covered, for almost 12 hours. These rituals were done as part of a two-year-long ritual in which Edward also engaged in what she describes as 'sacred sex work' to heal past painful and traumatic experiences she has had while doing 'profane sex work'.

Her experience of pain or danger however is mitigated by the creation of the ritual space as the totality of her own body and the allowance of liminality as the path to a place between or in in some cases without identity:

> I don't see it as danger. I see it as death, and transgression and confronting into my own mortality and this is also a part of Spiritualism or Voodoo where people honor the dead and this is part of this confrontation. I see this as destroying my own ego and confronting death. But also I don't feel the pain because I am in a trance and I am not thinking about that.
>
> (Angela Edwards Interview 3: 2016)

It may be this transcendence of pain and experience of miraculous 'painlessness' in the sacred space of the body which can create the healing moment of transition. This is the case with the rituals of sacred pain as discussed by Glucklich in which 'The correlation between pain and truth . . . is not punitive but "gnostic" (truth eliciting)' and 'The power of pain, or the miracle of insensitivity to pain, is attributed to divine oversight in the proceedings' (Glucklich, 2001: 20).

Edwards does not call herself a performance artist because she sees her work as real and actual, as 'non-acting' (Kirby, 1995) and not within the same tradition of other artists who explore similar themes. Artists like California performance artist Chris Burden, whose 'painful exercises were meant to transcend physical reality' (Goldberg, 1993: 159), are not on the same level of experience because they are not in the process of creating esoteric ritual, that is they are not creating the kind of ritual within a tradition of western esotericism which leads to a liminal space in which magickal consciousness is experienced. According to Edwards this is something which one must participate in and be fully oneself in, not as a performance of the self, but as the self in what is called performance because it is being watched by others. It is this self as oneself in the ritual space which creates meaning in the work of Angela Edwards. As well, to return to Douglas' assessment of the rituals and purity and danger, it is ritual itself which creates these special and specific meanings:

> Ritual . . . can permit knowledge of what would not be known at all. It does not merely externalise experience, bringing it out into the light of day, but it also modifies experience, in so expressing it. . . . There are some things we cannot experience without ritual.
>
> (Douglas, 1984: 65)

Orryelle Defenestrate-Bascule and the Metamorphic Ritual Theatre

Orryelle Defenestrate-Bascule is a prolific artist and performer, well known in Neo-Pagan and occult communities for his varied arts and mythic esoteric performances. He has not only created paintings, drawings and a tarot deck with imageries based in the western esoteric tradition, but has also produced, written and performed in ritual theatre projects with his Australian-based collective Metamorphic Ritual Theatre (Lingan, 2006). These performances have toured across Europe, the UK and the US, to audiences of mainly other Neo-Pagans and occultists. I have worked with Orryelle on several projects (*Parzival a Fool's Journey* 2005: Glastonbury, *Loom of Lila*, 2007: Edinburgh Fringe Festival, *Solve et Coagula* 2012: Brighton, *Alchemical Chess*: 2015 Exeter University) as a performer and deviser. I will be discussing the private and public aspects of his performances in sacred spaces which are mythic, transgressive and based in long term initiatory practices.

One of the ways Neo-Pagan communities create meaning in the context of having no accepted religious authority is to collectively devise new ritual texts (Bloch, 1998; Stone, 1978; Adler, 1979). Barbara Rensing explains that 'neo-pagan spirituality is flexible and personal' and how 'this personal spirituality is expressed in poetry' (Rensing, 2009: 184).[13] These ideas are reflected in the work of Metamorphic Ritual Theatre. Poetic plays such as *Parzival a Fool's Journey* (2005), which was performed outdoors in locations considered sacred (such as Glastonbury Tor), use integrated mythologies from various traditions.

This use of specific mythologies is linked to the style, direction and meaning of the ritual format of the play. Within the context of establishing a text for performance, the text will always be based on personal rituals created for the experience of magical consciousness, liminality, initiation and healing and yet at the same time, shared in a public and open space as part of a healing for a 'collective consciousness'. In the Metamorphic Ritual Theatre play *Oedipus Tyrannos* (London 2008):

> The whole idea of doing a rewrite of the Oedipus myth was based on a healing or transformational premise, not necessarily for myself or the cast but it seemed this myth mostly through theatre has made a very strong impression on the collective consciousness of humanity . . . and I felt this to be a primarily negative one. It is symptomatic of patriarchal 'heroism' mythology in that the central character is so proud of his 'wit' which 'defeats' the sphinx and yet he seems ignorant of his own subconscious and his own fate.
>
> (Orryelle Defenestrate-Bascule Interview 4: 2015)

Similar themes abound in other Metamorphic Ritual Theatre plays such as *Loom of Lila* (2007), a dance based piece in which the performers challenge their own ideas of fate through an embodiment of the goddess Kali (Figure 17.2), or in *The Choronzon Machine* (2003), a rock-opera in which ideas of initiation and

Figure 17.2 Metamorphic Ritual Theatre in performance of *Loom of Lila* (2007). Eleven dancers form the face and hands of Kali-Arachne, a composite form of the Mother Goddess.

Source: Orryelle Defenestrate-Bascule.

transformation were the main focus. This ritual framework is a way of delineating space which is esoteric from everyday normal space, the main difference between these being the level of liminality which is being experienced. In the ritual space an esotericist is creating a liminal zone where different types of transmutations can take place. Turner as mentioned has written of how liminality in ritual structures leads to various types of initiatory changes (Turner, 2009), while Van Gennep describes the 'transition' stage in a rite of passage as the place between identities (Van Gennep, 2004; Bell, 1997).

Where a ritual is public or private then is a matter of degrees, and making public what is private means in some way opening or expanding a traditionally and probably practically closed space (Bogdan, 2007). The public or open performance of rituals allows personal experiences to be communicated and in this way also changed or shaped; for the private ritual has now been compromised and a public form of it is taking place, leading to changes and transformations of the ritual. This is part of the fluidity of the open ritual as it is a step into the unknown, into the possibly chaotic event of performing and including strangers in a public display. This is true for all public esoteric rituals, from the open rituals of the Pagan Federation in London to the performance of esoteric theatre. Although in the context of esoteric theatre this chaotic process is refined and specialized.

In the *Metamorphic Ritual Theatre*, although the rehearsal process is private, and the creation individual, poetic and initiatory, containing a variety of rituals within its process, the ultimate 'Rite of Performance' is always in public. However these boundaries between private ritual and public and inside and outside of a ritual or liminal space are themselves in a state of change and flux as private overlaps with public:

> I find degrees of privacy in my work rather than it simply being either public or private . . . the (often ritualistic) process of its creation is a more private one. Yet a part of this process imbues what the viewer eventually receives. . . . So all ritual really has aspects of the private and the public, there are just different degrees, levels and means of interaction.
>
> (Orryelle Defenestrate-Bascule Interview 4: 2015)

At some point the public-making is itself a ritual and a point of transmutation since as the ritual transforms so may the performers and an unknown variant may add to the experience of liminality. Part of the meaning of esoteric theatre as public ritual performance is that through the opening of a doorway and the communication of a ritual in a chaotic or unknown fluid ritual, space creates a liminality extending back to the ritual boundaries and almost re-sealing the public space into a private one:

> The energetic feedback of the audiences can affect the performer as well as the audience, and this is part of the purpose of public ritual. In a way all ritual

> is a kind of performance, as you are establishing parameters of sacred time and space, and therefore become more conscious of your behaviour.
> (Orryelle Defenestrate-Bascule Interview 4: 2015)

The experience shared among all the people present becomes a new private space and the spectators are now transformed into participants, having shared in the esoteric knowledge or mytho-signifiers of the ritual makers. The work of the Metamorphic Ritual Theatre has for instance led to many audience members asking to be initiated into his 'Chaorder of the Silver Dusk', an occult order associated to his theatre company. Concurrent with the work of Foolish People, audience members are able to become initiated into or at the very least have a liminal experience of the western esoteric tradition.

Conclusion: insider perspectives on ritual practice and the rite of performance

As part of my work with esoteric theatre I have become focused on issues of the insider as a performer, researcher, experiencer and the necessity of continuing to develop appropriate insider research methodologies. This is because the creation of ritual in esoteric theatre is based on very specific meanings already known as well as codes and knowledge which the performers have already integrated.[14] As well, in terms of esoteric theatre, once the ritual space has been opened, all who are 'in' this space are in it as various types of insiders. Outsiders are the ones who are not present, not participating, not knowing. As has been expressed in this article, this ritual then leads to a liminal space wherein a new meaning, which can mean healing or a new form of identity, or some type of permanent magical change, occurs (Greenwood, 2005; Hanegraaff, 1998: 42–61).

There are many ways in which this insider perspective or collection of 'insider codes' has been expressed as part of my experience working with the groups in this article. While performing with Metamorphic Ritual Theatre for instance, I was first formally initiated into the occult order associated to the group and founded by leader Orryelle Defenestrate-Bascule. I was then asked to pull a card from a tarot deck to determine what my interaction with the group might entail; I pulled the card called 'Babalon' which in turn was a character I played in a number of productions. My playing of the character as well as my interaction with the performance in general were predicated on my experience in a formal initiation.

I feel that to study esoteric ritual and specifically esoteric performance is to participate in it on some level. This idea is echoed in the participatory mythogeography of Phil Smith (Smith, 2010) and other performers or practitioners of western esotericism who have written about their own practice (Evans, 2007; Letcher, 2001). There is a shared experience which is based on insider codes and which is able to be interpreted both communally and individually, hence how initiations, ideas or experiences may be passed on to the audience as Harrigan spoke of above.

This place 'between' identities or realities, this place where in magic as an active force is part of consciousness, leads into a definition of esoteric ritual in which one meaning creates or practices a ritual which creates or develops into a new meaning. For instance:

Meaning A = Ritual = Meaning B.

'Meaning A' is the previous self, set of circumstances and/or previous rituals which lead the individual to the creation of a new ritual. 'Meaning A' could be the self in a way which the individual wants to alter, to heal or to initiate into a new understanding of magic, the world, the self, the body or into a hierarchical magickal order.

The 'Ritual' in this model is all the actions which take place in whatever is being called the 'ritual space'. This space may be by its nature transgressive since it is outside of regular time and space; it 'transgresses' the normal boundaries of experience. Therefore if 'ritual' is the action, then 'ritual space' is the conceptual space of that action. In my work with Angela Edwards on *Death Shrine* I was able to enter in a liminal space which allowed me to stick over 50 needles into her skin, which is not something I had ever done before, but which as part of the ritual we engaged in was necessary; it was part of the code of the ritual. In this liminal space audience members were also invited to stick needles in and they did, thus becoming part of a transgressive act which was outside of the normal experiences between strangers.

'Meaning B' is the new identity being created by the ritual experience. The way in which the identity is altered depends on the ritual itself, its content and structure are all important aspects which influence the outcome and the new meaning being created. Also there is an element of the unknown, in which the new meaning may not be fully understood or may not be as expected.

I feel that similar to the aforementioned 'Rites of Passage', what is being practiced by esoteric performers may be called something like a 'rite of performance'. This possible rite of performance would be the creative process of esoteric theatre; from the inception of the idea to its performance in front of an audience. There is no form or specific content to a rite of performance but types of practice which draw together various elements of esotericism and theatre into a repeatable model of performance, comprising an extended set of rituals in rehearsals, performances and repeat performances. It is based not only on the body and identity of the actor but also on a greater text of mythologies and ideas which have a place in the history and culture of western civilisation. It is in this way an important part of western culture questioning western models of liminality, the interaction of private and public spaces, transgression and taboo in contemporary western spirituality, current models of ritual practice for healing studies, new insider research methodologies and ways in which the arts interact with ideas of the sacred and the profane, hopefully to be discussed as part of ongoing research.

This ritual is now closed
Its meanings have entered
Into the space of the reader
Let it create more meanings,
Wherever it goes.

Notes

1 During the occult revival as explored by Lingan (Lingan, 2014). Another example would be the occult plays of W.B Yates, who was a member of the Hermetic Order of the Golden Dawn (Macneill, 2013).

2 For the purposes of this article, western esotericism is defined as a world view or mode of thought with four essential components as proposed by Antoine Faivre (Faivre, 1994). These are: (1) Correspondences: 'Symbolic and real correspondences are said to exist among all part of the universe, both seen and unseen'; (2) Living Nature: 'Nature occupies an essential place'; (3) Imagination and Mediations: '. . . mediations of all kinds such as rituals, symbolic images, mandalas, intermediary spirits'; (4) Experience of Transmutation: '. . . understood also as "metamorphosis". It consists of allowing no separation between knowledge (gnosis) and inner experience' (Faivre 1994: 10–14).

3 See Innes and Schechner on various types of non-esoteric theatre practice using ritual and containing esoteric ideas but in which ultimately the performers are not aiming at the same kind of esoteric transmutation, nor do they uphold an esoteric worldview or claim to be western esotericists (Schechner, 1993; Innes, 1981).

4 There is currently a lack of academic writing on contemporary western esoteric ritual practice, with most studies of ritual being written in a historical context. Bodgan has written at length about rituals of initiation in western esotericism but limits this to the Freemasons and ends his study with Gardnerian Wicca era 1950 (Bogdan, 2007). Historian Ronald Hutton likewise has written about many aspects of Neo-Paganism but without a specific focus on the contemporary ritual experience or practice (Hutton, 2001).

5 As well, these participations themselves and the nature of participation and researcher engagement, insider research and insider lexicography can be looked at through the sociological and pedagogical methodologies of Adler and Adler (Adler and Adler, 1987) both of whom involve themselves in the 'core activities of group members' (Dwyer and Buckle, 2009: 55).

6 According to Greenwood the 'magical' state of consciousness is also something which 'must be experienced' in order to be understood and has an 'intrinsically subjective and sensory quality' (Greenwood, 2005: 7).

7 For a discussion of transgression as a part of western esoteric practice in general see Urban (2003).

8 Unlike much of the historical esoteric theatre, such as that of Rudolf Steiner and Florence Farr, contemporary esoteric theatre has a tendency towards these extreme experiences and has more in common with the 'Theatre of Cruelty' of Antonin Artaud, who believed that 'There is a mysterious identity of essence between the principal of alchemy and that of theatre' (Artaud, 1958: 48).

9 Interviews and experiences with these three groups are varied, and each performer uses slightly different language to describe similar events. For this reason some words will be used interchangably; esotericist/occultist/pagan/Neo-Pagan; initiation/transmutation/identity change.

10 Recent productions include *Virulent Experience* (Conway Hall, London, 2012), *The Woods Trapped at the Edge of Midnight* (Cornbury Park, Oxfordshire, 2015) and the feature film *Strange Factories* (2013).

11 Their esoteric practice is more akin to the flexible and individual spirituality of Neo-Paganism (Evans, 2007; Rensing, 2009; Hanegraaff, 1998) than to the more formal initiatory rituals of contemporary occult orders (Bogdan, 2007; Faivre, 1994).
12 See Peter Brook and his Artaud inspired Theatre of Cruelty Season at the RSC (1964–65), Jerzy Grotowski and his Paratheatre to Barba and the Odin Teatret, and more recently Nicholas Nunez and his work on 'Anthropocosmic Theatre' (Nunez, 1996). These are approaches which venture to some degree into the same territory as the ritual performances of Foolish People.
13 These elements accord with other definitions posited by Faivre and Hannegraft, both about the involvement with nature (Faivre,1994) and the 're-illusionment' of the individual through new age spiritualities (Hanegraaff, 2000).
14 According to practitioner-researcher Evans, 'participation within magical rituals requires compliance with tacit and often complex, and in some cases contradictory codes of conduct, plus some prior knowledge' (Evans, 2007: 61).

References

Adler, M. *Drawing Down the Moon*. New York: Viking Press, 1979.

Adler, P., and Adler, P. *Membership Roles in Field Research*. Newbury Park, CA: SAGE, 1987.

Armstrong, G. S. *Theatre and Consciousness: The Nature of Bio-Evolutionary Complexity in the Arts*. New York: Peter Lang, 2003.

Artaud, A. *The Theatre and Its Double*. Trans. M. C. Richards. New York: Grove Press, 1958.

Bell, C. *Ritual: Perspectives and Dimensions*. New York and Oxford: Oxford University Press, 1997. Originally published 1966.

Bergunder, M. "What Is Esotericism? Cultural Studies Approaches and the Problems of Definition in Religious Studies", *Method and Theory in the Study of Religion*, 22(1), 2010. Brill.

Bloch, J. P. *New Spirituality, Self, and Belonging: How New Agers and Neo-Pagans Talk About Themselves*. Westport, CT and London: Praeger, 1998.

Bodgan, H. *Western Esotericism and Rituals of Initiation*. Albany: State University of New York Press, 2007.

Crowley, A. *Rites of Eleusis as Performed at Caxton Hall*. South Cockerington, UK: Mandrake Press, 1990.

Denzin, N. K., Lincoln, Y. S., and Smith, L. T. (Eds.) *Handbook of Critical and Indigenous Methodologies*. Thousand Oaks, CA: SAGE, 2010.

Didion, J. *The Year of Magical Thinking*. New York: A.A. Knopf, 2005.

Douglas, M. *Purity and Danger: An Analysis of Concepts of Pollution and Taboo*, first published 1966. New York: Ark Edition, Routledge & Kegan Paul Ltd., 1984.

Dwyer, S. C., and Buckle, J. L. "The Space Between: On Being an Insider-Outsider in Qualitative Research", *International Journal of Qualitative Methods*, 8(1): 54–63, 2009.

Eliade, M. *Patterns in Comparative Religion*. London: Sheed & Ward, 1958. (Trans. from *Traite d'Histoire des Religions*, 1949).

Ellis, C. *The Ethnographic I: A Methodological Novel About Autoethnography*. Walnut Creek, CA: AltaMira Press, 2004.

Ellis, C. "Telling Secrets, Revealing Lives: Relational Ethics in Research With Intimate Others", *Qualitative Inquiry*, 13(1): 3–29, 2007.Evans, D. *The History of British Magick After Crowley: Kenneth Grant, Amado Crowley, Chaos Magic, Satanism, Lovecraft, the*

Left-Hand Path, Blasphemy and Magical Morality. Harpenden, UK: Hidden Publishing, 2007.

Faivre, A. *Access to Western Esotericism*. Albany, NY: State University of New York Press, 1994.

Farr, F. *The Serpents Path: The Magical Plays of Florence Farr*. London: Holmes Pub Group, 2006.

Garner, S. B. *Bodied Spaces: Phenomenology and Performance in Contemporary Drama*. Ithica, NY: Cornell University Press, 1994.

Glucklich, A. *Sacred Pain: Hurting the Body for the Sake of the Soul*. Oxford: Oxford University Press, 2001.

Gordon, A. *Ghostly Matters: Haunting and the Sociological Imagination* (2nd ed.). Minneapolis: University of Minnesota Press, 2008.

Goldberg, R. *Performance Art From Futurism to the Present*. Slovenia: Thames and Hudson, 1993.

Greenwood, S. *Magic, Witchcraft and the Otherworld: An Anthropology*. Oxford, UK: Berg Publishers, 2000.

Greenwood, S. *The Nature of Magic: An Anthropology of Consciousness*. Oxford, UK: Berg Publishers, 2005.Greenwood, S. *The Anthropology of Magic*. New York: Berg Publishers, 2009.

Greenwood, S., and Goodwyn, E. D. *Magical Consciousness: An Anthropological and Neurobiological Approach*. London, Routledge, 2015.

Grimes, R. L. *Rite Out of Place: Ritual, Media and the Arts*. Oxford: Oxford University Press, 2006.

Grosz, E. "Transgressive Bodies", in *Performance Analysis: An Introductory Coursebook*. Eds. C. Counsell and L. Wolf. Oxford and New York: Routledge, 2001: 140–153.

Hanegraaff, W. J. "Empirical Method and the Study of Esotericism", *Method and Theory in the Study of Religion* 7(2): 99–129, 1995.

Hanegraaff, W. J. *New Age Religion and Western Culture: Esotericism in the Mirror of Secular Thought*. New York: State University of New York Press, 1998.

Hanegraaff, W. J. "New Age Religion and Secularization", *Numen* 47(3): 288–312, 2000. Brill.

Hurley, E. *Theatre and Feeling*. New York: Palgrave Macmillan, 2010.

Husserl, E. *The Idea of Phenomenology*. Trans. W. P. Alston and G. Nakhnikian. The Hague, The Netherlands: Martinus Nijhoff, 1964.

Hutton, R. *The Triumph of the Moon: A History of Modern Pagan Witchcraft*. Oxford: Oxford University Press, 2001.

Innes, C. *Holy Theatre: Ritual and the Avant Garde*. London, New York and Melbourn: Cambridge University Press, 1981.

Jennings, S. *Theatre, Ritual and Transformation: The Senoi Temiars*. London and New York: Routledge, 1995.

Jennings, S. "The Healing Practices of the Senoi Temiar". Lecture at *Ritual Theatre Celebration* 2012.

Jennings, S., Cattanach, A., Mitchell, S., Chesner, A., Meldrum, B., and Mitchell Nfa, S. *The Handbook of Dramatherapy*. London: Routledge, 1994.

Kirby, M. "On Acting and Non Acting", in *Acting (Re) Considered: Theories and Practices*, Ed. P. B. Zarrilli. London and New York: Routledge, 1995: 43–58.

Kristeva, J. *Power of Horror: An Essay on Abjection*. Trans. L. S. Roudiez. New York: Columbia University Press, 1982.

Letcher, A. *Role of the Bard in Contemporary Pagan Movements*. PhD Thesis, King Alfred's College, University of Southampton, 2001.
Lingan, E. "Contemporary Forms of Occult Theatre", *Performance Art Journal*, PAJ 84: 23–38, 2006.
Lingan, E. *The Theatre of the Occult Revival: Alternative Spiritual Performance From 1875 to the Present*. New York: Palgrave Macmillan U.S., 2014.
McNeill, P. "The Alchemical Path: Esoteric Influence in the Works of Fernando Pessoa and W. B. Yeats", in *Fernando Pessoa's Modernity without Frontiers: Influences, Dialogues, Responses*, Ed. M. G. De Castro, NED – New edition ed. Suffolk, UK: Boydell and Brewer, 2013: 157–168.
Nunez, N. *Anthropocosmic Theatre: Rite in the Dynamics of Theatre*. Amsterdam, The Netherlands: Harwood Academic Publishers, 1996.
Pearson, J. "Going Native in Reverse: The Insider as Researcher in British Wicca", *Nova Religio: The Journal of Alternative and Emergent Religions* 5(1), 2001. University of California Press.
Plummer, K. *Documents of Life: An Introduction to the Problems and Literature of a Humanistic Method*. London: Unwin Hyman, 1983.
Rensing, B. "Individual Belief and Practice in Neopagan Spirituality", in *Vol. 21 Postmodern Spirituality*. Abo, Finland: Donmar Institute for Research in Religious and Cultural History, 2009: 182–195.
Schechner, R. *The Future of Ritual: Writings of Culture and Performance*. London and New York: Routledge, 1993.
Smith, P. *Mythogeography: A Guide to Walking Sideways*. Devon, UK: Triarchy Press, 2010.
Sontag, S., and Artaud, A. *Antonin Artaud, Selected Writings: Edited and with an Introduction by Susan Sontag*; translated from the French by Helen Weaver; notes by Susan Sontag and Don Eric Levine Farrar, Straus and Giroux New York 1976.
Steiner, R. *Four Mystery Dramas*, Trans. and Ed. H. Collinson, M.A. Oxon: S.M.K, 1925.
Stone, D. "New Religious Consciousness and Personal Religious Experience", *Sociological Analysis* 39, 123–134,1978.
Stuckrad, K. "Esoteric Discourse and European History of Religion", in *Western Esotericism*, Ed. T. Ahlbak. Abu, Finland: Donmar Institute for Research in Religious and Cultural History, 2007.
Turner, E. *Experiencing Ritual: A New Interpretation of African Healing*. Philadelphia, PA: University of Pennsylvania Press, 1992.
Turner, V. *Dramas, Fields, and Metaphors: Symbolic Action in Human Society*. Ithaca, NY and London: Cornell University Press, 1974.
Turner, V. *The Anthropology of Performance*. New York: PAJ Publications, 1988.
Turner, V. *The Ritual Process: Structure and Anti-Structure*. New Brunswick, NJ and London: Transaction Publishers, 2009.
Urban, H. B. "The Power of the Impure: Transgression, Violence and Secrecy in Bengali Śākta Tantra and Modern Western Magic", *Numen* 50(3): 269–308, 2003.
Van Gennep, A. *The Rites of Passage: An Anthropology and Ethnography*. Trans. M. B. Vizdom and G. L. Caffee. London and Henley: Routledge and Kegan Paul, 2004. Originally published 1909.
Versluis, A. *Esotericism, Art, and Imagination*. East Lansing: Michigan State University Press, 2008.
Walsh, R. *The World of Shamanism: New Views of an Ancient Tradition*. Woodbury, MN: Llewellyn Publications, 2007.

Whitehead, N. L. *Dark Shamans: Kanaima and the Poetics of Violent Death*. Durham, NC and London: Duke University Press, 2002.
Wyllie, A. "Philip Ridley and Memory", *Studies in Theatre and Performance* 33(1): 65–78, 2013.

Websites accessed

Angela Edwards Art, www.angelacarolinedwardsart.com. Last accessed 09/08/16.
Foolish People, www.foolishpeople.com. Last accessed 07/08/16.
Metamorphic Ritual Theatre, www.crossroads.wild.net.au/morph.htm. Last accessed 05/08/16.

Interviews

Angela Edwards, Interview 3. January 2016.
Orryelle Defenestrate-Bascule, Interviews 3–4. December 2015-January 2016.
John Harrigan, Interview 2. June 2016.

18 Reading three ways

Ask me how!

professor dusky purples

A reading by professor dusky purples

Preparation

Why are you here?
Why are you *really* here?
Are you doing your work?
Are you doing the work someone set you the task of doing?
Isn't that satisfying?
Is that satisfying?
How did you learn to read?
Do you remember it being painful?
What was the first book you read?
What was the first book from which someone read to you?
Were these words read over your little body?
Were you a little boy or a little girl?
Were there witches?
Were there evil spirits?
Was there Jesus?
Was there war?
Did we win?
Were there prophets?
Did they see it coming?
Were their faces clear?
Could they see you?
Could you hear them?
Did you hide from them?
Did you think about them in the dark?
Do you pass these stories on?
Do they frighten you?
Do they bring another feeling?

Get to know that feeling, sit with it, let it sit with you.

Orientation

What are you asking?
Or are you seeking?
Is it yes, a confirmation?
Is it no, a negation?
Is it neither?
Do you have something – an object, a talisman, a totem, a stone – at hand?
Can you take it in hand?
Can you put the question there?
Can you inquire?
Can you ask it in a good way?
Can you anticipate an answer?
Can you find out where the question wants to be put now?
Is there a right way to place it?
What are the elements of this way?
Do they have the quality of air, water, earth, or fire?
Are they fixed, cardinal, or mutable?
How do these elements relate to each other?
Are they harmonious, tense, oblique, acute, distant, or near?
If the question lives in your body, where is it most comfortable?
If the question lives through your body, what are its points of entry and exit?
If the question itself has a body, what is your posture toward it?
How can you face it?
Does it unmask you?
Does it speak with one voice?
Are you at home with the question?
Is it traveling with you?
Is it orbiting you?

Pull at the thread of the question and, when you reach a snag, an uncertainty, take a chance.

Cherished Reader,

Since this is a piece about where we have been and where we might now be going, let me begin by asking you to travel. Move inward from the vitreous humor that suspends your eye. Release the worldly seer. Make contact with your bodily ancient. Forgo the demands of faciality. Take off the mask. Flow inward through the teary mineral headwaters of your overprivileged oculus. Cross the apparent border of your sebum coated skin. Glide through the cerebrospinal barrier that encases your grey brain. Let the fungus of experience decompose your self-consciousness. Through the sticky cerumen yellow residue that gathers to protect you from aural inculcation, drift carefully across the endolymph lubricated passages of your inner ear. All sound is a soft touch. Swallow saliva, treasure of language. Gently sniff/slip through the mucoidal deposits of your sinus. Let your pleural fluid flow as you take in the substantiations of air. Churn and swish through gastric middle earth, your other brain. And, as you come into yourself, in whatever way, with whatever fluid, feel that breath which makes your not yet red blood flow. We begin from this place of in-folded containment, this bodily present/tense. We meet here, at dusk. We work in the purple twilight of the idea that theory comes from nowhere, has no biography. We greet the night sure that the world is subject to other influences, is not One, is comprehensible neither through force nor through field.

Now that we are a bit closer, I should tell you why I am here and where I've been. My last appearance, at the 2015 Emotional Geographies (#EmoGeo) conference in Edinburgh, Scotland, was prompted by an invitation from a longtime colleague and friend, Toby Sharp, '"tool" for urban change' (McLean, 2016, p. 39). Sharp reworks tired readings of regenerative/regenerating urban landscapes – the realm of the fabled creative class and its boosters – to show us how feminist, queer, and anti-racist performances walk the tightrope between neoliberal appropriation and creative subversion. Though I had long ago left behind the abandoned lots of urban creativity to join an autonomous feminist separatist commune in the last remaining swamplands of the Po River Delta, near Marina Romea, Italy, I could not resist the invitation to reflect on my life as a 'dirty, sweaty other' of urban theory, the field in which I had formally been trained as an akademik. So I convened a collectivized reading ritual in Edinburgh which culminated in taking a look at Henri Lefebvre's astrological birth chart, during which we discovered that he has a distinct lack of fire in his chart. Fire is the element associated with spirit, with sacred transformation, with warmth. So, we tried to bring some to him. I had an array of implements with which to work: rocks, smells, fabric. In the end, we held the circle and we danced to Buffy St. Marie's 'Keeper of the Fire,' from her 1969 masterpiece *Illuminations*.[1] And then we hiked up Arthur's Seat, an ancient dormant volcano just behind the Edinburgh conference center where we all met. We recalled its ancient fire and drank cheap spirits.

In akademia, such forms of conviviality, let alone conjuring, tend to be kept until after the end, in the drift space between scheduled events. Because I arrived to the conference after a long absence from the akademy, it felt easier to draw on a deeper embodiment of my femme-inity and to reconnect with two reading practices which had long ago characterized my life as a teenage witch: tarot and astrology. Both are conversation starters; they put us into relation with images, arrangements, constellations, orientations, chance (not so much fate, not for me, anyway), and, sometimes, they put us in relation to each other. Tarot deals with archetypal and situational dramas, casting the pieces of the present tense into new geometries according to the disposition of a querent and their encounter with a deck of cards, a particular question, and a reader, who is sometimes the same person as the querent. Astrology, on the other hand, renders the sky under which we were born as a stage with the planets as actors. The stories of their interactions, the tone and tenor of their relationships, is related to differing patterns and tendencies borne out in the life of the person whose chart is being cast. Astrology has many iterations, its own literatures, conferences, controversies, and charismatic figures.

It is not apart from this world. Astrology is like meteorology for everyday life; it is a guide, an indicator, a storytelling device.

I learned to read tarot and astrology through filters of gendered detachment and degendered attachment. To read, I had to access parts of myself not worn on my body while paying both more and less attention to those parts which are found there. Tarot and astrology attempt to re-link biography and geography over and above the dominant western figure of territory (see Povinelli, 2016, pp. 77–78). Popular conceptions suggest that tarot or astrology might be matters of belief. So many friends have asked me, just before they ask me to read their cards or their chart, 'Do you really *believe* in all that?' As practices long abjected into a vague realm of 'spirituality,' they have fallen out of conversation with dominant forms of contemporary literacy. They arrive from an 'enchanted' pre-capitalist realm (see Federici, 2004); their endurance in the present remains a sign of the incomplete, yet insistent, colonization of everyday life and expropriation of bodies and communal practices by the dominant order. Skeptics tend to overstate the extent to which tarot or astrology is absolutist, fatalistic, deterministic, a so-called pseudo-science or superstition. But it's not about that. Or, at least, that's not why I am here. It's about storytelling. The cards or the chart cannot tell you what you do not already know, but they can help you understand what you may know better. They might help you shoulder that burden of being who you are. They might help you carry that which you can neither put down nor push forward. They ask us to release what is no longer useful or what keeps us from learning more about what might be useful. Both practices rest on the notion that a querent, a person, someone with a question and a trajectory, can be understood in relation to both localized events, like drawing a card, and cosmic patterns, like the movement of planets and celestial bodies. We come to reading from both places. More than meaning, these modes question motive, stoke capacities for change in perspective and behavior, and reveal tendencies in what the Freudo-Lacanian mode might call 'the unconscious.' They are what we already know, but just a bit too deeply to claim with rationality.

Why conjure up tarot and astrology for the conference? During one of our routine letter exchanges, Toby informed me of the emergence of a framework in critical geography called planetary urbanization (Brenner, 2014). Proponents of the framework return to Henri Lefebvre's (2003) work in *The Urban Revolution* to substantiate the claim that the process of (urbanizing) capitalism must be understood as encompassing the entire surface of our beloved Earth. I was curious, so I took a look through the public archive of the project. Among the many rich cartographic and theoretical representations of this condition realized by the proponents of the framework, one apparently peripheral rendering stood out to me. Namely, that urbanization extends itself into the extra-atmospheric, insofar as the telecommunications and surveillance satellites orbiting the earth form a geocentric visualization prosthesis. As a by-product of this extension of vision and 'remote sensing' into orbit, the Earth is also a center of gravity for a proliferating cloud of 'space junk' (Figures 18.1 and 18.2).

Figure 18.1 professor dusky purples in Edinburgh, 2015. Source: Heather McLean.

Looking at this image, it occurred to me that the discipline of geography is still struggling to come to terms with even the most basic tenets of its own cosmological, if not cosmopolitical, orientation. The eye-like image of satellites and other *stuff* orbiting the earth appears, in aggregate, to form just another, larger Eye. This fundament of sensors sees without visualizing – or, better, without clearly acknowledging – its own prophetic aspirations. If this particular horizon of geography, is located 'out there,' what are we to make of its reliance on techniques and technologies which recursively cover the surface of the earth with the thought of urbanization, weaving it into the very fabric of everyday existence – human and not, living and not? Efforts to understand the planetary dynamics of urbanization stake out an enunciative position which may or may not *necessarily negate* the theoretical positions which preceded. Nonetheless, it seems to me that analytics organized around these dynamics tend toward an apparently singular approach to apprehending capitalism's survival in and as *this* world. How might the proposed priority of urbanization over other forms of cohabiting with and on Earth circumscribe the transformative potential rooted in other ways of knowing/sensing/interpreting both the origins and destiny of the planet? What are the modes of knowing and struggling together adequate to the present tense? Insofar as the epistemological orientation of planetary urbanization identifies a trajectory, a direction in which the planet might be headed, how does the architecture of its way of knowing constrain us to read the surface of the Earth? What difference does it make that this particular epistemology is the product of an akademik process?

As I made the journey North to the conference – where, in July, the sun rises at 4am – I asked myself these questions. I saw them all floating around the practice at the core of the reproduction of akademik knowledge: Reading. As I sped across

Figure 18.2 The back page of a U.S. Passport. Source: Darren Patrick.

Brexit territory, I thought about the perils of reading in only one way. I thought about divination, that maligned and minor realm of reading. I thought about divided nations and the ostensible retrenchment of the decadent imperial powers into their own miserable territories of hoped for homogeneity. I wondered about the signs of the times and the system/s adequate to interpreting those signs.

What better place than a speeding train to confront the dilemma of epistemic location? What is the cost of *going there from here*? And how might being in that *there* invite a challenge to our sense of what it is to *be here now*? So much aka-demic knowledge is narrowly concerned with the object of epistemology. That's fine, that's okay. There *are* objects. But, might other ways of reading – of divination, even – give us a better sense of the movement/s of the moment? Perhaps they will leave only a bitter sense. Addressing precisely these issues, Manulani Aluli Meyer (2010) draws on/unfolds Hawaiian Indigenous teachings (2001) to present what she calls holographic epistemology (2014). Meyer shows the ecological unity of three parts: knowledge rooted in experience, knowledge which is floating, and *aloha*, loosely translated, knowledge learned in the practice of loving and in service to others. She constellates epistemology as the trinity of knowledge, knowing, and understanding; noun, verb, and liberating practice, respectively. Together, these three modes give rise to questions which Meyer (2010) poses:

> What do you want to understand in your life? [. . .] What brings you meaning? What gives you meaning? When you do that in the practice of deep self-inquiry, then you hit the bottom of your own regenerative spirit. We truly know that true wealth is about giving; collaboration is more of an enduring practice, and truth telling is a higher frequency than the accumulation of facts . . . What is the difference between knowledge, knowing, and understanding?

While Meyer also demonstrates the wide array of cognate formulations in philosophy, metaphysics, spirituality and, yes, some religions, the distinction of her presentation seems to rest on the apparently ephemeral mode of understanding. As those of us who have been schooled – forcibly or otherwise – in dominant Western ideas of epistemology could no doubt attest, the pieces that Meyer calls collaboration, service, relation, loving, and truth telling are often mishandled, misunderstood, derided, or destroyed as we attempt to *know more*. These pieces are so often the seeds of understanding. Such tendencies toward neglect are correlated to the elevation of the so-called life of the mind, the mind scanned and mapped in MRIs and with radioactive isotopes, the mind ramified in the institution of the uni-versity, the mind which devises borders and property lines to steal and to criminalize and to outlaw. Though the uni-versity is becoming a corporation, and some states think of corporations as people, the uni-versity is not a person. It does not, as far as we know, have the capacity for embodied experience. And so it incorporates, developing new tools for epistemological validation and extraction; it is a factory, born of a guild. The uni-versity is a place where crude opinion is formed into transferrable knowledge by Learned Men.

Like all institutions, the uni-versity is both **a** formation and **in** formation; we can leave its **re**-formation for another day. It may not be a person, but it does relate, or acts as a relay. As **a** formation, it aspires to secular modes of uni-versality. Lately, it has been struggling, at least in North America, with diversity and inclusion, a signal that it is still **in** formation. It wants to know more about particular struggles without necessarily being changed by them or allowing itself to be the site of such struggles. It wants to know more about *you* and *your relations*. Because, if the uni-veristy is to become something else, and therefore to survive the great transformation, it will likely not be because of the Learned Men. It will be because of you and your relations. Here, perhaps, we shift from epistemology to ontology; we might enter and be together in the uni-versity, albeit with some struggle. We could dither about struggle against or within, we could quibble about struggle for power over or power to, these are certainly moments that hang between knowing and being. But, once we have crossed that bridge from epistemology to ontology, we have, at the root, struggle. Whether we persist or fail, whether we transform or remain rigid, each of these is a mode of struggle and has, as disciplinary geography will no doubt attest from its colonialist history and present, much to do with where we are born, how we remain there, or, if we move from that place, the conditions under which that happened and the losses and reorderings immanent to that displacement.

How do we come to whatever might be next from here?

This is also a political question.

How we navigate that question brings us closer to cosmology. A cosmology unfolds in the conversations we have with the night sky, rocks and rivers, magnetic fields, clouds and birds, these are conversations we can have with each other, sometimes without words. Cosmology is how we locate ourselves in the mesh of images, signs, stories which make us human and social. It is how we make sense out of the simultaneity of partial vision and an uncanny sense of wholeness. There are as many ways to locate cosmologies as there are vantage points from which to look up, out, in, and through. I do not claim to describe every possible cosmology, I could not! Besides, city lights are too bright. Instead I ask a question: How is it that the same geo-graphic/earth-writing apparatus that brings us space junk has so little to tell us about cosmologies? Are we reading correctly? When what we know or how we struggle to know it appears apart from, or without understanding or acknowledging where and how we have come to wonder, we are in danger. That danger may, in some sense, be symptomatic, especially if we are talking about highly specific akademik debates confined to particular disciplinary formations. Nevertheless, let's step over that threshold to consider, instead, what happens when we enter our reading practice not from the standpoint of epistemological subjecthood – unsure of who we know ourselves to be – nor from the immanence of ontological struggle – unsure of how we might be other/wise else/where. Let's begin again from a grounded spiritual cosmo-logic, a place which refuses and stands a-part from the Enlightenment vanquishment of spirit and the sacred, a place where we are reminded that, to be in a world, we must locate ourselves, our persistence, our ancestry, our lost memories, our dead ends.

Living in *a* world is not the same as living in *the* world. Even our neighbor may live in *a* world; distinct from, but adjacent to, ours.

Worlds too touch.

Though we are living, breathing beings we find ourselves subject to certain and defined logics. Or, at least, logics that tell themselves and us they are certain and defined. Like interstate highways, twelve hour clocks, Google Maps, and other networks of ordination and navigation. Some of these logics proclaim their universality, their applicability to all questions at all times and in all places. They wear their one-ish-ness too lightly and too seriously all at the same time. They can take you anywhere. The surety of these ways of knowing and reading the world is a patrilineal bond. It tends to order knowledge vertically, eclipsing story. Story asks us to *listen*, to *understand*, to *stand with* and draw lines of connection. In this way, we see that the patrilineal uni-versal is but a series of ramified (mis)alignments and premature unities.

From here to there through this.

From you to me through this.

By turning to different reading practices, we loosen the bind of that bond a bit. We do not try to save face in front of the baffling mysteries. We do not assume that the newest instrument is the best instrument. We do not operate in a mode of falsification. I am not for you and I am not against you. Instead, we begin to assess an agenda, a posture, we query unarticulated desires and sate unacknowledged needs while running a tired finger along lines marking paths of arrival and departure. Tired because falsifiability remains the reality of much of akademik knowledge production. Falsifiability and verifiability, specie and doubloon of crude empiricism. If you can do it here and there and elsewhere, so it is done to all. That is how laws are made. Crude empiricism is but a market whose trades ride the promissory rails of proof positive. Progress *will* be made. (We do indeed have a problem of *science literacy* – for which we may all be STEM'd to death – but I'll leave that to the scientifically literate to talk about; I am addressing the deficiency in sacred literacy. We don't need to choose between them, we need to see where they converge and where they complement each other.) We do not put our faith in the breakdown because we do not use a constituent model. We try to understand how the present configuration matters, how it might tell a story, point us in a direction, bring a prophecy, heed a warning. If only we could learn to read again . . .

Did akademiks forget to how to read? Perhaps not. But, who and what gets read in the contemporary uni-versity cannot be understood apart from the economy of citations,[2] itself a prosthesis of enduring heteropatriarchal masculinity. In the mode of falsification/verification, one must first recite the genealogical ancestry of Men quite apart from the man himself. The man's bio+geo+mytho+graphic location is not part of the metadata, it cannot easily be admitted. The man does not belong everywhere, and so he stakes his claim and travels motionlessly across an imaginary unified space, making laws for every body. Now it is flat, now it is round. Now we are central, now we are not. I should say, since I anticipate criticism for this point: I do not oppose the presence of some men in the academy, no more than I oppose science, no more than I refute the Copernican revolution.

Despite all of this, my decision to take that indefinite sabbatical was hurried forward by the particularly masculine insistence on taking up more space, being more well-funded, and remaining more entrenched than everybody and everything else.

From my vantage point: It is well past time to de-masculinize the process of knowledge production.[3] Such an operation cannot be prescribed, but it can easily be imagined. Why we must have read 'x' (xy) Theorist♂ to understand 'y' (Y?) event-horizon? Perhaps they had something interesting to say. Perhaps they just said what somebody else said. Maybe so. How does this situation demand our attention, our intention, and, maybe, our reply? Let's say that discernment grows when turning away from economies of citation *as such*. Look a way! I don't mean to say we should institute a boy-cott – citation is also a way of paying respect, of honoring origins, and of passing stories. It is a system of currency and so divestment assumes we are also gathering in other ways, sharing ideas in other ways, and honoring those gatherings and sharings beyond the matrix of equal value or differential value. Though I have been separatist, I know there is also a time and place for mutual acknowledgment. Let's be as concerned with how, where, and when we read the situation because our collective survival depends on rediscovering that ability to maintain a foot in other worlds, to hold fast to intuition, to cultivate rigorous belief, to honor collective knowing and ancient wisdom, to assemble with clear intention. All of these modes may only be ballast against the bleakness of the new formations of uni-versity knowledge, but still, we need them and we need them now (see Harney and Moten, 2013). We must continuously read differently, read more, and read better. We must get an education in spite of the uni-versity (see Kelley, 2016).

I don't just cite you, I see you.

I am not leasing you, I am listening to you.

I am not trading you on the marketplace of ideas, I am sitting with you.

I am not just processing your words, I am asking where you have been and how you have reached me here.

We meet each other in the way that water meets land.

To be sighted, to be at sea; to wade from the shallows to the depth, to fear the tide and, still, to confront what it brings ashore.

Come to the edge, one foot on the soft boundary of unblown broken glass, the sands of understanding.

Here, we learn to read.

After all that, I returned to the swamplands and, eventually, back to North Amerika, where the reading and ritual documented here was undertaken. This is a reading which asks: What does the un-masking/de-masculinization of geography look like? How can we prepare the space? I offer you some ritually produced maps (Figures 18.3 and 18.4) documenting responses to these queries and calling in various writers, thinkers, and storytellers. These maps are guides toward a reconstructed femme-inist reading practice. They are a mutation, an adaptation, an attempt at salvaging something from the pieces. For the tarot reading, I used a spread called the Path of Balance, which Angeles Arrien (1997) describes as 'an opportunity for us to see how balance is present in six areas of our nature:

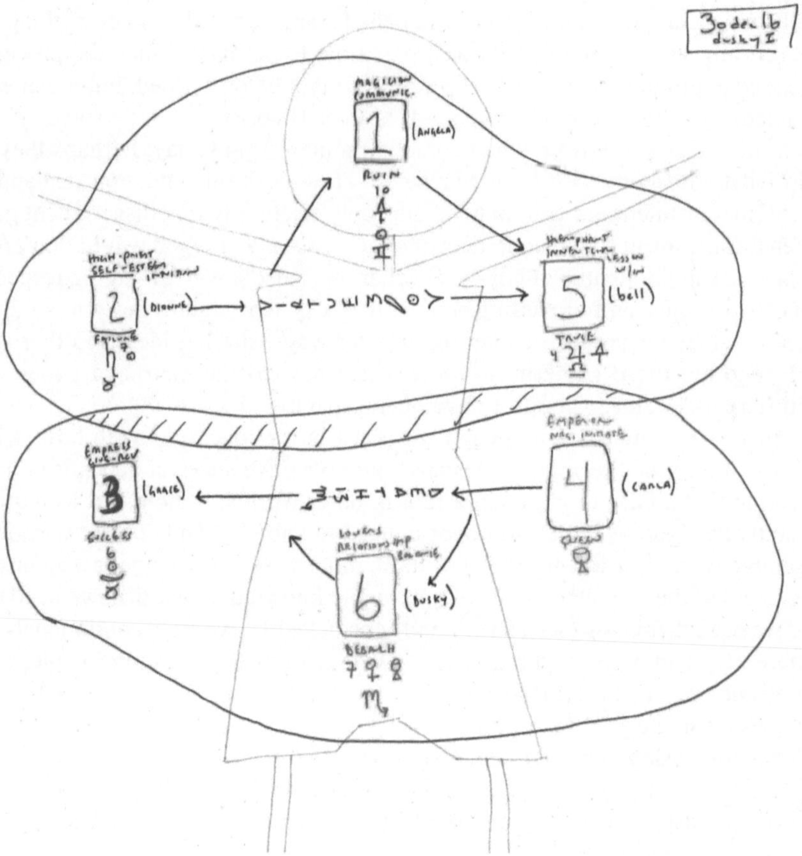

Figure 18.3 Map of the Path of Balance reading conducted on 30 December 2016.
Source: Darren Patrick.

in our self-esteem and self-trust; in our ability to give love and extend love; in what we are learning and teaching; in use of power and leadership; and in our relationships' (p. 260). I modified the reading with additional cards to aid in interpretation. I used the Thoth tarot deck because of its astrological inclinations, and in spite of its cloudy past. The tarot reading preceded and provided a framework for the effigy ritual, which you will find documented in the photo series at the end of this piece (Figures 18.5–18.12).

As a result of these efforts, I have been left with some visions of what I want to call unreconstructed masculinity. Masculinity in pieces. Now that these pieces are on the table (or, in this case, the floor): How will we decide which to keep, which to repurpose, and which to consign to the compost bin? How will we keep reading? How will we purpose reading? How will we consign ourselves to purpose?

Figure 18.4 Photo of the Path of Balance reading conducted on 30 December 2016.
Source: Darren Patrick.

Each piece of the ritual is, like anything worth repeating and working with ritually, borrowed or taught in friendship, siblinghood, comradely struggle, magick. That my sense of reading comes from elsewhere must be acknowledged, even if that acknowledgment is only ever a partial reflection of how I am still learning to read. Taken together, the reading is a testimony of my process to re-cover and to re-lease stories. The enclosed photos were made with the enwitching assistance of a poet and dear Spiritual SuperSheroe friend, Deidre 'D-Lishus' Walton, who also suggested I read the work of Joseph Roach (1996). The subtitle – Ask Me How! – was suggested by Lauren Berlant during the Q&A in Edinburgh.

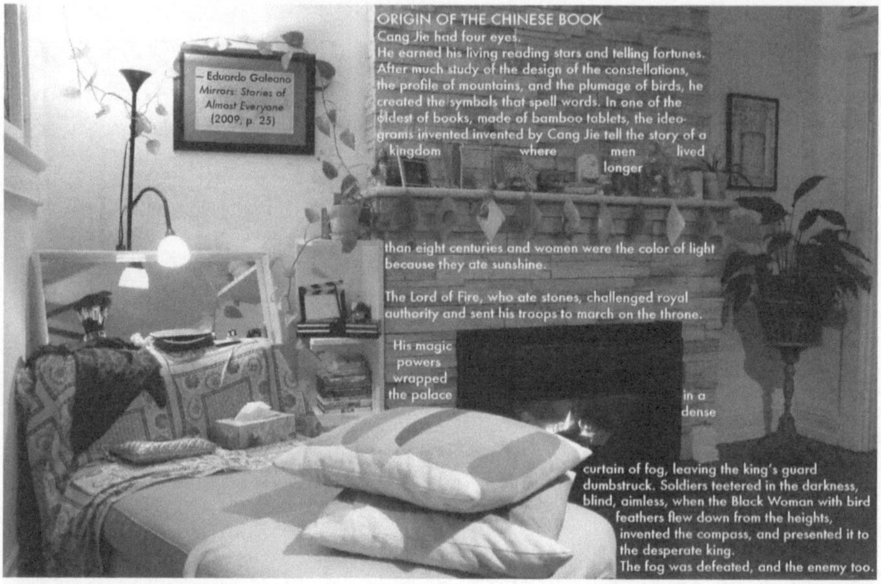

Figure 18.5 Ritual.

Source: professor dusky purples. All quotations as attributed.

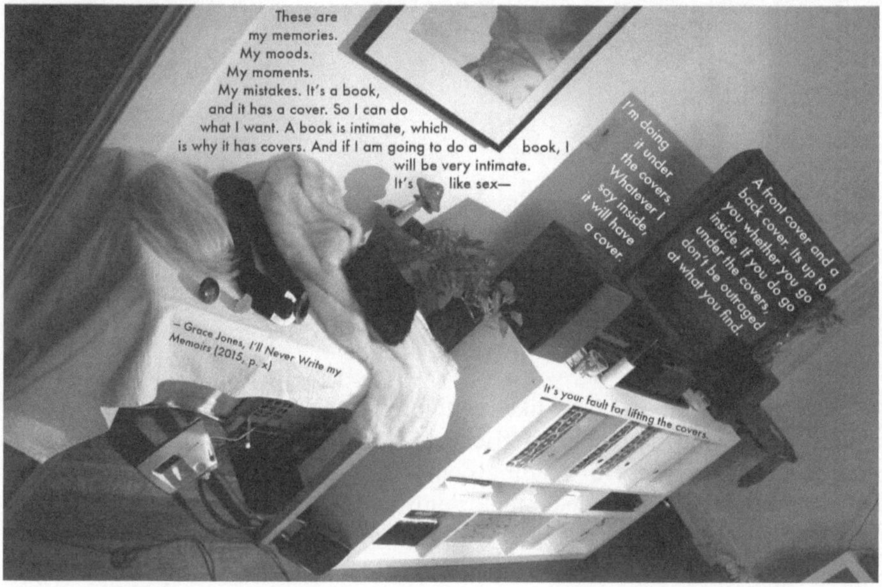

Figure 18.6 Ritual.

Source: professor dusky purples. All quotations as attributed.

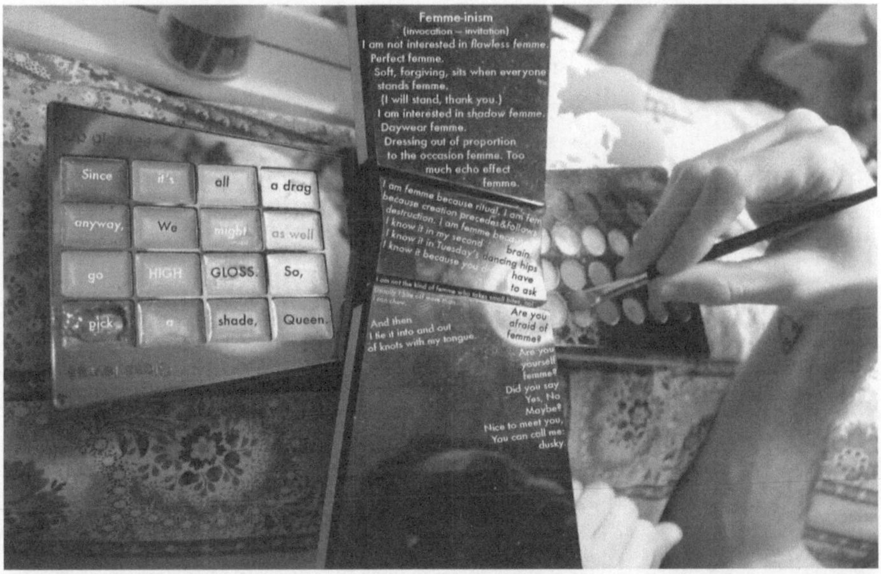

Figure 18.7 Ritual.

Source: professor dusky purples. All quotations as attributed.

Figure 18.8 Ritual.

Source: professor dusky purples. All quotations as attributed.

Figure 18.9 Ritual.

Source: professor dusky purples. All quotations as attributed.

Figure 18.10 Ritual.

Source: professor dusky purples. All quotations as attributed.

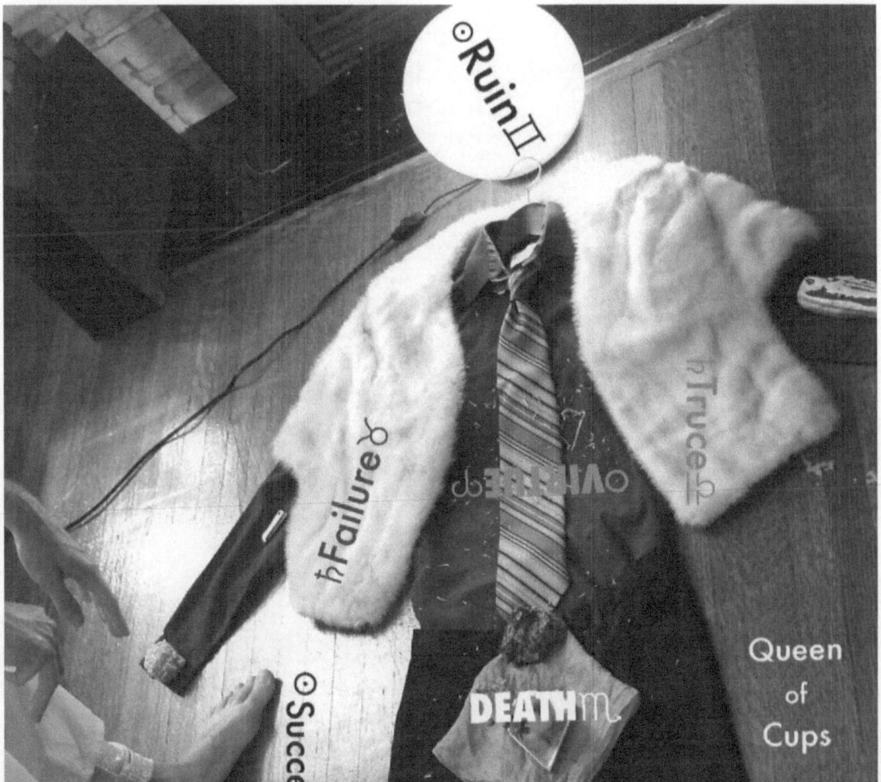

Figure 18.11 Ritual.

Source: professor dusky purples. All quotations as attributed.

Figure 18.12 Ritual.

Source: professor dusky purples. All quotations as attributed.

The ritual itself was performed on Turtle Island, in a place that has been home to humans for more than 10,000 years. Drawing on stories, teachings, acknowledgments, and writings from indigenous educators (see for example Mehot, 2012) and non-indigenous educators, I can tell you that the place now called Toronto has been home to the Haudenosaunee, Huron-Wendat and Petun First Nations, the Seneca, and, most recently, the Mississaugas of the Credit River. The land is subject to the Dish With One Spoon Wampum Belt Covenant between the Haudenosaunee Confederacy and the Confederacy of the Ojibwe. The covenant is a significant referent for U.S.-Amerikan democracy, appropriated uncited. The place where I live in Toronto used to be underwater, lending to one of the ways it is known: 'where the trees stood in water' (see Bambitchell, 2015). Today, this place is still home to many indigenous peoples from across Turtle Island, and to refugees, settlers, and new arrivals. I have learned and been taught that living on this land means we are all treaty people and, though we must walk our own path, we who live here all have obligations to this land and to resurgence (Simpson, 2011); we who live on Earth must continuously push toward decoloniality (Mignolo, 2000).

I offer the ritual reading, in part, as a modest expression of gratitude for the welcome, teaching, understanding, and friendship that I have received here; I offer it to friends and relations who continue doing the hard work of holding on and remaking relations. Misreads and half-steps are my own, everything else is for you. I ask only that you take this reading and make it your own. Make your own sense from it and, if things don't seem headed in that direction, ask for help. Where is that feeling which you asked to sit with you? Where has that thread led you? All is not lost, even if we are not quite sure where it has been stowed for safekeeping. Keep reading.

Yours in femme-inist struggle and siblinghood,
d purples/dp

Notes

1 The album, while critically panned upon release, is doubtless one of the most forward thinking of its generation. On the cover, Sainte-Marie appears in a pose similar to the Rider-Waite tarot's depiction of The Magician, which is the first numbered card in the Major Arcana.

2 This phrase comes from the SomMovimento NazioAnale [Natio-Anal Uprising] (Acquistapace *et al.*, 2015), a network of Italian trans*feminist-queer collectives with which I had the immense fortune of working while in Italy.

3 This is another term/praxis I learned in Italy, from Laboratorio Smaschieramenti [Laboratory for De-mask-ulinization]. For more on the fate of this autonomous collective project, see '#AtlantideOvunque: Statement of Solidarity for the Evicted Trans*-Feminist-Queer-Punk Space,' at https://atlantideresiste.noblogs.org/post/2015/10/11/atlantide ovunque-statement-of-solidarity/ (The statement is available in Italian, French, and English.)

References

Acquistapace, A., *et al.*, 2015. Transfeminist scholars on the verge of a nervous breakdown. *Feminismos*, 3(1), pp. 62–70.

Arrien, A., 1997. *The Tarot Handbook: Practical Applications of Ancient Visual Symbols*, New York: Jeremy P. Tarcher/Penguin.

Bambitchell, 2015. Where the trees stood in water. *Undercurrents*, 19, pp. 62–66. Available at: http://currents.journals.yorku.ca/index.php/currents/article/view/37295/36069.

Brand, D., 2001. *A Map to the Door of No Return: Notes to Belonging*, Toronto: Vintage Canada.

Brenner, N. ed., 2014. *Implosions/Explosions: Towards a Study of Planetary Urbanization*, Berlin: Jovis.

Federici, S., 2004. *Caliban and the Witch*, Brooklyn: Autonomedia.

Galeano, E., 2010. *Mirrors: Stories of Almost Everyone*, New York: Nation Books.

Harney, S. and Moten, F., 2013. *The Undercommons: Fugitive Planning and Black Study*, Wivenhoe: Minor Compositions/Autonomedia.

Jones, G., 2015. *I'll Never Write My Memoirs*, New York: Gallery Books.

Kelley, R.D.G., 2016. Black study, black struggle. *Boston Review*. Available at: http://bostonreview.net/forum/robin-d-g-kelley-black-study-black-struggle [Accessed May 1, 2017].

Lefebvre, H., 2003. *The Urban Revolution*, Minneapolis, MN: University of Minnesota Press.

McLean, H., 2016. Hos in the garden: Staging and resisting neoliberal creativity. *Environment and Planning D: Society and Space*, 35(1), pp. 38–56.

Mehot, S., 2012. "Toronto" is an Iroquois word. *Dragonfly Consulting Services Canada*. Available at: http://dragonflycanada.ca/toronto-is-an-iroquois-word/ [Accessed May 1, 2017].

Meyer, M.A., 2001. Our own liberation: Reflections on Hawaiian epistemology. *The Contemporary Pacific*, 13(1), pp. 124–148.

Meyer, M.A., 2010. *An introduction to "Indigenous Epistemology"*. Available at: www.youtube.com/watch?v=lmJJi1iBdz [Accessed May 1, 2017].

Meyer, M.A., 2014. Holographic epistemology: Native common sense. In C. Smith, ed. *Encyclopedia of Global Archaeology*. New York: Springer, pp. 3435–3443.

Mignolo, W., 2000. *Local Histories/Global Designs: Coloniality, Subaltern Knowledges, and Border Thinking*, Princeton, NJ: Princeton University Press.

Povinelli, E., 2016. *Geontologies: A Requiem to Late Liberalism*, Durham, NC: Duke University Press.

Roach, J., 1996. *Cities of the Dead: Circum-Atlantic Performance*, New York: Columbia University Press.

Simpson, L., 2011. *Dancing on Our Turtle's Back: Stories of Nishnaabeg Re-Creation, Resurgence, and a New Emergence*, Winnipeg: Arbeiter Ring.

Index